Next-Gen Semiconductor Device Design

Next-Gen Semiconductor Device Design

Prof. Dr. R.K. Singh

ARCLER
PRESS

www.arclerpress.com

Next-Gen Semiconductor Device Design

Prof. Dr. R.K. Singh

Arcler Press
4164 Lakeshore Road
Burlington, ON L7L 1A4
Canada
www.arclerpress.com
Email: orders@arclereducation.com

ISBN: 978-1-77956-973-8 (Hardcover)

ABOUT THE AUTHOR

Prof. (Dr.) R. K. Singh is a highly respected academician and administrator with over twenty-eight years of experience in teaching, research, academic leadership, and consultancy. He is a Professor of Electronics Engineering and has previously served as the Vice Chancellor of Patliputra University, Patna, where he contributed significantly to institutional growth, academic governance, and policy implementation at the state university level.

Born on 11 June 1969, Prof. Singh has maintained an outstanding academic record throughout his career. He earned his Ph.D. in Electronics and Communication Engineering from Allahabad University and completed his Master of Engineering from BITS Pilani, one of India's premier technical institutions. He began his professional journey at BITS Pilani as a Teaching Assistant and later worked as a Research Associate and Scientist at CEERI, Pilani (CSIR), gaining strong research and technical experience early in his career.

His academic career has been closely associated with B. T. Kumaon Institute of Technology (BTKIT), Dwarahat, where he progressed through the ranks from Lecturer to Professor, attaining professorship at a notably young age. In addition to teaching and research, he has held several key administrative roles, including Director of BTKIT, Registrar, Controller of Examinations, and Officer on Special Duty at Uttarakhand Technical University.

Prof. Singh is an accomplished researcher who has guided 22 Ph.D. scholars and more than 15 M.Tech students, published over 120 research papers in reputed journals, and authored eight academic books. He is a member of leading professional bodies such as IEEE and the Institution of Engineers (India).

TABLE OF CONTENTS

List of Figures ... *xi*

Preface ... *xvii*

Chapter 1 Introduction to Semiconductor Device Design ... 1

1.1. Historical Evolution of Semiconductor Devices ... 6

1.2. Fundamentals of Semiconductor Physics ... 11

1.3. Design Principles of Semiconductor Devices ... 16

1.4. Applications of Semiconductor Devices ... 22

1.5. Challenges in Semiconductor Device Design ... 27

1.6. Emerging Trends in Semiconductor Technology ... 31

Summary ... 36

References ... 39

Chapter 2 Semiconductor Material Science ... 47

Introduction ... 47

2.1. Crystal Structures of Semiconductors ... 51

2.2. Doping in Semiconductors ... 56

2.3. Defects in Semiconductor Materials ... 61

2.4. Bandgap Engineering ... 66

2.5. Wide-Bandgap Materials ... 71

2.6. Two-Dimensional Materials ... 76

Summary ... 81

References ... 83

Chapter 3 Carrier Transport Mechanisms ... 89

Introduction ... 89

3.1. Drift Mechanism ... 93

3.2. Diffusion Mechanism ... 98

3.3. Recombination and Generation 103
3.4. High-Field Effects 108
3.5. Ballistic Transport 113
3.6. Quantum Confinement 118
Summary 124
References 127

Chapter 4 MOSFET Fundamentals and Scaling Challenges 133
Introduction 133
4.1. Basic Principles of Metal-Oxide-Semiconductor Field-Effect Transistor (MOSFET) Operation 138
4.2. Short-Channel Effects 144
4.3. Scaling of Metal-Oxide-Semiconductor Field-Effect Transistors (MOSFETs) 149
4.4. Advanced Metal-Oxide-Semiconductor Field-Effect Transistor (MOSFET) Architectures 154
4.5. Subthreshold Slope and Leakage Control 159
4.6. Future Trends in Metal-Oxide-Semiconductor Field-Effect Transistor (MOSFET) Technology 164
Summary 169
References 172

Chapter 5 Advanced Transistor Architectures 177
Introduction 177
5.1. Historical Context 182
5.2. Tunnel Field-Effect Transistors (TFETs) 188
5.3. Negative Capacitance Field-Effect Transistors (NC-FETs) 192
5.4. Vertical Transport Devices 197
5.5. Emerging Trends in Transistor Architectures 203
5.6. Challenges and Limitations 207
Summary 212
References 214

Chapter 6 High-Power and High-Frequency Devices 221
Introduction 221
6.1. Types of High-Power Devices 226
6.2. Materials Used in High-Frequency Devices 232

6.3. Applications of High-Power Devices 236
6.4. Thermal Management Strategies 241
6.5. Challenges in High-Power and High-Frequency Device Design 247
6.6. Future Trends in High-Power and High-Frequency Devices 252
Summary 256
References 259

Chapter 7 Spintronics and Magnetic Devices 267
Introduction 267
7.1. Fundamentals of Spintronics 272
7.2. Spin Transport and Spin Torque Transfer 278
7.3. Magnetic Random Access Memory (MRAM) 282
7.4. Spin Field-Effect Transistor (Spin-FET) Design 288
7.5. Prospects for Neuromorphic Computing 293
7.6. Current Research and Innovations 298
Summary 302
References 305

Chapter 8 Photonic and Optoelectronic Devices 311
Introduction 311
8.1. Fundamentals of Photonics 316
8.2. Photonic Integrated Circuits (PICs) 321
8.3. Quantum Dot (QD) Lasers 326
8.4. Photodetectors 331
8.5. Silicon Photonics 336
8.6. Challenges in Photonic and Optoelectronic Devices 341
Summary 345
References 348

Chapter 9 Quantum and Single-Electron Devices 353
Introduction 353
9.1. Fundamentals of Quantum Mechanics 356
9.2. Qubits in Quantum Computing 361
9.3. Single-Electron Transistors (SETs) 366
9.4. Quantum Dots and Their Applications 371

9.5. Cryogenic Electronics Design Considerations 376
9.6. Current Research and Developments 381
Summary 386
References 389

Chapter 10 Fabrication Techniques for Next-Gen Devices 395
Introduction 395
10.1. Advanced Lithography Techniques 400
10.2. Atomic Layer Deposition (ALD) 405
10.3. Integration of Heterogeneous Materials 409
10.4. 3D Printing in Device Fabrication 414
10.5. Nanofabrication Techniques 419
10.6. Quality Control and Testing Methods 424
Summary 429
References 431

Index 437

LIST OF FIGURES

Figure 1.1. Transformative aspect of semiconductor materials
Figure 1.2. Evolution of semiconductor technologies
Figure 1.3. Semiconductor process technology
Figure 1.4. The integration of semiconductor physics
Figure 1.5. The evolution of band theory and energy levels
Figure 1.6. Charge carriers are essential for enabling the flow of electrical current in semiconductors
Figure 1.7. The landscape of semiconductor device design
Figure 1.8. Semiconductor devices from classical architectures to next-generation designs
Figure 1.9. Semiconductor devices classical architectures
Figure 1.10. The development of semiconductor devices
Figure 1.11. Artificial intelligence (AI) transformation in semiconductor industry
Figure 1.12. Challenges in the behavior of semiconductor devices based on fundamental physics
Figure 1.13. The advancement in semiconductor technology
Figure 2.1. The specification of semiconductor material science
Figure 2.2. Crystal structure and behavior of semiconductor
Figure 2.3. The analysis of crystal structures in semiconductor material
Figure 2.4. The manipulation of doping in semiconductor properties
Figure 2.5. Fundamental doped process in semiconductor material science
Figure 2.6. The crystal structures and defects in semiconductor materials
Figure 2.7. The semiconductor technical approach to defect pattern analysis
Figure 2.8. The analysis bandgap energy
Figure 2.9. Revisiting the optical bandgap of semiconductors
Figure 2.10. Wide bandgap semiconductors
Figure 2.11. The exploration of wide-bandgap semiconductors

Figure 2.12. Two-dimensional semiconductors and transistors for future integrated circuits

Figure 2.13. Important advancements in semiconductor material science

Figure 3.1. Different types of current transport mechanism in metal/semiconductor Schottky diodes

Figure 3.2. The interplay between drift and diffusion processes

Figure 3.3. A comprehensive examination of drift in semiconductor

Figure 3.4. The mechanism of diffusion

Figure 3.5. The diffusion mechanisms in semiconductor materials

Figure 3.6. Fick's first law of diffusion

Figure 3.7. Generation, recombination of charge carriers

Figure 3.8: Direct generation/recombination process

Figure 3.9. The high-field effects of semiconductor physics

Figure 3.10. Understanding of carrier transport mechanisms in high-field environments

Figure 3.11. Ballistic transport in action

Figure 3.12. Ballistic transport in transistor channel

Figure 3.13. Schematic representation of the quantum confinement effects

Figure 3.14. Illustration of quantum confinement effect

Figure 3.15. Charge carrier transport in a nanofilm metal-semiconductor structure

Figure 4.1. Function of metal-oxide-semiconductor field-effect transistor (MOSFET)

Figure 4.2. Principle of metal-oxide-semiconductor field-effect transistor (MOSFET)

Figure 4.3. Structure of metal-oxide-semiconductor field-effect transistor (MOSFET)

Figure 4.4. (a) Metal-oxide semiconductor field effect transistor (MOSFET), and (b) Schematic describing scaling laws applied to MOSFET where scaling down by a factor λ implies increasing the density of devices on IC by a factor of λ 2 and decreasing the gate-length or CD by 1/λ. The cost reduction is a function of scaling factor, decreasing with the increase in λ

Figure 4.5. Short channel effects in a metal-oxide-semiconductor field-effect transistor (MOSFET)

Figure 4.6. The miniaturization of short-channel effects

Figure 4.7. Schematic illustration of the scaling of a Si metal-oxide-semiconductor field-effect transistor (MOSFET)

Figure 4.8. Semiconductor miniaturization: limitations of metal-oxide-semiconductor field-effect transistor (MOSFET) scaling

Figure 4.9. Evolution of metal-oxide-semiconductor field-effect transistor (MOSFET) from classical to nanowire FETs

Figure 4.10. Fin field-effect transistor

Figure 4.11. Planar transistors vs. finFETs vs. gate-all-around

Figure 4.12. Subthreshold Slope in FDSOI FD-SOI has a more vertical subthreshold slope (SS) than bulk CMOS (Vth) and requires less subthreshold voltage to forward bias the transistor in Thus, FD-SOI can match bulk CMOS performance with less voltage

Figure 4.13. The enhancement property of metal-oxide-semiconductor field-effect transistor (MOSFET)

Figure 4.14. Performance improvements in metal-oxide-semiconductor field-effect transistor (MOSFET) technology

Figure 5.1. The technical advancements in transistor architectures

Figure 5.2. The evolution of advanced transistor architectures

Figure 5.3. The Evolution of Transistors: From PLANAR MOSFETs to FINFETs to NANOSHEET FETs

Figure 5.4. The historical development in transistor

Figure 5.5. Interplay between bipolar junction and field effect transistors (BJTs and FETs)

Figure 5.6. Tunnel field-effect transistor (TFET): At a sufficient gate bias a tunnel current flows across the device

Figure 5.7. Schematic structure of a typical tunnel field-effect transistor (TFET)

Figure 5.8. Negative capacitance field effect transistor (NC-FETs)

Figure 5.9. The negative capacitance (NC) transistor (FET) based on monolayer black phosphorus (ML-BP)

Figure 5.10. The working principle of vertical transport devices

Figure 5.11. The structure of vertical transportation system ⋆

Figure 5.12. The technical advancements in transistor

Figure 5.13. Specific advanced transistor challenges in the formation of fin, spacer, gate stack and junction

Figure 6.1. The modern landscape of high-frequency devices

Figure 6.2. Commercially available high-power semiconductor devices

Figure 6.3. High electron mobility transistor (HEMT)

Figure 6.4. Schematic cross-section of a laterally-diffused metal-oxide-semiconductor (LDMOS)

Figure 6.5. The advancements in IGBT technology

Figure 6.6. Types of materials used in high-frequency devices

Figure 6.7. Possible applications of gallium nitride (GaN) power devices as a function of the voltage. The range of application of silicon carbide (SiC) devices is also shown for comparison

Figure 6.8. Power electronics in modern power transmission systems and its increasing applications in future energy systems
Figure 6.9. Personal thermal management
Figure 6.10. A comprehensive evaluation on thermal management of electronic devices |
Figure 6.11. Techniques for cooling power and other electronic devices -
Figure 6.12. The design of high-frequency devices
Figure 6.13. The future demand of high power devices
Figure 7.1. The base of semiconductor architecture
Figure 7.2. Magnets and electronics: The close relationship driving modern devices
Figure 7.3. Future developments in spintronics
Figure 7.4. The mechanism between spin and charge in electrons
Figure 7.5. Mechanisms for spin transport
Figure 7.6. Spin transfer torque
Figure 7.7. Spin transport in graphene/transition
Figure 7.8. Basic features of magnetic random access memory (MRAM)
Figure 7.9. The operational activity of MRAM (Magnetic random access memory)
Figure 7.10. The spin field-effect transistor (Spin-FET) schematic design
Figure 7.11. The interplay between metal-oxide-semiconductor field-effect transistor (MOSFET) (left) and a spin field-effect transistor (Spin-FET) (right)
Figure 7.12. The integration of advancement in neuromorphic computing
Figure 7.13. Neuromorphic computing architecture inspired by a human brain
Figure 7.14. Innovations in the field of spintronics
Figure 7.15. The contemporary advancements in the field of spintronic materials
Figure 8.1. Utilization of photonic devices in modern technology
Figure 8.2. Types of optoelectronic devices
Figure 8.3. Principles behind photonic
Figure 8.4. Basic principles of fiber optics series: refraction
Figure 8.5. Photonic integrated circuit (PIC)
Figure 8.6. Structure of photonic integrated circuits (PICs)
Figure 8.7. Quantum dot (QD) laser technology
Figure 8.8. Design and application of photodetectors
Figure 8.9. Function and mechanism of photodetectors
Figure 8.10. Silicon photonics: Designing and prototyping
Figure 8.11. The integration of photonic devices onto silicon platforms

Figure 8.12. Representative photonic and optoelectronic devices using CNTs
Figure 9.1. Method of quantum mechanics
Figure 9.2. The working principle of quantum mechanics
Figure 9.3. Basic features of qubit
Figure 9.4. The pathway of superconducting qubits
Figure 9.5. Schematic of a basic single-electron transistors (SET) and its internal electrical components
Figure 9.6. The manipulation of single-electron transistors (SETs)
Figure 9.7. Applications of quantum dots
Figure 9.8. Structure of a quantum dot
Figure 9.9. Cryogenic applications: from high-performance computing and deep space electronics, to quantum computing
Figure 9.10. (a) Structure of a hexagonal transition metal dichalcogenide (TMD) monolayer. M atoms are in black and X atoms are in yellow. (b) A hexagonal TMD monolayer
Figure 9.11. The future advancement in quantum device technology
Figure 10.1. Characteristics of next-gen technology
Figure 10.2. Working cycle of atomic layer deposition (ALD)
Figure 10.3. Characteristics and process of atomic layer deposition (ALD)
Figure 10.4. The principle of atomic layer deposition (ALD)
Figure 10.5. The implementation of heterogeneous integration
Figure 10.6. Technology building blocks for heterogeneous integration
Figure 10.7. The pattern of 3D printing
Figure 10.8. Application of 3D printing technology
Figure 10.9. The methods of nanofabrication techniques
Figure 10.10. Purpose of quality control

PREFACE

Electronics is revolutionizing at an unprecedented rate, driven by demands for faster, smaller, and lower-power devices. Semiconductor technology is the foundation of all things in modern life—communications and transportation, as well as medicine and solar energy. The challenge for engineers and researchers is not only to improve current device structures but to develop completely new paradigms that will enable the exponential scaling of computing and sensing uses. The transition from classical designs to next-generation architectures requires an understanding of fundamental principles, new materials, and new device physics far beyond classical limits.

As materials become smaller, to the nanometer range, and are being called upon to function under increasingly rigorous conditions, boundaries between material science, quantum mechanics, and engineering break down. Concepts previously emerging only in theory—such as quantum confinement, spintronics, and single-electron transport—are now viable design and fabrication factors. Such transformation underscores the need for large materials that span elementary knowledge with breakthroughs reaching forward.

Next-Gen Semiconductor Device Design meets this demand by presenting a detailed, systematic coverage of basic as well as frontier topics in semiconductor devices. The book bridges the gap between materials foundation and carrier transport and high-end architectures, high-performance applications, and cutting-edge techniques such as spintronics, photonics, and quantum devices. It not only charts the existing technologies but also calls forth the barriers that must be broken if the next jump in performance and efficiency is to occur. Balance between theory and practice, it is a blueprint presented to students, researchers, and practitioners engaged in designing the future of semiconductor technology.

Chapter 1 presents semiconductor device design, establishing the foundation upon which the ideas to be cultivated in the book are built. Chapter 1 provides a historical context, describes the place of semiconductors in today's world, and outlines the fundamental principles that govern the design of devices. Chapter 1 also demonstrates how trends in scaling, power constraints, and performance have shaped the development of device structures over time.

Semiconductor material science is dealt with in Chapter 2, a fundamental subject for understanding the properties governing the behavior of devices. Band structures, doping, crystal growth techniques, and novel materials such as wide-bandgap semiconductors and two-dimensional systems are taught to the reader. The analysis identifies atomic-scale tuning of materials as the means to achieve breakthroughs in mobility, thermal stability, and efficiency required for future devices.

Chapter 3 addresses carrier transport mechanisms, the phenomena that control the way electrons and holes move themselves within semiconductor devices. The chapter synopsizes drift, diffusion, tunneling, and ballistic transport, connecting microscopic physical principles with macroscopic device performance. Transportation in nanostructures is addressed in particular and how quantum mechanics modifies the classical models to provide insights into ultra-scaled device performance limits.

Chapter 4 expands the fundamentals of MOSFETs and the scaling issues that they will face as dimensions shrink. The review considers threshold voltage control, short-channel behavior, leakage current, and variability issues. Readers will also gain an appreciation of why conventional MOSFET scaling is reaching its limits and how such limits drive the search for novel device architectures.

Chapter 5 discusses the next-generation transistor architectures that overcome the drawbacks of traditional planar MOSFETs. Topics discussed are FinFETs, gate-all-around transistors, and nanosheet transistors. These architectures are not just explained in terms of design concepts but also in terms of fabrication viability, performance benefit, and adoption in the commercial world. The chapter presents a vision-based perspective on the future of transistor technology.

Chapter 6 demonstrates high-power and high-frequency devices, which are crucial in telecommunication, power electronics, and automotive applications. The chapter addresses wide-bandgap materials like GaN and SiC, their use towards increasing efficiency, and design approaches to dealing with thermal and reliability problems. High-frequency operation principles are also addressed, correlating material characteristics with circuit-level performance.

Chapter 7 discusses spintronics and magnetic devices, a new technology that uses the spin of electrons as a new degree of freedom for information processing. From magnetic tunnel junctions to spin-transfer torque devices, the chapter describes how spin-based technologies promise to revolutionize non-volatile memory, low-power logic, and neuromorphic computing. It shows how magnetism and quantum spin phenomena are being integrated into semiconductor design.

Chapter 8 covers photonic and optoelectronic devices, which convert electrical signals into light and vice versa. Some of the subjects covered are LEDs, laser diodes, solar cells, and photodetectors, with an emphasis on how optical interactions with semiconductors enable new functionalities in communication, sensing, and renewable energy. Hybrid electronic-photonic integration is also covered, a path towards faster and more efficient data processing.

Chapter 9 focuses on quantum and single-electron devices, where quantum mechanics rather than classical laws is controlling operation. Discussion includes quantum dots, tunneling devices, and issues surrounding quantum computing. Problems of practical interest such as decoherence and scalability are covered as well, describing both the promise and challenges of incorporating quantum devices into mainstream use.

Chapter 10 describes the next-generation device fabrication techniques, with emphasis on how enhancements in lithography, etching, deposition, and self-assembly contribute to nanoscale advancements. In addition to usual CMOS processes, the chapter introduces advanced techniques such as atomic layer deposition, extreme ultraviolet lithography, and emerging patterning techniques. The chapter illustrates how device ideas must be supported by fabrication techniques in order to arrive at the next generation of semiconductor technology.

This book aims at a general readership including graduate students, research-oriented scholars, and practitioners. For students, the book is intended to provide a well-organized step-by-step treatment from the fundamentals to the advanced, handy both as a course material and as a study guide. For researchers, it is a work that brings together theoretical ideas with advances, making cross-disciplinary interactions between physics, materials science, and engineering easier. For working professionals, it shares details on both established and emerging device technologies, which is enough to allow them to deal with actual world problems of design, fabrication, and utilization. Whether new to semiconductor science or moving towards cutting-edge innovations, this book aims to offer clarity, depth, and usefulness.

CHAPTER 1

INTRODUCTION TO SEMICONDUCTOR DEVICE DESIGN

Contents

1.1. Historical Evolution of Semiconductor Devices......6
1.2. Fundamentals of Semiconductor Physics......11
1.3. Design Principles of Semiconductor Devices......16
1.4. Applications of Semiconductor Devices......22
1.5. Challenges in Semiconductor Device Design......27
1.6. Emerging Trends in Semiconductor Technology......31
Summary......36
References......39

INTRODUCTION

The semiconductor device world has undergone revolutionary evolution from conventional designs to complex next-generation solutions that drive technological advancements in a variety of fields. Semiconductors were initially only utilized in simple electronic devices such as rectification and amplification. With the transistor still decreasing in size based on Moores Law, however, the focus is now switched to construction while maintaining energy efficiency. The advent of ultra-wide bandgap (UWBG) and wide bandgap (WBG) materials has brought new horizons for device performance enhancement through the reduction of conduction and switching losses. This is particularly important since it directs the entire design goals toward enhancing the efficiency of semiconductor devices in a wide variety of applications, paving the way for new functionalities in AI, IoT, and high-performance computing (Porter et al., 2024). In recent years, semiconductor devices have occupied key positions in emerging technologies such as artificial intelligence and the Internet of Things (IoT) because they can be versatile and adaptive. They serve as the basis for sophisticated systems, enabling data processing and interconnectivity that create the essence of smart devices and autonomous systems. Quantum computer developments also reflect this tendency; for instance, companies like Intel leverage semiconductor manufacturing know-how to develop scalable spin-qubit systems. The fabrication of high-quality quantum dots illustrates how advancements in mainstream semiconductor technology can be transferred to new applications, which significantly impact computational efficiency and speed (George et al., 2024). This spectrum of application points out the pivotal role played by semiconductor devices to sustain digital transformation. As semiconductor technology continues to improve, it is confronted with scaling limits as it goes beyond Moore in pushing the boundaries of material and device performance. Component miniaturization, for all its benefits, creates problems in thermal dissipation, operational reliability, and power use. Emerging paradigms like integrated photonics and 2D materials are being worked on to surmount these constraints to achieve enhanced performance without relying solely on scaling. Secondly, new manufacturing technology such as extreme ultraviolet lithography (EUV) is enabling quicker and smaller production of highly complex silicon chips. It is essential that researchers and engineers possess this overall picture of semiconductor device design since it paints a picture of the future environment and thus shapes the future of technology in many industries.

Definition and Functionality

The relentless evolution of semiconductor devices has compelled innovation from traditional to next-gen technology, impacting all industries significantly. As devices evolve, their capability has increased in terms of complexities, enabling multifarious uses ranging from AI to the IoT. Expansion into such new-age technologies demands that there is an exploration of new materials and structures that can enable the requirements of high-end computing. Enhanced considerations of processing speed, power consumption, and thermal management are crucial with devices shrinking in size and becoming more functional. The development not only increases the functional range but also provides the capability of processing more complicated tasks that will be required in future technology. Welding in such progress permits an instantaneous reaction to shifting application requirements, hence securing the position of semiconductor devices at the forefront of technology advancement. Only by extremely close observation of scaling limits and beyond-Moore trends could one understand semiconductor operation complexities. With transistors getting close to atomic size, traditional scaling practices, encapsulated by Moores Law, are faced with significant challenges. These challenges are among others such as increased out power density, thermal management, and reduced quantum effects, all of which affect device performance. Consequently, new methods outside the normal scaling models are becoming apparent, and they are focusing on 2D materials, quantum computing, and heterogeneous integration. The outreaching to these beyond-Moore trends not only presents ways to overcome the constraints brought about by physical scaling but also enables enhanced functionalities and performances in various applications. This research underscores the versatility of semiconductor design and supports the imperative of innovation in addressing the demands of today and tomorrow's technology. Specific breakthroughs, for instance, FSDOI physical design methods, illustrate the functionality enhancement needed for specialized use like automotive systems. These methods concentrate on the optimization of thermal dissipation and mechanical stress tolerance in semiconductor chips, critical in high-temperature and high-mechanical-stress environments. Research has indicated that outstanding advancements are evident, with FSDOI-optimized chips exhibiting better reliability and durability, essential characteristics for modern automotive applications requiring high performance and safety. Furthermore, sophisticated techniques such as Finite Element Analysis (FEA) allow engineers to simulate a chips reaction to various environmental conditions and thereby improve design parameters for optimal functionality. In addition, reverse engineering methods have also exhibited innovation possibilities in already established technologies, as presented in the instance of ultrasonic saw blade cutters that aim to improve semiconductor material production procedures for substances like

The innovation is aimed at reinforcing the essential role of design enhancement to achieve the continuous advancement of semiconductor device functionality in fulfilling evolving requirements by various high-tech markets.

Types of Semiconductor Materials

Evolution and improvement in semiconductor materials have spurred the transition from traditional electronics to sophisticated next-generation devices, which have profoundly impacted many industries. The conventional silicon-based semiconductors, with their reliability and highly conductive properties, have dominated the electronic domain for several decades. With the current breakneck pace of developments in the areas of AI, the IoT, and quantum computing, materials that can offer improved operational performances and sustain high-performance computing are needed. Advances in compound semiconductors such as gallium nitride (GaN) and silicon carbide (SiC) are being researched more and more for their high electron mobility and thermal conductivity, paving the way for devices with higher operating frequencies and efficiencies and therefore enabling their uses to go beyond conventional boundaries. For this purpose, mixing various types of materials is needed to keep up with the ever-growing demand for efficiency and performance in semiconductor technology. A comparable transformative attribute of semiconductor materials is in being able to keep pace with scaling problems in modern electronic design (Figure 1.1).

Figure 1.1. *Transformative aspect of semiconductor materials.*

Source: https://www.elprocus.com/wp-content/uploads/Semiconductor-Materials.jpg.

Systematic miniaturization of device elements has led to a point of saturation known as the scaling limit, where reductions in size become progressively burdensome in terms of technical challenges such as heat dissipation and quantum effects. In order to overcome such barriers, researchers are looking into beyond-Moore trends, which are based on the use of new materials rather than physically scaling. For instance, two-dimensional (2D) materials like graphene and transition metal dichalcogenides are emerging as strong competitors. Their unique electronic properties enable them to design transistors that occupy much less space but continue to be very efficient. Thus, the transition to these materials could signal a paradigm in overcoming the boundaries that presently constrain silicon technologies, which will ultimately lead to more efficient and speedier electronic devices. In organic semiconductor materials, there has been great success in enhancing electronic transport properties, crucial to the development of many electronic and optoelectronic devices. The synthesis of n-type organic semiconductors, as outlined in recent reports, is a significant accomplishment, particularly for application in organic field-effect transistors (OFETs). The development of novel conjugated copolymers, such as PDPPTT-2Tz, is illustrative of how slight molecular-level changes are adequate to produce high-electron-mobility materials that are essential for successful logic circuit operation. Electron-deficient frameworks offer a platform for efficient charge conduction, opening the way for novel applications in organic electronics for complementing traditional semiconductor applications. Beyond providing alternatives for traditional materials, these advances show that semiconductor technology is adaptable to meeting requirements of modern developments, ranging from flexible electronics to sustainable green energy technology (Ren et al., 2023).

Importance in Modern Technology

In today's technologically advanced world, the evolution from traditional semiconductor devices to next-generation innovations is a milestone that opens doors to new avenues. The conventional semiconductor materials employed, including silicon, have predominantly dictated the limits of electronic circuits in regards to performance and efficiency. But the trend towards heterogeneous integration and high-capability materials engineering is provoking the development of devices capable of meeting the needs of application in the IoT and AI. Such is evidenced by the development of resistive gas sensors, which have been optimized for high sensitivity and selectivity using new material design and device architecture. These sensors demonstrate the manner in which semiconductor technology develops to address emerging challenges in diagnostics and monitoring, underlining the critical contribution of semiconductor device design to the future of networked technologies (Wang et al., 2025). Semiconductor design plays a significant role in high-performance computing and quantum technology, areas that require unprecedented processing power and efficiency. With devices becoming smaller, physical constraints have arisen, and attention has turned toward beyond-Moore approaches that consider alternatives to scaling. 2.5D and 3D integration solutions have appeared as promising solutions, enhancing performance without linear shrinking otherwise inherent in Moores

Law. The 3D SRAM innovation, for instance, shows the prowess of staggered pillar structures in bandwidth optimization with reduced space, demonstrating a monumental leap in semiconductor capabilities (Swarnkar et al., 2025). The new design allows for increased computational density and acceleration, which are key requirements for applications that enhance real-time processing and complex data handling, areas critical for AI and quantum computing applications. Besides, the development of smart systems is basically revolutionizing the semiconductor world, where machine learning algorithms optimize the function and operation dependability of devices. AI use in semiconductor device design optimizes not just material selection but also how multisensory functions are carried out. Integration facilitates signal processing and data integrity enhancement, particularly in complex environments where precision is paramount. Since the semiconductor industry does embrace these trends, the continuous integration of smart technologies ensures future devices not just compact and energy-efficient but also more capable of real-time information interpretation and decision-making processes. This is a paradigm shift that will demonstrate how important semiconductor device design is in pushing technological advancements and addressing the challenges of industries today.

1.1. HISTORICAL EVOLUTION OF SEMICONDUCTOR DEVICES

The history of the evolution of semiconductor devices is marked by an unprecedented transition from conventional technologies to next-generation, high-tech devices. The earlier breakthroughs in semiconductor material have acted as a key force to make electronic revolution accessible. Vacuum tubes formed the backbones of electronic devices until the invention of the transistor in 1947, which opened the door towards compact, more reliable, and cost-effective devices. This was followed by the development of integrated circuits during the 1960s, which further minimized the size of electronic devices and led to exponential increases in processing power. Each progress has not only improved the functionality of typical electronic applications but laid the groundwork for complex systems, such as artificial intelligence (AI) and Internet of Things (IoT) functions. As these technologies continue to develop, the position of semiconductor devices at the forefront of societal infrastructure is only going to become more robust. The application of semiconductor devices has grown exponentially, particularly in the field of high-performance computing and quantum technology. The past several years have seen the integration of semiconductor technology create possibilities for AI advancements through the aid of sophisticated data processing algorithms and enormous parallel processing capability (Figure 1.2).

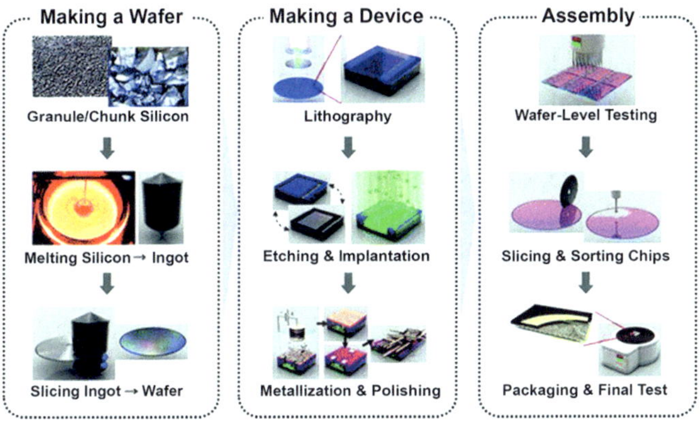

Figure 1.2. *Evolution of semiconductor technologies.*

Source: https://media.springernature.com/lw685/springer-static/image/chp%3A10.1007%2F978-94-017-8768-0_3/MediaObjects/273504_1_En_3_Fig13_HTML.gif

Similarly, the history of quantum computing, with computers that leverage quantum mechanics principles, relies heavily on semiconductor materials for qubits. The history of superconducting qubits, built on semiconductor foundations, is a good case in point. The integration of matter with light at the quantum level has opened new paths in circuit quantum electrodynamics (QED), thus opening a new area of study in semiconductor science focused on quantum information processing (Blais et al., 2021). Such a highly advanced reliance of semiconductor growth on emerging technologies shows how pivotal it is to various disciplines. Still, semiconductor devices' trajectory is not without challenges, especially considering scaling limits and beyond-Moore revolutions. As semiconductor technology approaches its economic and physical limits, traditional scaling methods are becoming increasingly impractical. The industry is turning to novel architectures such as heterogenous integration and 3D stacking in order to maintain the performance trajectory. Reconfigurable intelligent surfaces (RISs) constitute a potential feature of future networks enhancing wireless communications by means of intelligent manipulation of the propagation environment (Liu et al., 2021). Furthermore, this evolving environment calls for synergistic approaches, employing machine learning techniques to optimize performance under evolving environments. With scientists entering this phase of transition, the historical underpinning of semiconductor devices is at once foundation and filter for reading and designing future advancements.

1.1.1. Development of the Transistor

Semiconductor technology has had its turning points, most significantly perhaps with the discovery of the transistor. Early equipment focused on vacuum tubes, which were heavy and power consuming. The introduction of the transistor was a revolutionary change, offering a smaller alternative that, besides being more energy efficient, also made electronic components smaller. This change opened the way to the production of integrated circuits, continuing to fuel consumer electronics and computational device innovations. The pioneering transistor design principles work established by the fundamental transistor development is still an influencer of modern device engineering and forms the core of classical as well as upcoming computing and communication technology inventions (Nguyen et al., 2023). The application of transistor technology has expanded immensely during recent times, particularly in artificial intelligence (AI) and the Internet of Things (IoT). Such applications require devices with the ability to process large volumes of data effectively, which mandates the use of advanced transistor designs that can effectively meet the demands of high-performance computing systems. The development of specialized transistors such as those used in neural processing and ultra-low-power consumption is a demonstration of how semiconductor technology is able to meet contemporary challenges of computing. As the need for greater computational power increases, the focus is increasingly on how to get transistors to function better to enhance machine learning programs and networks of connected devices (Nguyen et al., 2023). Transistor development has thrived, but it is being constrained by tremendous barriers to scaling boundaries. The traditional Moores Law, forecasting doubling in the density of transistors on integrated circuits approximately every two years or so, is being challenged by physical limits. These are caused by issues such as quantum tunneling and short-channel effects that detract from performance with decreasing devices. Beyond-Moore trends are starting to surface, utilizing heterogeneous integration and novel materials to improve functionality and efficiency without relying solely on transistor scaling. With technologies such as spintronics and photonic integration, the future of semiconductor device design is on the cusp of revolutionary changes that can potentially redefine computing paradigms (Crawford, 2023).

1.1.2. Transition from Classical to Next-Gen Devices

Evolution from classical semiconductor devices to next-generation devices is not only an increase in performance but also a revolutionary shift in architecture and application. This is most driven by the requirements of modern technologies such as artificial intelligence (AI), Internet of Things (IoT), and high-performance computing, which necessitate devices capable of efficiently managing a large amount of data. As classical equipment, founded to a great extent on bulk silicon technology, approaches its scaling limit, the semiconductor community is forced to innovate on a constant basis. While classical aesthetics have filled basics, they cannot respond to the growing power and performance requirements called for by these new technologies. Therefore, emerging architectures like FinFET and Gate-All-Around (GAA) devices are being explored to address both energy efficiency and scalability challenges, which open the way to a new generation in semiconductors. New uses place high value on the imperative requirement for such next-generation devices. In artificial intelligence, machine learning algorithms need hardware supporting parallel processing efficiently, which conventional transistor architectures cannot provide reliably. Similarly, IoT devices require small energy-efficient components that are capable of operating even under harsh power limits without disconnecting. High-performance computing applications value utilization of next-generation semiconductor technology, which enables higher rates of computation and larger memory capacities (Figure 1.3).

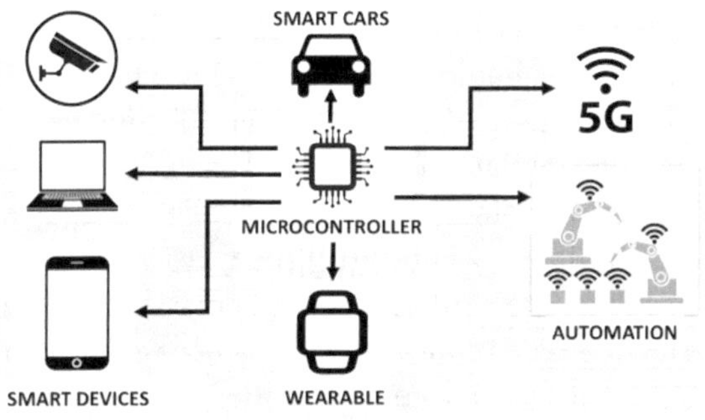

Figure 1.3. *Semiconductor process technology.*

Source: https://www.renesas.com/sites/default/files/media/images/microcontroller-applications-en.png.

The new designs' flexibility also allows for the testing of new materials outside silicon, which is central to overcoming the limitations imposed by traditional approaches. Thus, with the nature of the technology landscape shifting, so too should the founding principles of semiconductor design, calling for a shift towards devices appropriate for today and tomorrow's needs. Along this transition, research in scaling limits and beyond-Moore trends gives crucial insight into the future of semiconductor device design. As increments in performance with traditional scaling decrease, researchers seek alternatives increasing the range of semiconductor technology. Approaches like

3D integration and heterogeneous integration offer promising paths to enhance device performance above relying on shrinking geometries. Furthermore, the study of quantum computing introduces radically different operation paradigms that are likely to transform computational capabilities. The adoption of advanced architectures, including the emergent RibbonFET technology featured in recent studies, attests to these trends and their implications on circuit design, particularly in analog use as documented in Gaddipati (2025). These technologies signal the leap beyond conventional devices and also lay the foundation for performance levels, efficiency, and functionality never hitherto achieved in next-generation semiconductor applications as discussed in Arsid (2025).

1.1.3. Key Milestones in Semiconductor Technology

During the evolution of semiconductor technology, there have been some key milestones that have altered the very fabric of device design. The transition to next-generation semiconductor devices from the traditional is a significant milestone that makes it possible to integrate new materials and innovative architectures to enhance performance characteristics. Most importantly, the introduction of FinFET technology represented a paradigm shift in transistor design because it overcame the constraints of vintage planar transistors by ensuring better electrostatic control and reduced leakage current. This innovation not only facilitated continuous scaling but also ushered in the unrelenting quest for efficiency in performance, essential for application across different fields of labor like artificial intelligence and the Internet of Things (IoT). With these technologies converging, they necessitate calibration of design techniques to realize optimal device capabilities, consequently paving the way for the next wave of semiconductor design advancements. As the demand for larger amounts of computing power grows ever larger, the technology of high-performance computing (HPC) is testament to the knowledge that comes from breakthroughs in semiconductor technology. The application of specialized processing blocks, such as graphics processing units (GPUs) and tensor processing units (TPUs), serves to underscore the importance of custom architecture in reducing performance. All these innovations facilitate record computer speeds, particularly for artificial intelligence where tremendous parallel computing capacity is required. Furthermore, these dedicated chips are profoundly impacting the future of semiconductor design paradigms since designers must specialize in creating flexible systems that can efficiently perform multiple computer chores. In this case, an awareness of the milestones in semiconductor technology is essential to researchers and engineers who want to develop novel devices that can cope with future computational demands. The transition into the beyond-Moore era encapsulates the semiconductor device design challenges and opportunities today. Scaling limitations have encouraged researchers to pursue new materials and future technologies beyond traditional silicon-based approaches. Technologies such as neuromorphic computing and quantum computing depict a trend towards new architectures that are promising further increases in processing power. Moreover, employing digital twins, a virtual representation of physical assets, offers a new way for real-time performance monitoring and design improvement (Rasheed et al., 2020). It is crucial to find the necessity of cooperative multidisciplinary challenges to overcome challenges of new paradigms. By examinations of these advances, scholars can build an elaborate picture of how the technology of semiconductors can advance further to support differing applications without becoming obsolete in an ever-more complex and interdependent world.

1.2. FUNDAMENTALS OF SEMICONDUCTOR PHYSICS

Due to the continuously changing technology, semiconductor device development has shifted from conventional techniques to even more sophisticated designs that are developed for next-generation applications.

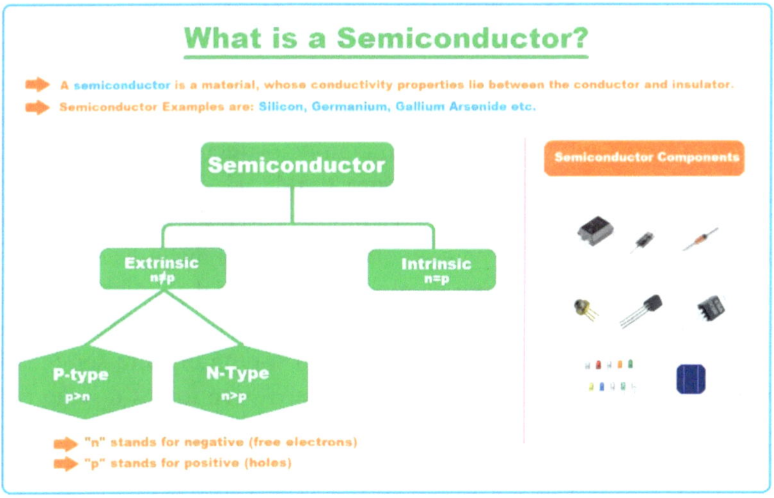

Figure 1.4. *The integration of semiconductor physics.*

Source: https://images.theengineeringprojects.com/image/webp/2020/08/Semiconductor-1.jpg.webp?ssl=1.

This is driven by the increased demand for high-performance computing devices that support the booming fields of artificial intelligence and the Internet of Things (IoT). The problems of consumer electronics, including power and thermal management, need new solutions that are outside the conventional semiconductor paradigms. To this end, researchers are focusing on enhancing material properties in one way or another, such as the fabrication of organic semiconductors that can offer flexibility and lower production cost compared to conventional materials of silicon (Figure 1.4).

Incorporating these novel materials into device structures is a critical step to achieving operating efficiencies required in modern semiconductor design. Moreover, ongoing research on quantum behavior in semiconductors holds tremendous potential for advancing computational functions further. Technologies such as strain engineering are being researched to manage electronic and optical nature of semiconductor materials to enhance their performance profiles significantly. Such a strategy allows for the application of a tailor-made band structure to improve carrier mobility and lower the effective electron mass. Methods like those outlined in Tsymbalov et al. (2021) show how machine learning algorithms can facilitate the identification of optimal strain paths, paving the way for new device applications. These are combinations of computational techniques and actual device design that put researchers in a position to sufficiently address the limitations of conventional scaling methods. As performance needs evolve, so do the means employed to produce next-generation semiconductor devices with the capability of being competitive on a consistent basis in an increasingly dynamic environment. In venturing into the scaling limits of semiconductor technology, the industry is witnessing a shift towards beyond-Moore trends where emphasis is laid on performance rather than just transistor counts. Three-dimensional integration and the use of new materials—namely organic semiconductors as mentioned in Kumagai et al. (2022)—are being employed as potential solutions to transcend traditional limits. The search for improved carrier mobilities in organic thin-film transistors (OTFTs), along with innovations in n-type organic semiconductor design, is just one of the expressions of the innovative forces driving the field forward. By harnessing a synergy of material science and device physics insight, the semiconductor world is opening up new territory that not only enhances performance but also creates the potential for applications across the full breadth of industries from telecommunications to medicine. Remaining at the forefront of such trends will be critical in order to keep up with the pace of technology development in the semiconductor industry.

1.2.1. Band Theory and Energy Levels

Development of semiconductor devices has necessitated a detailed understanding of band theory and energy levels since they are the foundation of operation for a vast range of applications, from Artificial Intelligence (AI) to the Internet of Things (IoT). As technology evolves to transition away from conventional silicon geometries toward next-generation possibilities, control of the energy bands takes center stage. Electronic properties imposed by band structures determine device performance, especially as dimensions keep shrinking; reduction of barriers and enhancement of conductivity become absolutely critical to achieving desired functionalities. This systematic approach toward understanding how energy levels are in relation with each other in materials allows scientists to design semiconductor properties that meet the evolving needs of new technologies. Without a profound appreciation for these underlying principles, the creation of integrated circuits and computing systems with high performance may be faced with insurmountable challenges. With regards to recent advances, the study of doped semiconductors illustrates the decisive impact of band theory on material property optimization. For instance, studies on copper-doped zinc oxide (CZO) show that donor atom doping can lead to an appreciable decrease in band gap of the materials, to enhance conductivity and to make potential applications available in optoelectronic devices (Soussi et al., 2023) (Figure 1.5).

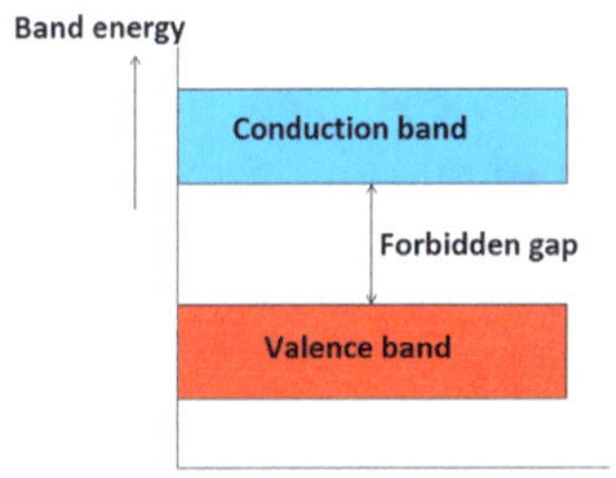

Figure 1.5. *The evolution of band theory and energy levels.*

Source: https://www.physics-and-radio-electronics.com/electronic-devices-and-circuits/introduction/images/energy-band-theory-in-solids.png.

The shift of the Fermi level toward the conduction band signifies a shift toward metallic behavior which is the prerequisite of future devices. Moreover, the application of nanoscale materials to semiconductor technology introduces an additional phase, where scientists can create and tailor specific electronic and optical properties that are essential in high-technology applications. Band theory is thus a pivotal framework for understanding such innovation in semiconductor technology. According to this, it is necessary to address the implications of band theory in the formation of low-dimensional materials and their electronic properties. Recent research on monolayers, such as MoS_2, has shown that the introduction of functional groups holds the potential to fundamentally modify electronic functionality, such that semiconductor behavior transforms into metallic behavior (Pan et al., 2019). This alteration is particularly important because the demand for efficient electron transport keeps on rising, particularly in the area of high-performance computing. Through tools such as density functional theory, researchers are able to predict and even optimize the performance of such materials under various conditions. As we proceed further to transcend the constraints of conventional scaling, understanding the energy levels and band structures will be essential in tapping the potential of future semiconductor devices in a changing technology landscape.

1.2.2. Charge Carriers: Electrons and Holes

The path from conventional semiconductor devices to future technologies has been one of increasing sophistication in understanding charge carriers, particularly electrons and holes. Catalyst for devices such as transistors and diodes, these charge carriers play an integral role in allowing the transfer of electrical current through semiconductors. With shrinking device size, control over these charge carriers becomes even more important. Progress now relies on more efficient electron-hole pair generation and separation for better performance in small-scale devices. New developments in material science, especially in the fabrication of substitute semiconductors, offset some of the challenges concerning charge recombination and mobility of electrons.

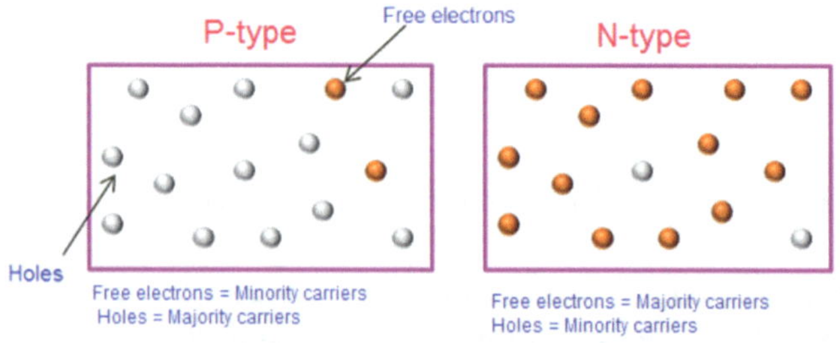

Figure 1.6. *Charge carriers are essential for enabling the flow of electrical current in semiconductors.*

Source: https://www.physics-and-radio-electronics.com/electronic-devices-and-circuits/semiconductor/images/majorityandminoritychargecarriers.png.

This advancement has far-reaching consequences for application in artificial intelligence (AI) and the Internet of Things (IoT), where improved semiconductor operation directly affects computation rate and efficiency.

In seeking applications that benefit from charge carriers, high-performance computing and quantum devices are vastly improved. The intricate manipulation of electrons and holes allows for increased rates of data processing and energy efficiency, hallmarks of modern models of computing (Figure 1.6).

Devices are being more and more designed to take advantage of the unique features of these charge carriers in new materials, such as two-dimensional semiconductors, which possess remarkable electrical characteristics. Through the focus on electron mobility enhancement and hole recombination inhibition, researchers are creating systems that can overcome the traditional limits of silicon technology. As newer methods for semiconductor structure development are explored, opportunities for revolutionary application in AI and quantum technologies look hopeful, positioning these future devices as important building blocks in the future of computational development. The boundaries faced by existing semiconductor technologies bring about the necessity for intense examination of both scaling barriers and beyond-Moore directions, namely charge carrier-related ones. Conventional scaling has given rise to severe difficulties, including increasing power density and thermal management issues, which compromise the purity of microelectronics. Addressing these issues necessitates thinking anew about the generation, manipulation, and utilization of charge carriers in devices. For instance, more recent trends towards novel architectures, such as neuromorphic and quantum computing devices, target specifically the utilization of electron-hole dynamics for information processing akin to biology. Furthermore, existing work explores how tunable interface properties and defect states can lead to improved charge carrier lifetimes, as in the example of progress in materials such as TiO_2, where surface-specific properties are tuned for the purpose of increasing photocatalytic activity. Innovations that maximize charge carrier efficiency not only enhance performance but also pave the way for sustainable device design in a world of finite resources, bridging the gap between theory and practical applications in semiconductor technology.

1.2.3. Doping and Its Effects on Conductivity

Theevolutionofdopinginsemiconductormaterialsisarevolutionarybreakthrough in device technology, pushing the boundaries of conductivity and performance. Historically, doping impurity atoms into silicon and other semiconductors gave the ability to produce p-type and n-type materials that improved their electrical characteristics by orders of magnitude. This initial work laid the groundwork for subsequent generations of devices that exploit advanced doping techniques, such as atomic layer deposition (ALD), with which dopant incorporation can be implemented with high accuracy. Importantly, the specific modification of carrier mobilities and densities creates customized electronic properties, which are essential for current uses like artificial intelligence (AI), the Internet of Things (IoT), and quantum computing. With advancements in semiconductor technology, understanding doping mechanisms becomes more valuable in an effort to implement concepts that meet the requirements of high-performance computing and other emerging fields. In the context of such progress, doping is an important role in overcoming scaling barriers that traditional silicon-based technology is limited by. Device miniaturization calls for materials that not only exhibit great conductivity but also offer better performance when miniaturized. Recent research has demonstrated that applying conductive and semiconducting oxides can give rise to better electrical and optical properties to a large extent due to new methods of doping. These compounds have found widespread applications in solar cells and display technology, paving the way towards next-generation devices tailored to a broad spectrum of applications. The controlled doping levels of transparent conductive oxides, for instance, can lead to improved efficiency in electronic and optoelectronic devices, demonstrating how doping can transcend its conventional bounds and redefine the semiconductor device engineering frontier (Macco et al., 2022). Nonetheless, the doping challenges in organic semiconductors indicate a complex situation that requires continued exploration. The doping conundrum, where molecular dopants have a detrimental effect on the structural stability and properties of organic semiconductors, points out the interplay between material performance and doping efficiency. Improvements are needed in an effort to enhance n-doping efficiency, particularly because the development of n-type molecular dopants remains much behind p-type counterparts. Through optimization of molecular designs and charge transport insights, researchers are beginning to overcome these challenges and expand n-doped organic material applications. This ongoing study is not only an advancement of existing technology but a big step toward future innovation in high-performance semiconductor devices that are anticipated to spearhead computing and energy solution industries (Lu et al., 2021).

1.3. DESIGN PRINCIPLES OF SEMICONDUCTOR DEVICES

Semiconductor device design has evolved greatly from the conventional structures to new innovative next-generation devices. This evolution is driven by increasing demands for improved performance and efficiency in many areas of technology. As conventional scaling methods approach their physical limitations, such as short-channel effects and power dissipation, designers are compelled to look for creative ways to enhance the performance of devices. Emerging materials and device architectures, such as FinFETs and quantum-dot transistors, exemplify this shift towards more sophisticated approaches. Each advancement exhibits greater understanding of semiconductor physics and is aimed at creating devices that meet the stringent requirements of artificial intelligence (AI), Internet of Things (IoT), and high-performance computing, where computational capability is exponentially increasing, demanding new paradigms for design. In applied technologies, semiconductor devices play a central role in facilitating breakthroughs in AI and IoT. For instance, the rapidly developing fields of machine learning and edge computing demand chipsets with high-speed processing capacity for massive volumes of data while consuming low energy. Here, the design methodology of semiconductor devices must not only look at performance parameters but also focus on low power consumption and heat dissipation methods (Figure 1.7).

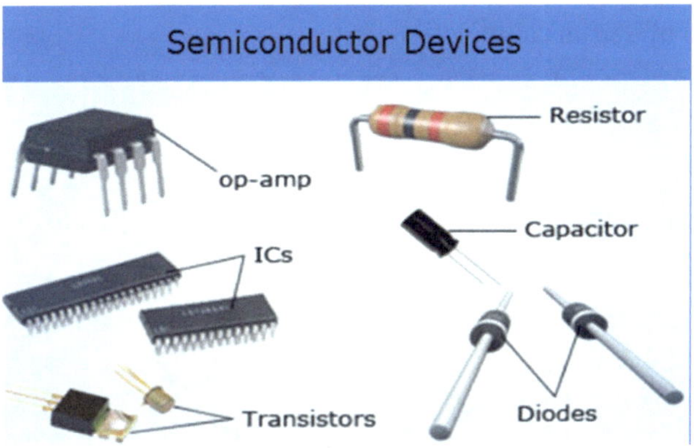

Figure 1.7. *The landscape of semiconductor device design.*

Source: https://www.elprocus.com/wp-content/uploads/2016/06/2016-06-10_11-41-46.jpg.

Existing innovations, such as the integration of CMOS comparators of low power dissipation and high level, precisely meet such needs, following the implicit principles presented in the literature (Zeng, 2024). Designers can enhance the performance necessary for use in the current world by optimizing device structure and materials, thus keeping semiconductor devices in line with technological advancements. In the future, it is crucial to understand the scaling limits of semiconductor devices and the emerging trends outside the traditional Moores Law. With decreasing size, quantum effects and increasing leakage currents preclude performance increase. To overcome these constraints, researchers are studying new approaches, such as elastic strain engineering and machine learning techniques for device parameter optimization (Tsymbalov et al., 2021). These methods provide an unprecedented degree of ability to tailor the electrical and optical performance of semiconductors using advanced computer algorithms. Through examining the interaction between strain and band structure, engineers are able to find optimal configurations that enhance device performance. This convergence of physics and high-performance computation not only defines semiconductor device design principles but also positions them for innovation in the rapidly evolving realm of technology, the gateway to quantum computing breakthroughs and further ones to follow.

1.3.1. Device Architecture and Layout

The development of semiconductor devices from old-fashioned architectures towards new-generation devices is a stark indication of such significant technological improvements, driven mainly by demands for more performance in a multitude of various applications.

Figure 1.8. *Semiconductor devices from classical architectures to next-generation designs.*

Source: https://tilos.ai/wp-content/uploads/2024/10/OpenROAD.jpg.

The transition has introduced complex device architectures such as FinFETs and more recently RibbonFETs that provide better electrostatic control and performance improvements necessary for today's high-performance computing and artificial intelligence applications. Such new transistor architectures allow for further scaling, which is necessary as manufacturers reach the limits of traditional technologies. The finding of new materials and structures, particularly in the area of IoT and quantum computing, is an urgent need to transcend the shortcomings of conventional architecture as well as to fit into various operation environments and requirements (Gaddipati, 2025). Inherently linked with the research of device architecture is the layout concept, which has the key significance in shaping the device performance and manufacturability. The topology also encompasses various design aspects such as parasitic capacitance, inductance, and heat dissipation, all of which have significant implications on signal integrity and power dissipation. With various industries such as AI and high-performance computing demanding more efficiency and less latency, semiconductor design methodologies need to adapt accordingly (Figure 1.8).

Scaling techniques must ensure effective scaling underlying the physical and electrical properties' inherent constraints. Analog integrated circuit design is where this is most evident, with precise matching and parasitic control paramount to signal processing reliability and accuracy (Nagaraj, 2025). Moreover, beyond-Moore trends need to be investigated in enabling semiconductor design innovation. As integration levels rise, the rising device architecture complexity requires adaptive solutions in design implementation and verification, especially regarding energy efficiency and performance. Innovative methods such as Dynamic Voltage and Frequency Scaling (DVFS) and power gating have emerged to address the pressing power management issues that arise with greater functionality in semiconductor devices. A focus on low-power design verification methods ensures that these adaptive architectures meet performance levels required without compromising efficiency. Research in these contemporary challenges also seeks to push the semiconductor technology further, opening avenues for upcoming advancements that will set the direction for electronics and computation (Nagaraj, 2025).

1.3.2. Simulation and Modeling Techniques

The intricate topography of the development of semiconductor devices has witnessed an incredible shift, from the conventional to the very brink of emerging technologies. This shift is not merely a function of advances in materials science but also depends on sophisticated simulation and modeling techniques. As the devices are becoming increasingly complex, particularly in terms of artificial intelligence (AI), the Internet of Things (IoT), and quantum computing, the demand for reliable modeling tools has never been higher. They allow designers to predict how newly designed devices will behave under various operating conditions, essentially bridging the gap between theory and practical use. Effective modeling, on the other hand, allows innovation to the point where it is possible for scientists to explore new architectures and materials that will be able to deliver the needed performance for tomorrow's demanding applications. Special challenges happen to face thermal and electrical dynamics in leading-edge semiconductor devices, necessitating the application of advanced simulation techniques. For instance, the operation of high-performance devices, such as high electron mobility transistors (HEMTs), must be modeled accurately in order to understand intermodulation distortion and dynamic large-signal response (Figure 1.9).

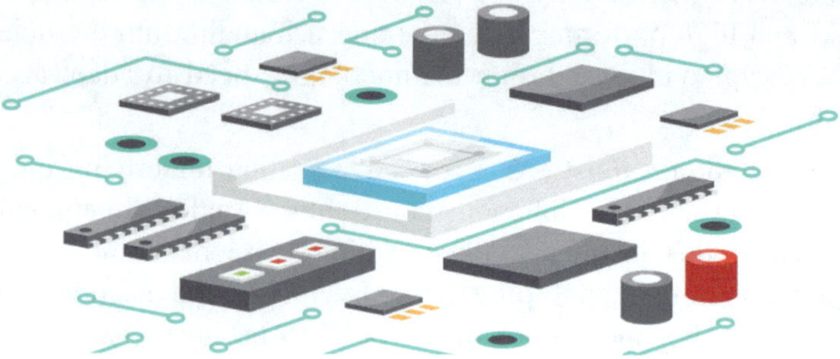

Figure 1.9. *Semiconductor devices classical architectures.*

Source: https://srmap.edu.in/wp-content/uploads/2023/03/Semiconductor-Design-Engineer.jpg.

With the aid of sophisticated device-circuit interaction models, engineers can gain insight into the internal physical behavior under realistic operating conditions, as demonstrated by recent work (Miles et al., 2001). Besides, the combination of computer-aided design (CAD) software with circuit-oriented modeling techniques has been demonstrated to enhance prediction fidelity and accuracy of device performance. Of key importance is the inclusion of thermal dynamics in the model framework since neglecting thermal nonlinearity tends to introduce deviations in operating behavior or responsiveness, ultimately affecting device performance and reliability (Bonani et al., 2008). With the semiconductor technology trending towards beyond-Moore technologies, the challenges in traditional scaling methods necessitate a paradigm change in modeling and simulation. Researchers are more inclined to come up with hybrid model techniques to accurately simulate the quantum mechanical and classical behavior of semiconductor devices. This process is essential to understanding how new device structures and materials can be made to work at reduced size and increased functions, particularly in high-performance computing devices. By extending the simulation capability to accommodate these new structures, engineers can now forecast performance metrics aligned with the demands of new technology. Technological advancements in model-making techniques will ultimately play a vital role in defining the future path of semiconductor device design, enabling quantum computing and advanced AI systems to emerge.

1.3.3. Fabrication Processes and Techniques

The development history of semiconductor devices has witnessed groundbreaking fabrication processes with far-reaching influence on device performance and integration capability. The shift from conventional semiconductor materials characterized by bulk silicon to new materials like perovskites is the most significant shift towards maximizing efficiency and functional ability. The heart of this shift is the quest for maximizing material pairs and employing sophisticated fabrication techniques that can accommodate the unique attributes of new materials. To this extent, the use of machine learning and physics modeling provides scientists with a breakthrough that makes solar cell and other semiconductor device design easier. The use of these technologies lowers the complexity of getting the best performance out of devices, illustrating a route to more efficient and economically viable solutions across industries, from renewable energy to consumer electronics (Chakar et al., 2023). With increased need for high-performance semiconductor devices in applications such as artificial intelligence (AI) and Internet of Things (IoT), the exploration of new fabrication technologies is essential. Thin-film microscale light-emitting diodes (LEDs) are a prime example of how new fabrication technologies can lead to compact, efficient devices compatible for integration with biomedical devices. Techniques such as device lift-off and transfer printing enable the fabrication of microscale LEDs that can function in various environments, even those requiring biocompatibility.

Such manufacturing techniques not only enable complex light sources integration but also offer implantation and wireless power delivery solutions, enabling the semiconductor device to open new possibilities for use in more than the traditional sense to other areas like medicine (Zhang et al., 2022). Further, the trajectory towards next-generation semiconductor devices has been marked by bridging intrinsic scaling obstacles and embracing beyond-Moore principles. As devices shrink, the limitations of traditional scaling methods render it inevitable to seek alternative fabrication techniques that bring improved performance without compromising functionality. Innovative approaches such as 3D integration and heterogenous materials assembly provide avenues to achieve improved device geometries. By moving in this direction, researchers can design even more advanced semiconductor devices that have the potential to enable next-generation technology within quantum computers and high-performance applications. The interaction of the process of fabrication with materials science underscores the central role that such advanced techniques play in shaping next-generation directions for semiconductor device technology, a shift by which miniaturization is reconciled with enhanced performance capabilities required in today's electronic demands.

1.4. APPLICATIONS OF SEMICONDUCTOR DEVICES

The evolution of semiconductor devices has stepped forward progressively from conventional frameworks to advanced configurations that respond favorably to contemporary challenges. Innovations have led to the development of high-performance semiconductors that enable rapid computation and efficient energy usage, driving applications in artificial intelligence (AI) and the Internet of Things (IoT). As device complexity increases, the physical scaling boundaries have dropped increasingly in sight and presented an engineering problem and force for advancement. Indeed, such scaling boundaries, established by the physics of conventional silicon technology, have spurred researchers to consider beyond Moores Law, e.g., to develop new materials and designs that enhance performance without compromising efficiency and functionality. Semiconductor devices play critical roles in the area of power electronics in different industries, from renewable energy systems to electric vehicles. Power semiconductors are the backbone of many systems, enabling effective energy conversion and management (Figure 1.10).

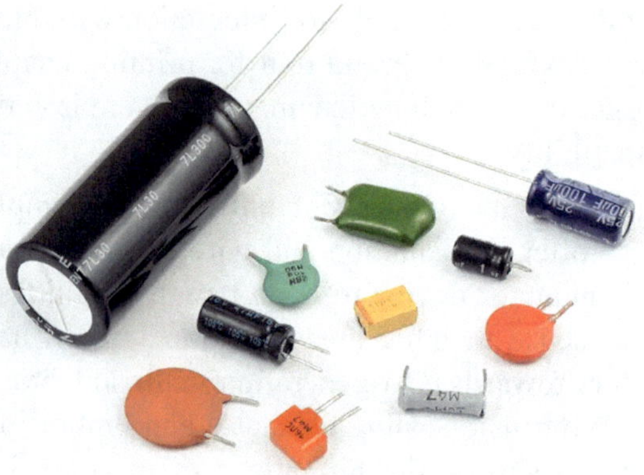

Figure 1.10. *The development of semiconductor devices.*

Source: https://www.agsdevices.com/wp-content/uploads/2024/06/what-are-semi-conductor-types.jpg.webp.

Some of the main advancements include the introduction of power semiconductor databases where the most suitable devices for a given application can be selected and utilized, and in return, it can lead to improved system design. Efficient system design and simulation, such as that of three-phase inverters using specially selected power devices, illustrate how these advancements can lead to efficient and reliable electrical power delivery systems (Shaheen et al., 2000).

This intersection of application and design demonstrates that semiconductor devices are still current today to the development of technology in various sectors. Beyond classic applications, semiconductor devices are being more and more designed for novel technologies like quantum computing and photonics. For instance, two-photon absorption photoconductivity in semiconductor microcavities demonstrates the potential for significantly enhanced nonlinear optical responses, which has interesting potential for low-intensity, high-speed devices like autocorrelation and optical demultiplexing. These advancements are particularly significant in applications that require immediate data processing and transmission, ideally suited to the demands of future generation computing architectures. With additional research continuing to debunk the promise held by these advanced devices, their role towards enabling revolutionary technologies presents a vision of semiconductor applications that is not only vast but also integral to future endeavors in how we approach and utilize technology (Barry et al., 2002).

1.4.1. Role in Artificial Intelligence (AI)

The evolution of semiconductor devices has been at the forefront in shaping the future of AI and its associated applications. From traditional to future-generation devices, the importance of high-quality materials and innovative designs has taken center stage. Silicon-based traditional technology has limitations when it comes to performance and efficiency due to the expanding computational demands with data-centric tasks prevalent in AI and machine learning. By integrating emerging semiconductor devices such as memcapacitors, scientists are opening up the gates to effective neuromorphic computing, enabling better processing of complicated AI algorithms, as illustrated in recent studies developing high accuracy in classification tasks (Singh et al., 2025). This innovation not only enhances the operation of AI systems but also impacts design aspects of emerging semiconductor devices. Applications of next-generation semiconductor technology are not just limited to AI but encompass elite fields like the Internet of Things (IoT), quantum computing, and high-performance computing (HPC). As more and more devices get networked, the demand for power-efficient, low-power solutions looms large. Semiconductor breakthroughs through new ways of processing and new materials are enabling the realization of the scalability required in IoT networks. For example, emerging innovations involving stacked complementary field-effect transistors (CFETs) and emerging two-dimensional materials are expected to significantly improve device performance while minimizing energy consumption. Furthermore, as industries start exploring potential for quantum computing, the ability of the semiconductor industry to design and manufacture unique devices exclusive to quantum applications presents an exciting future direction for research and development (Mii, 2024). This crossover of different applications and semiconductor design will certainly drive smart system innovations (Figure 1.11).

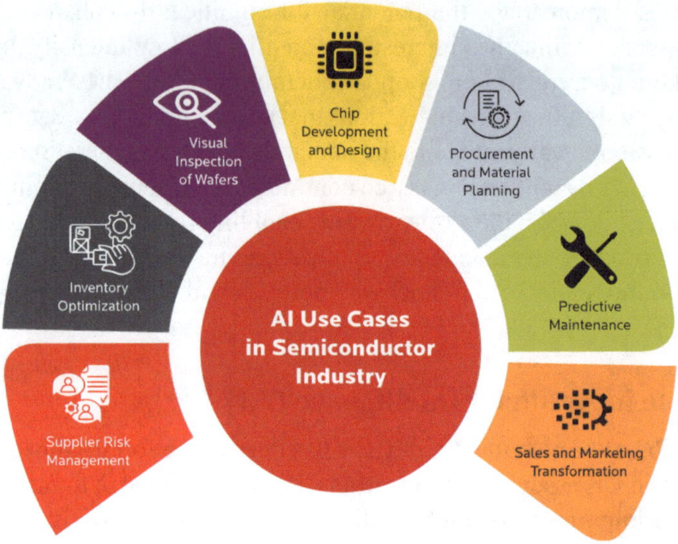

Figure 1.11. *Artificial intelligence (AI) transformation in semiconductor industry.*

Source: https://www.birlasoft.com/articles/ai-semiconductor-industry-use-cases-and-benefits.

With the background in mind, scaling boundaries and beyond-Moore directions are at the center stage as extremely critical impetuses of semiconductor device design. The technology scaling paradigm has long driven innovations in the semiconductor industry, but physical and economic constraints now necessitate rethinking of traditional approaches. New strategies revolve around integrating next-generation packaging strategies and exploring new architectures to enhance overall system performance. Methods like design-technology co-optimization (DTCO) help overcome the challenges by enabling enhanced synergies between fabrication and design operations. As the demand for higher performance in AI and other future high-tech applications grows, it will be critical to understand and overcome these scaling constraints to propel the next semiconductor devices (Mii, 2024). This convergence of ideas shall play a key role in promoting intelligent technologies which are suitable for the needs of an ever-evolving digital world.

1.4.2. Impact on Internet of Things (IoT)

Increasing IoT application complexity calls for better semiconductor device design, and thus encourages innovations along various dimensions. As the need for connected devices increases, particularly in smart home solutions, health care, and industrial automation, semiconductor companies are repurposing their business models to meet the special needs of IoT systems. Embedded systems are being designed to provide greater connectivity features with the capability to work within low-power modes, as IoT devices are often used in battery-powered or remote locations. The miniaturization of semiconductor components has become the need to meet the growing demand for small-device products to facilitate the integration of technology into daily products smoothly. This technology advancement in semiconductors not just facilitates greater functionality but also facilitates greater efficiency and performance in IoT, hence transforming market forces and opening up new innovation opportunities in the sector of semiconductors (Tareq et al., 2024). The intersection of the newest technologies such as Narrowband Internet of Things (NB-IoT) is a classic example of the intersection of semiconductor design and use of IoT, bringing into focus the strategic importance of low-power wide area network technology. With network performance and energy efficiency in consideration, NB-IoT semiconductors utilize features such as Power Saving Mode (PSM) and long Discontinuous Reception (eDRX) to provide excellent battery life. This is especially crucial for applications that need to operate continuously without constant recharging, thereby increasing the applicability of IoT solutions in different industries. The investigation into power saving techniques in NB-IoT not only highlights the importance of semiconductor technology in IoT but also mirrors wider trends, including the demand for sustainable technology solutions that reduce resource utilization. In this regard, semiconductor innovations play a crucial role in optimal IoT deployments, taking into consideration the operational economy and sustainability of the environment (Al-Sammak et al., 2024). Moreover, the trajectory of semiconductor device design is also directly connected with the continuous evolution of the IoT ecosystem, ranging from artificial intelligence applications to high-performance computing. As the technology of semiconductors progresses toward extending the scaling boundaries and moves into beyond-Moore trends, it becomes more feasible to integrate AI capability into IoT devices. Enhanced processing and improved power management functionality enable the development of smart devices that can process data in real time and decide, allowing the technologies to blend together to offer smarter and more responsive interactions between humans and devices and hence a smarter and more responsive IoT environment. Through that, advancements in semiconductor design not only enhance existing applications of IoT but also drive innovation in a wide array of industries and create an avenue for transformative changes in the ways technology is impacting everyday life and industry operations (Tareq et al., 2024).

1.4.3. Use in Quantum Computing and High-Performance Computing

The face of computer technology has undergone a revolutionary shift, from the classical systems based on classical semiconductor principles to futuristic devices capable of computing sophisticated computations at record speeds. This transition has been felt most prominently in the fields of artificial intelligence (AI), the Internet of Things (IoT), and above all, quantum computing. In quantum computing, qubit design scalability and coherence issues call for new semiconductor device engineering innovation. Emerging materials such as two-dimensional transition metal dichalcogenides have shown potential to achieve efficient hybrid structures that can improve qubit performance in terms of improved charge transport and thermal stability, both essential to sustaining quantum superposition states, which are indispensable to moving forward with quantum algorithms and applications (Wang et al., 2023). In high-performance computing (HPC), the integration of advanced semiconductor devices is at the center of providing exceptional processing power. HPC computers need ongoing developments in speed, power, and capacity to process vast amounts of data. To this end, semiconductor device design has evolved to capture computational power through architectural improvements, such as stacking levels of processors and using advanced cooling techniques. Furthermore, the incorporation of nanophotonics is critical to expand the communication bandwidth between processors, which can significantly address latency—a prime concern for real-time data processing in numerous applications, from climate modeling to financial forecasting. The focus on optimizing the conversion of electrical-to-optical signals ties into the integration of photonic devices with semiconductor technology, underscoring their significance in next-generation HPC environments (Dalton et al., 2023). As the semiconductor ecosystem transforms to address these new challenges, it should be taken into consideration the limitations imposed by traditional scaling methods, which are usually referred to as Moores Law. While chip size and power dissipation limitations are already on the horizon, researchers begin to look beyond Moore methods that incorporate novel materials and device structures in order to sustain computational performance progress. The marriage of semiconductor technology and quantum computing is particularly promising, as researchers investigate the properties of hybrid structures comprised of organic and inorganic materials that make stable, high-performance devices. Hybridization not only opens up new pathways for quantum bit control but also renews the science of high-performance computing through improved device efficiency through elaborate material combinations. The assurance of going beyond current limitations makes this research a frontier of semiconductor device design (Dalton et al., 2023; Wang et al., 2023).

1.5. CHALLENGES IN SEMICONDUCTOR DEVICE DESIGN

The transition from conventional semiconductor devices to emerging technologies presents monumental challenges that designers must overcome. One is the increasing complexity of device structures, often with innovative solutions to maintain performance and efficiency while keeping power consumption in check. As semiconductor products continue to evolve to address advanced applications in artificial intelligence (AI), the Internet of Things (IoT), and quantum computing, there is the daunting challenge of combining different functions into a compact form factor.

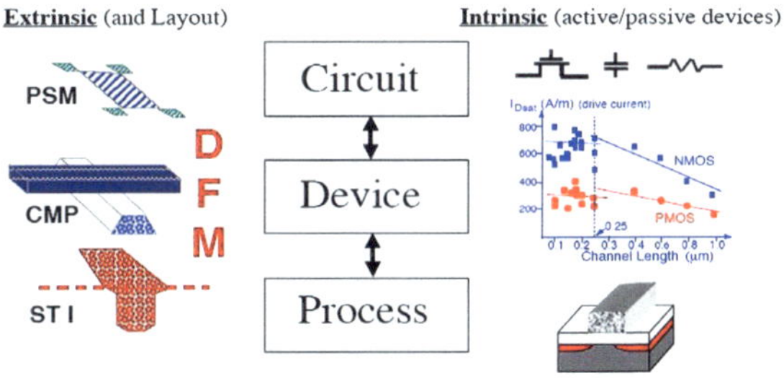

Figure 1.12. *Challenges in the behavior of semiconductor devices based on fundamental physics.*

Source: https://upload.wikimedia.org/wikipedia/en/5/5a/TCAD-overview.JPG.

Miniaturization produces intricate design issues with the requirement of thermal management and reliable interconnects. Growing demand for high-performance computing compounds these issues, with designs having to optimize for speed as well as data handling capabilities. Addressing these requirements requires a design approach to semiconductors that not only encourages innovation but also interprets the trends that will define the future direction of the industry. Addressing the limitations imposed by conventional scaling techniques, semiconductor designers are increasingly turning to beyond-

Moore approaches. These approaches seek to enhance device performance without relying on the outdated scaling laws, which have begun to approach their limits. One of the largest challenges here is handling the vast amounts of data produced via the design and simulation processes (Figure 1.12).

NanoCMOS project work highlights pertinent issues in multi-TByte data volumes, which will probably be enormous hurdles to proper data management and analysis (Asenov et al., 2009). Thus, semiconductor design needs sophisticated software solutions to make this selection of these datasets with specific care, particularly when they are structured together across institutions with varied expertise. With increasing necessity for inter-institutional collaboration, there is an increased need for imparting secure and effective operating mechanisms that minimize data handling without compromising on design integrity. With semiconductor technology venturing into new frontiers, inter-disciplinary collaboration is a prerequisite for overcoming device design-related issues. Recent developments, such as the electronic vibration bracelet for the visually and hearing impaired, illustrate the potential of new semiconductor applications to adapt and the requirement for modular design (Ababei et al., 2019). Such a project emphasizes the importance of developing systems that not only perform a purpose but also can provide room for future upgrades. Adapting is crucial as designers try to deal with evolving consumer demands and technological advancements. Consequently, as the technology further develops, fostering a culture of collaboration among engineers, researchers, and developers will be instrumental in overcoming the numerous challenges faced in semiconductor device design and propelling innovation under competitive conditions.

1.5.1. Scaling Limits and Physical Constraints

The relentless quest for miniaturization defines the development of semiconductor devices from conventional design to intricate architectures optimized for emerging applications. As transistors scale down to atomic dimensions, scaling limits become more obvious due to the physical limits of materials and quantum phenomena. Some of these effects include short-channel effects and quantum tunneling, which make it harder to miniaturize further and need innovative design methods. The trend is not only of academic significance but also directly impacts actual applications in artificial intelligence, the Internet of Things (IoT), quantum computing, and high-performance computing, where small yet powerful devices are of highest importance. The transition to models accommodating these challenges will require rethinking device structures and materials to enable the preservation of key performance characteristics, such as speed and power efficiency, in light of these constraints. The scaling bounds challenge is not merely a matter of transistor size but entails going into the field of materials science as well. Semiconductor material properties in nanoscale dimensions can lead to device performance-compromising effects like increased leakage currents and reduced channel control. Beyond-Moore technologies have emerged as a solution to these limitations, capitalizing on alternatives like quantum bits in quantum computing and low-energy neuromorphic computing devices for artificial intelligence systems. These structures leverage novel materials, such as 2D materials and organic semiconductors, to transcend traditional silicon constraints to fulfill the particular demands of today's technology. With the semiconductor design landscape unfolding, researchers and engineers face a challenge of developing architectures that can function within novel physical limits, initiating a renaissance in device innovation. Achieving these scaling limits and physical constraints is an arena of various complexities and interdependencies. Multidisciplinary collaborative work among computer science, material science, and electrical engineers is critical in formulating sustainable solutions. Project nanoCMOS is a case in point of how this is attained, not only focusing on device performance but also security features demanded by the sophisticated ecosystems encircling next-generation electronics (Asenov et al., 2008). In the progress of such projects, they present insightful observations regarding the handling of the vast amounts of data in simulations and designs. Besides, the utilization of sophisticated security frameworks highlights the multidisciplinary aspects of today's semiconductor development, implying that fulfilling the requirements of future applications is a holistic process with both physical and operational undertones (Green et al., 2012).

1.5.2. Thermal Management and Reliability Issues

The intricate relationship of thermal management and the reliability of semiconductor chips is of utmost significance, particularly as these devices evolve from conventional structures to future-generation chips. The rapid progress of applications in areas such as artificial intelligence, Internet of Things (IoT), quantum computing, and high-performance computing necessitates a rethinking of thermal management paradigms. As devices shrink, thermal dissipation becomes the main concern because, most of the time, smaller geometries mean higher power densities. In these scenarios, suitable thermal management solutions have to be there in order to be able to maintain performance as well as prevent cataclysmic failure by way of overheating, which tends to reduce device reliability. Thus, solutions toward better thermal performance must be integrated into the design process so that the new technologies can withstand demanding operation conditions without stretching beyond thermal limits. In addition, semiconductor product reliability also relies on thermal parameters, whose utility has gained more attention as the industry seeks to beyond-Moore advances. As highlighted, the mutual relationship of thermal management techniques, such as junction temperature estimation and thermal regulation, has a direct influence on device lifespan and performance stability. Traditional reliability evaluation techniques are likely to overlook these thermal issues, while the economic cost thereof might be incurred through device failures. Emergent geometries and materials, such as gallium oxide power devices, illustrate the pressing need for specialized thermal management strategies that can address existing needs in the semiconductor sector. In particular, the methodologies offered to optimize gallium oxide device performance specifically respond to thermal management issues, as a greater industry shift toward building thermally robust devices aimed at combating performance degradation over time (Ren, 2024). Moreover, examination of the evolving topography of thermal network models gives us a glimpse into shortcomings and prospects of current reliability improvement techniques. A sharp examination reveals that despite numerous means having been developed to enhance device reliability through thermal parameters, there exist gaps in applying and leveraging them. Current thermal models help predict how temperature variations influence device performance but rely primarily on the desired application and operational conditions. Better predictive models and better integration of thermal management techniques also contribute to overall device reliability. As semiconductor devices further impact various fields, the search for new cooling technologies and creation of better reliability assessment frameworks will be critical to counteract the thermal management issue and deliver long-term performance in future applications (Shen et al., 2024).

1.5.3. Economic and Environmental Considerations

As semiconductor technology continues to advance, the economic and environmental implications of device design become increasingly important. Over the last few years, the shift away from conventional devices toward next-generation devices has highlighted the need for cost-effective manufacturing methods that include minimal waste and energy consumption. The rising need for better applications in Internet of Things (IoT) and artificial intelligence (AI) necessitates a careful examination of the impact of semiconductor production on both economic efficiency and sustainable environment. With the emphasis on producing novel materials and manufacturing techniques, manufacturers are able to potentially reduce their carbon footprint without sacrificing performance requirements of high-performance computing. This dual focus on economic efficiency and sustainability will be the determining factor in shaping the direction of semiconductor device design in the future. Moreover, as growing numbers of industries rely on high-performance computing and new technology such as quantum computing, energy-efficient and scalable semiconductor architectures become a necessity. Traditional scaling techniques are hampered by physical limits, and future trends need to explore beyond-Moore that not just boost performance but also reduce energy consumption and operational costs. Among these approaches are introducing low-power design practices, as envisioned in Nagaraj (2025), where there is immense reduction in power wastage in verification, thus more environmentally sustainable device operation. These practices not only counterbalance the economic burden of increasing energy needs but also support overall environmental goals, driving the semiconductor industry toward green initiatives. The region of intersection between the environment and economics in semiconductor design ultimately enforces the responsibility upon engineers and scientists to innovate responsibly. By incorporating eco-friendly practices such as seawater electrolysis for energy needs as quoted by Wang et al. (2025), the semiconductor industry can work towards meeting global demands for energy-saving devices and minimizing its impact on the environment. Sustainability focus must extend beyond initial design choices to the whole life cycle of semiconductors, from production to disposal. This negrated approach ensures that the updates in semiconductor technology not only enhance economic growth but also enrich the efforts in environmental conservation, opening doors to future development in the industry.

1.6. EMERGING TRENDS IN SEMICONDUCTOR TECHNOLOGY

Accelerated development of semiconductor technology has been characteristic of a fundamental change from conventional to next-generation devices, consistent with the rapidly evolving character of the technology. Semiconductor devices have evolved conventionally by scaling down transistor size, leading to enhanced performance and reduced power consumption. But as these devices approach physical boundaries, the industry is shifting toward more extreme materials and architectures outside traditional scaling dogmas. For instance, novel materials such as graphene are being explored for their ability to alter electrical properties and performance parameters, particularly for high-frequency applications. Similarly, process technology advancements allow for more complex device structures to be produced with high heat dissipation capabilities and enhanced electron mobility, enabling the creation of smaller, faster, and more efficient devices to address the growing demands of modern technology (Lixue et al., 2018). Artificial Intelligence (AI), Internet of Things (IoT), quantum computing, and high-performance computing are influencing the way semiconductors are being designed significantly. The proliferation of many AI applications has given rise to chips targeted at machine learning algorithms, with parallel processing and power efficiency emerging as essential. Also, IoT devices, characterized by variability in functionality and sensitivity of sensors, require semiconductors that are able to deliver low power consumption but high performance across a large number of devices (Figure 1.13).

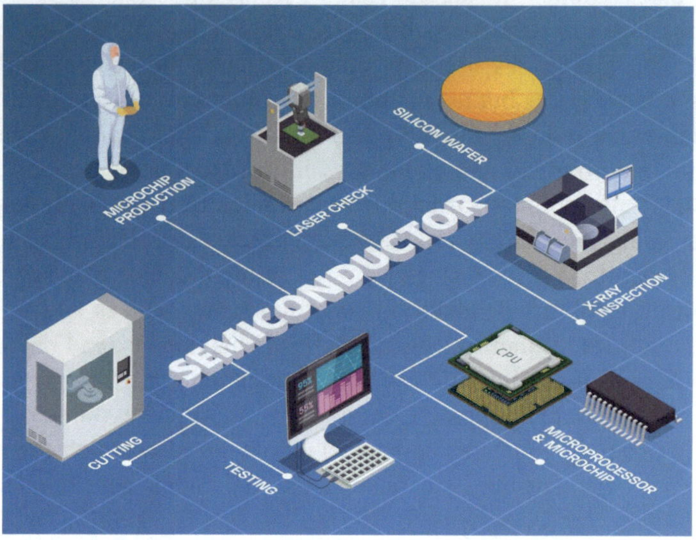

Figure 1.13. *The advancement in semiconductor technology.*

Source: https://inc42.com/cdn-cgi/image/quality=75/https://asset.inc42.com/2024/02/Untitled-design-95.png.

Quantum computing is a paradigm shift, beyond what can be achieved by conventional semiconductors and requiring new designs that leverage quantum effects to support improved computing. These tendencies indicate a transition not only to more powerful devices but to specialized semiconductor solutions to meet specific requirements of various advanced technologies (He et al., 2017). With advances in semiconductor technology research and development, there is a strong trend towards investigating beyond-Moore trends that target the constraints of conventional scaling approaches. Methods like 3D integration, heterogeneous integration, and neuromorphic computing are rising to prominence, threatening conventional paradigms and encouraging a more complete system approach to device design. These technologies support enhanced performance and capability, combining different types of materials and components to form more advanced systems-on-chip (SoCs). Additionally, with energy efficiency becoming more significant, especially in an increasingly sustainability-conscious age, the semiconductor design is transforming to minimize power consumption and maximize output. This convergence of materials science, engineering, and computer science is essential to the semiconductor industry's success in sustaining its success in an era characterized by rapidly advancing technology and growing demands for applications (He et al., 2017).

1.6.1. Beyond-Moore Trends and Innovations

The semiconductor industry has undergone a revolutionary shift, from conventional transistor design toward future-oriented devices that are progressively in the vanguard of meeting the demands of advanced applications. This environment requires new solutions in device architecture and material composition from technologies such as quantum computing, artificial intelligence (AI), and the Internet of Things (IoT). With traditional scaling of CMOS transistors approaching physical boundaries, there is a growing need for new designs. This environment has driven the innovation of concepts such as heterogeneous integration and three-dimensional (3D) packaging that allow different technologies to coexist and function together. These methods maximize performance without detracting from energy efficiency, therefore addressing the more sophisticated requirements of modern computing systems, particularly where rapid processing and data handling are critical in applications well beyond ordinary computer function. Realization of next-generation semiconductor technologies must go hand-in-hand with steps that most advantageously take advantage of their unique strengths. Modularity, open innovation, and platform-based innovation are among the key business strategies identified for mapping this intricate terrain (Sturgeon, 2019). Through exploiting tested fundamentals and then integrating them with innovative device concepts and materials, firms can place themselves in an optimal position for success in evolving markets. New semiconductor technologies are not only going to enable the computational demands of machine learning and AI but also enable the deployment of intelligent devices and systems in the IoT space. With these technologies taking to the skies, the effect on manufacturing and resource allocation becomes pivotal, and it necessitates a revisit of the existing paradigms in semiconductor manufacturing and designing. The strategic options made possible by these innovations are a foundation on which industries can transform and thrive despite rapidly changing technology. Independent of the achievement realized through continuous adherence to Moore's Law, the semiconductor industry must cope with increasing complexity of next-generation devices and implications on performance. As previous 10 years of scaling strategies have demonstrated, device structure and materials progress are necessary to achieve improvements in capability and efficiency (Bohr et al., 2017). This places exciting frontiers such as spintronics and neuromorphic computing, which entail new operation paradigms that could substitute traditional CMOS paradigms, on the agenda. By exploring these post-Moore technologies, semiconductor designers can take advantage of new design paradigms that go beyond the growth in transistor density, leading to innovations that redefine expectations of performance. As the boundaries of semiconductor technology continue to expand, an understanding of the interplay between new-generation applications and new design paradigms will be necessary for the maintenance of momentum in this vital industry.

1.6.2. Integration of New Materials and Technologies

History has shown semiconductor devices traversing a remarkable journey from legacy architectures to increasingly sophisticated next-generation designs. Much of this has resulted from the introduction of new materials and advanced technologies facilitating increased performance and efficiency. For instance, the trend towards gate-all-around (GAA) architectures and future complementary field-effect transistors (CFET) promises spectacular improvements in device scalability and performance. Such nascent architectures take advantage of the unique properties of emerging materials, providing a blueprint for going beyond existing scaling roadblocks within the confines of the stringent specifications of advanced semiconductor processing technology. Additionally, advancements in design technology co-optimization (DTCO) and integration engineering are imperative to address the issues of complexity associated with these materials to guarantee that the transition to new architectures maintains the target reliability and efficiency and increases device capability horizons (Lamb, 2025). Within the realm of emerging applications such as artificial intelligence (AI), Internet of Things (IoT), and high-performance computing, the application of new materials and technologies is the central theme to producing devices that possess the capability to deal with the increasing needs of these fields. The capability of semiconductor devices to maintain effective performance at decreasing sizes relies on advances in material science, particularly in the development of two-dimensional materials and other emerging composites. These materials also have better electronic properties, including higher charge mobility and reduced power consumption, and are thus the prospective building blocks of future transistors. The promise of materials like graphene and transition metal dichalcogenides does not just provide a means of improvement in performance but also provides a path for ultralow latencies needed in applications from real-time data processing to machine learning algorithms (Long et al., 2022). As market participants look for the embrace of newer technology, they encounter challenges and opportunities that will shape the future of semiconductor design. The push towards heterogenous integration systems is a good example of how traditional semiconductor manufacturing methods are being adapted to accommodate multiple functionalities within a compact architecture. Through integration of display, sensing, communication, and computing elements into unit systems, the semiconductor market is on the cusp of revolutionizing with performance and efficiency mantra. However, this vision calls for innovative solutions to device miniaturization and maturity in advanced interconnection technologies. As the industry wrestles with such demands, attention should also be given to developing materials that will be able to enable such integration and satisfy increasingly stringent performance needs (Long et al., 2022). Thus, the merging of new material and new technology is simultaneously a response to and stimulus for the changing nature of semiconductor device design so that future advances are capable of sustaining the rising pace of technological innovation.

1.6.3. Future Directions in Device Design

The future of semiconductor device design itself is in radical transition with directions towards future devices that break traditional scaling paradigms. With the industry contending with challenges that come with Moores Law, fresh remedies are being explored to expand functionality as room is being created for the exploding needs of applications in artificial intelligence (AI), Internet of Things (IoT), and quantum computing. Researchers are researching material and architecture beyond silicon, such as 2D materials and nanostructures, that have the potential to release enhanced performance. These advances are triggered by the need for devices that will handle and process vast amounts of data in real time, critical to AI applications. The advent of such technologies is a paradigm shift in device capability, bringing to the table possibilities previously considered impossible, and thus remapping semiconductor devices as they are conceived and designed. In addition to new materials, device fabrication is now fueled by concepts such as high-performance computing and beyond-Moore paradigms. High-performance uses intensify the need for more energy-efficient devices that still deliver unmatched computing capability. New technologies like three-dimensional (3D) stacking and heterogeneous building block integration open new routes to surmount traditional limitations of planar architectures. Furthermore, the integration of advanced functionalities like sensing and communication on the same chip is a significant future direction for device design. As highlighted by recent industry trends, the combination of various technologies in a single semiconductor architecture not only improves performance but also paves the way to wiser, more responsive devices. The trend is in line with ongoing efforts at making systems not only faster but greener. Not less important is the emphasis on people-centric design principles in the development of semiconductor devices. With the increase in device and application sophistication, considerations of ergonomic design to device usability have become imperative. Incorporation of cognitive ergonomics as evidenced in efforts to reduce cognitive load facilitates improved user interaction with advanced systems, ultimately leading to maximum operational efficiency. Wearable technologies, which are advancing at high speeds, offer a good case in point of this trend toward user-centric designs that are health- and well-being-centric, such as hydration monitoring devices (Belabbaci et al., 2024). The future of device design will largely be a combination of technological advancement with the requirement of humanity, with the understanding that as devices become more powerful, they still have to be used and useful to a growing number of users, eventually driving the next revolution in semiconductor technology.

SUMMARY

The quest for semiconductor device design has undergone great transformation, particularly as we transition from conventional implementations to next-generation devices as part of contemporary technological advances. This evolution is characterized by the ongoing pursuit of increased efficiency, miniaturization, and scaling of performance, which are essential to uses in AI, IoT, quantum computing, and high-performance computing. Usage in these technologies calls for devices and systems that can operate at record-breaking speeds and with enhanced capabilities, thus pushing the boundaries of semiconductor technology. Understanding the meaning of this change not only advances our appreciation of device structure but also sets the stage for dealing with the challenges that are brought by the limitations of current scaling technologies, particularly in energy consumption and thermal management. Limitations in semiconductor device scaling have become more apparent as the world moves into a post-Moores Law paradigm. Historically, Moores Law had dictated the exponential increase in transistor density on integrated circuits, but today's innovations indicate that new methods have to be pursued in order to continue performance enhancement. Although traditional scaling methods comply with thermal and electrostatic limitations, the integration of beyond-Moore technologies offers new pathways. The new developments include using high-tech materials, three-dimensional structures, and novel device architectures that tap quantum effects to develop enhanced functionality. These techniques not only aspire to address the performance bottlenecks encountered in devices that are miniaturized but also revolutionize the world of semiconductor design, allowing for more complex and efficient electronic systems. An understanding that is appreciative of these innovations and bottlenecks suggests towards the necessity for designers to remain agile with their approach. Because semiconductor technology continues to advance, it is essential to incorporate new concepts from other sciences, such as physics and materials science, to properly solve design problems. Knowledge of effects like single-electron effects, which have a significant impact on the low-frequency noise behavior of small-area MOSFETs, is essential for circuit designers who want to maximize performance (Hoekstra et al., 2007). Thus, the future of semiconductor device design is to an integrated method that not only embraces new materials and architecture but also leverages new technologies to carry out practical solution-finding. This end-to-end insight can be a blueprint for constructing efficient and faultless devices in a speeding digital world (Green et al., 2012).

Throughout the history of semiconductor devices, the evolution from legacy technologies to next-generation designs has been marked by state-of-the-art approaches and methodologies. As semiconductor technology advances, performance enhancement has been top of mind along with energy optimization. Low-power design verification has become more dominant in this new era, particularly in application domains such as AI and HPC. By applying concepts like Design for Test (DFT) and the use of power gating techniques, designers are able to create semiconductor devices that meet the high performance and functional specifications of contemporary applications. Such focus on energy efficiency serves not only to optimize the performance of semiconductor devices but also supports greater goals in the realm of sustainability in technology development, thereby securing the importance of verification processes in the design process (Nagaraj, 2025). With the emergence of different application domains such as the IoT and quantum computing, semiconductor design has shifted increasingly to discover new paradigms that diverge from traditional scaling limits.

As applications require increasing processing speed and improved energy efficiency, semiconductor designs must employ innovative approaches to tackle these. The research into post-Moore trends is significant in this context, as it entails advanced materials and novel transistor schemes that have the capability to surpass the scaling boundaries of the planar technology. Integrating techniques that enhance the functionality of the device, including multi-gate transistors and 2D materials, enables designers to stretch performance boundaries without reducing power levels. This complex interplay of design techniques and emerging technology directions is at the center of the future of semiconductor devices (Yang et al., 2024). In summary, the domain of semiconductor design is going through a revolutionary phase that involves the union of sophisticated technologies and techniques domains of special significance such as power efficiency, low-power design verification, and beyond-Moore strategies are of utmost importance to address the growing needs of modern applications. As AI, IoT, and HPC technology evolve, it is increasingly important for semiconductor architects to embrace such new approaches. Each step toward semiconductor architecture optimization must also include sustainable strategies, creating a future wherein designs are not only energy-efficient in use but contribute positively to overall environmental objectives. The ongoing emphasis on enhancing semiconductor device design underlines the critical role of this field in shaping technological innovation and addressing the evolving needs of the global marketplace (Nagaraj, 2025).

REFERENCES

1. Ababei, C., Conley, K., D'Souza, J., et al. (2019). Vibration alert bracelet for notification of the visually and hearing impaired. https://core.ac.uk/download/322826217.pdf (accessed on 27th September 2025).
2. Al Tareq, A., Rana, M. J., Mostofa, M. R., & Rahman, M. S. (2024). Impact of IoT and embedded system on semiconductor industry: A case study. *Control Systems and Optimization Letters*. https://www.semanticscholar.org/paper/73865add8fb002f730d1c48eb46a33aee599db62 (accessed on 27th September 2025).
3. Al-Sammak, K., Al-Gburi, S., Marghescu, I., Dragulinescu, A., Marghescu, C., & Al-Sammak, N. A. H. (2024). An experimental study of power consumption in narrowband IoT devices. *2024 15th International Conference on Communications (COMM)*. https://www.semanticscholar.org/paper/8165dc21f3a8ded0bf5575d72987584682b0cdba (accessed on 27th September 2025).
4. Arsid, R. (2025). Evolution of transistor architecture: From FinFET to gate-all-around technology. *International Journal of Advanced Research in Science, Communication and Technology*. https://www.semanticscholar.org/paper/ab9c1c1ad70fe536eff213e3955611a13e60f984 (accessed on 27th September 2025).
5. Asenov, A., Bayliss, C., Davenhall, C., et al. (2008). Integrating security solutions to support nanoCMOS electronics research. https://core.ac.uk/download/1397145.pdf (accessed on 27th September 2025).
6. Asenov, A., Davenhall, C., Harbulot, B., et al. (2009). Data management of nanometre-scale CMOS device simulations. https://core.ac.uk/download/1397183.pdf (accessed on 27th September 2025).
7. Barry, L. P., Bradley, A. L., Donegan, J. F., et al. (2002). Two-photon-induced photoconductivity enhancement in semiconductor microcavities: A theoretical investigation. https://core.ac.uk/download/11308033.pdf (accessed on 27th September 2025).
8. Belabbaci, N. A., Anaadumba, R., & Ul Alam, M. A. (2024). Recent advancements in wearable hydration-monitoring technologies: Scoping review of sensors, trends, and future directions. *JMIR mHealth and uHealth, 13*. https://www.semanticscholar.org/paper/a620b5e843aac48422483877a28ce0fb99513f86 (accessed on 27th September 2025).
9. Blais, A., Grimsmo, A. L., Girvin, S. M., & Wallraff, A. (2021). Circuit quantum electrodynamics. *Reviews of Modern Physics, 93*. https://doi.org/10.1103/revmodphys.93.025005 (accessed on 27th September 2025).

10. Bohr, M., & Young, I. A. (2017). CMOS scaling trends and beyond. *IEEE Micro, 37*, 20–29. https://doi.org/10.1109/mm.2017.4241347 (accessed on 27th September 2025).
11. Bonani, F., Camarchia, V., Cappelluti, F., et al. (2008). When self-consistency makes a difference. https://core.ac.uk/download/pdf/11399348.pdf (accessed on 27th September 2025).
12. Chakar, J., Julien, A., Medjoubi, K., Posada, J., Guillemoles, J., Puel, J. B., & Bonnassieux, Y. (2023). Advanced characterization and degradation analysis of perovskite solar cells using machine learning and Bayesian optimization. *2023 IEEE 50th Photovoltaic Specialists Conference (PVSC)*. https://www.semanticscholar.org/paper/286afd60353fa08386fb339902c7cb156e793714 (accessed on 27th September 2025).
13. Crawford, G. P. (2023). *Flexible Flat Panel Displays*. Wiley eBooks. https://doi.org/10.1002/9781118751077 (accessed on 27th September 2025).
14. Dalton, L. R., Leuthold, J., Robinson, B. H., Haffner, C., Elder, D. L., Johnson, L. E., Hammond, S. R., et al. (2023). Perspective: Nanophotonic electro-optics enabling THz bandwidths, exceptional modulation and energy efficiencies, and compact device footprints. *APL Materials, 11*. https://doi.org/10.1063/5.0145212 (accessed on 27th September 2025).
15. Freund, R., Zaremba, O., Arnauts, G., Ameloot, R., Skorupskii, G., Dincă, M., Bavykina, A., et al. (2021). The current status of MOF and COF applications. *Angewandte Chemie International Edition, 60*, 23975–24001. https://doi.org/10.1002/anie.202106259 (accessed on 27th September 2025).
16. Gaddipati, K. (2025). Navigating the architectural shift: RibbonFET implementation strategies for next-generation analog integrated circuits. *Journal of Computer Science and Technology Studies*. https://www.semanticscholar.org/paper/81538e86c735410bf25223a17f7a957aec5bb329 (accessed on 27th September 2025).
17. George, H. C., Mądzik, M., Henry, E., Wagner, A. J., Islam, M. M., Borjans, F., Connors, E. J., et al. (2024). 12-spin-qubit arrays fabricated on a 300 mm semiconductor manufacturing line. *Nano Letters, 25*, 793–799. https://www.semanticscholar.org/paper/a4e515bc6239a0b3911d0188957117ef794ad3d6 (accessed on 27th September 2025).
18. Green, T. C., & Mitcheson, P. D. (2012). Maximum effectiveness of electrostatic energy harvesters when coupled to interface circuits.

19. He, J., & Tritt, T. M. (2017). Advances in thermoelectric materials research: Looking back and moving forward. *Science, 357*. https://doi.org/10.1126/science.aak9997 (accessed on 27th September 2025).

20. Hilmi, A., Abdul Hamid, A. R., & Wan Ibrahim, W. A. R. A. (2024). Human-centered ergonomics: Advancements, challenges, and future directions in industrial and occupational settings. *Malaysian Journal of Ergonomics (MJEr)*. https://www.semanticscholar.org/paper/7ec03a5026e519676d19222ebf411c40fcb9b7a0 (accessed on 27th September 2025).

21. Hoekstra, E., Klumperink, E. A. M., Kolhatkar, J. S., et al. (2007). Low-frequency noise phenomena in switched MOSFETs. https://core.ac.uk/download/pdf/11461852.pdf (accessed on 27th September 2025).

22. Kumagai, S., Ishii, H., Watanabe, G., Yu, C. P., Watanabe, S., Takeya, J., & Okamoto, T. (2022). Nitrogen-containing perylene diimides: Molecular design, robust aggregated structures, and advances in n-type organic semiconductors. *Accounts of Chemical Research, 55*, 660–672. https://doi.org/10.1021/acs.accounts.1c00548 (accessed on 27th September 2025).

23. Kurniawan, R., Chen, S., Teng, H., Ko, T., Han, P., Kim, G., & Kiswanto, G. (2025). Reverse engineering of commercial ultrasonic saw blade cutter for cutting semiconductor material. *Seminar Nasional Tahunan Teknik Mesin XXII 2024*. https://www.semanticscholar.org/paper/6668f7b142c3f8ce3b4e963c1dc80aaf331afa54 (accessed on 27th September 2025).

24. Lamb, J. E. (2025). An integration engineering approach to material design and development. *13428*, 1342803–1342803-8. https://www.semanticscholar.org/paper/19c27747a1c3ca4095e0585a214d321a23866bf1 (accessed on 27th September 2025).

25. Liu, Y., Liu, X., Mu, X., Hou, T., Xu, J., Di Renzo, M., & Al-Dhahir, N. (2021). Reconfigurable intelligent surfaces: Principles and opportunities. *IEEE Communications Surveys & Tutorials, 23*, 1546–1577. https://doi.org/10.1109/comst.2021.3077737 (accessed on 27th September 2025).

26. Lixue, Z., Wang, L., Wu, Y., Ma, C., Yu, S., & Liu, X. (2018). Trends analysis of graphene research and development. *Journal of Data and Information Science, 3*, 82–100. https://doi.org/10.2478/jdis-2018-0005 (accessed on 27th September 2025).

27. Long, H., Zhang, L., Ma, H., Li, J., Xia, J., Zhang, Y., & Chen, J. (2022). Heterogeneous integration system in display (HiSID) for next-generation terminal device. *IEEE Transactions on Components, Packaging and Manufacturing Technology, 12*, 731–739. https://www.semanticscholar.org/paper/6ebcde12d06b5024acc3886bfcee88b4e7d88d5b (accessed on 27th September 2025).

28. Lu, Y., Wang, J. Y., & Pei, J. (2021). Achieving efficient n-doping of conjugated polymers by molecular dopants. *Accounts of Chemical Research, 54*, 2871–2883. https://doi.org/10.1021/acs.accounts.1c00223 (accessed on 27th September 2025).

29. Macco, B., & Kessels, W. M. M. (2022). Atomic layer deposition of conductive and semiconductive oxides. *Applied Physics Reviews, 9*. https://doi.org/10.1063/5.0116732 (accessed on 27th September 2025).

30. Mii, Y. J. (2024). Semiconductor industry outlook and new technology frontiers. *2024 IEEE International Electron Devices Meeting (IEDM)*. https://www.semanticscholar.org/paper/2c79f51a8df8028287ea1dc761875b39d00a8849 (accessed on 27th September 2025).

31. Miles, R. E., Rudge, P. J., Snowden, C. M., et al. (2001). Investigation into intermodulation distortion in HEMTs using a quasi-2-D physical model. https://core.ac.uk/download/58699.pdf (accessed on 27th September 2025).

32. Nagaraj, V. (2025). Ensuring low-power design verification in semiconductor architectures. *Journal of Information Systems Engineering and Management*. https://www.semanticscholar.org/paper/83424cdea7faf23da0b37dc4f93a611c210a976c (accessed on 27th September 2025).

33. Navidpour, A. H., Abbasi, S., Li, D., Mojiri, A., & Zhou, J. L. (2023). Investigation of advanced oxidation process in the presence of TiO_2 semiconductor as photocatalyst: Property, principle, kinetic analysis, and photocatalytic activity. *Catalysts*. https://www.semanticscholar.org/paper/a570d05f84b64efd07e21b8c74015a514b125054 (accessed on 27th September 2025).

34. Nguyen, T. T. H., Nguyen, M., Huynh, M. A., Vu, H. H., Nguyen, T. K., & Nguyen, N. T. (2023). Field effect transistor based wearable biosensors for healthcare monitoring. *Journal of Nanobiotechnology, 21*. https://doi.org/10.1186/s12951-023-02153-1 (accessed on 27th September 2025).

35. Piskun, G., Alexeev, V. F., Stsepchankou, A. V., Popov, A. N., Belikov, A. N., & Rybakov, D. G. (2023). Effect of the configuration and shape of external ribs of sealed enclosures of electronic devices on heat removal efficiency. *Journal of the Russian Universities. Radio Electronics*. https://www.semanticscholar.org/paper/71eacdcf3c86749f482a852c473c8696e401e2c2 (accessed on 27th September 2025).

36. Porter, M., Yang, X., Gong, H., Wang, B., Yang, Z., & Zhang, Y. (2024). Switching figure-of-merit, optimal design, and power loss limit of (ultra-) wide bandgap power devices: A perspective. *Applied Physics Letters*. https://www.semanticscholar.org/paper/

bdbf4840907bd351f4e9eb0de4b109c8d678d401 (accessed on 27th September 2025).

37. Rasheed, A., San, O., & Kvamsdal, T. (2020). Digital twin: Values, challenges and enablers from a modeling perspective. *IEEE Access, 8*, 21980–22012. https://doi.org/10.1109/access.2020.2970143 (accessed on 27th September 2025).
38. Ren, S. (2024). Research on reliability issues of gallium oxide power devices. *2024 3rd International Symposium on Semiconductor and Electronic Technology (ISSET)*. https://www.semanticscholar.org/paper/08080f829752f1e6deddf1c6d8092cefb3f4a378 (accessed on 27th September 2025).
39. Ren, S., Zhang, W., Wang, Z., Yassar, A., Liao, Z., & Yi, Z. (2023). Synergistic use of all-acceptor strategies for the preparation of an organic semiconductor and the realization of high electron transport properties in organic field-effect transistors. *Polymers, 15*. https://www.semanticscholar.org/paper/975182c32bdb4c40bf7d0b25df130ae073e49db6 (accessed on 27th September 2025).
40. Shaheen, G. M. (2000). Development of power semiconductor devices database for knowledge-based systems. https://core.ac.uk/download/43000491.pdf (accessed on 27th September 2025).
41. Shen, H., Zhang, J., Sun, P., & Ma, X. (2024). Review on the thermal parameter's applications in the reliability of power semiconductor device. *2024 IEEE 7th International Electrical and Energy Conference (CIEEC)*. https://www.semanticscholar.org/paper/cf7016451981a362477da86f67dd82738ef7fa4b (accessed on 27th September 2025).
42. Singh, A., Kim, D., & Lee, B. (2025). Analysis of a memcapacitor-based online learning neural network accelerator framework. *Advanced Intelligent Systems*. https://www.semanticscholar.org/paper/7438fbf9b65b17df3226f1d963177d8773c62382 (accessed on 27th September 2025).
43. Soussi, A., Haounati, R., Ait Hssi, A., Taoufiq, M., Asbayou, A., Elfanaoui, A., Markazi, R., et al. (2023). First principal study of structural, electronic, optical properties of co-doped ZnO. *Journal of Composites Science*. https://www.semanticscholar.org/paper/ca5bba6f52ce240de1f7c21acbdaf79ddcf87107 (accessed on 27th September 2025).
44. Sturgeon, T. J. (2019). Upgrading strategies for the digital economy. *Global Strategy Journal, 11*, 34–57. https://doi.org/10.1002/gsj.1364 (accessed on 27th September 2025).

45. Swarnkar, A., Abdi, D., Kumari, B., Venugopal, P., Pantano, N., Hellings, G., Kulkarni, J. P., et al. (2025). Design technology co-optimization of 3D SRAM macro in nanosheet technology for high-bandwidth applications. *2025 IEEE International Memory Workshop (IMW)*. https://www.semanticscholar.org/paper/dca96d455d798febae3778ae47d49e66e6c237ba (accessed on 27th September 2025).
46. Tsymbalov, E., Shi, Z., Dao, M., Suresh, S., Li, J., & Shapeev, A. (2021). Machine learning for deep elastic strain engineering of semiconductor electronic band structure and effective mass. *npj Computational Materials, 7*. https://www.semanticscholar.org/paper/035c9bb707080f3d4e11e7140 39de5cb8bf0c70f (accessed on 27th September 2025).
47. Wang, C., Cusin, L., Ma, C., Unsal, E., Wang, H., Girelli Consolaro, V., Montes-García, V., et al. (2023). Enhancing the carrier transport in monolayer MoS_2 through interlayer coupling with 2D covalent organic frameworks. *Advanced Materials, 36*. https://doi.org/10.1002/adma.202305882 (accessed on 27th September 2025).
48. Wang, P., Xu, S., Shi, X., Zhu, J., Xiong, H., & Wen, H. (2025). Recent advances in resistive gas sensors: Fundamentals, material and device design, and intelligent applications. *Chemosensory*. https://www.semanticscholar.org/paper/a286109699127e6d4bfe6d8914f3159aa74964a0 (accessed on 27th September 2025).
49. Wang, T. (2025). Research progress and optimization strategy of trans-calcite solar cells. *Applied and Computational Engineering*. https://www.semanticscholar.org/paper/9e40fc0720f67aae9f0ad8205e507d5e971914 5e (accessed on 27th September 2025).
50. Wang, Y., Li, F., Zhao, L., Wang, Y., Yang, G., Tian, J., Heng, S., et al. (2025). Strategies for industrial-grade seawater electrolysis: From electrocatalysts and device design to techno-economic analysis. *Nanoscale*. https://www.semanticscholar.org/paper/efac9bfda309274b801a144757493a860e8cfa42 (accessed on 27th September 2025).
51. Yang, Z., Sun, L., Li, Z., Chen, X., Xie, Y., & Liang, S. (2024). Design of V-band impedance tuner for parameter measurement of microwave semiconductor devices. *2024 International Conference on Microwave and Millimeter Wave Technology (ICMMT), 1*, 01–03. https://www.semanticscholar.org/paper/23e36d2a536157aecb7803abe3f702a16aeaf ea9 (accessed on 27th September 2025).
52. Zeng, Z. (2024). A comprehensive analysis on CMOS comparators: Principles, design optimizations, and applications. *Highlights in Science,*

Engineering and Technology. https://www.semanticscholar.org/paper/6cbd7f1822f701dbc8ac59a0a325efc2e083a1e1 (accessed on 27th September 2025).

53. Zhang, H., Peng, Y., Zhang, N. W., Yang, J., Wang, Y., & Ding, H. (2022). Emerging optoelectronic devices based on microscale LEDs and their use as implantable biomedical applications. *Micromachines, 13*. https://www.semanticscholar.org/paper/753f1259d40d68e02f82b581cec3ac0e191c6b41 (accessed on 27th September 2025).

54. Zhang, J., Wang, Y., Lu, C., Zheng, K., Li, R., Zhao, J., Yin, R., et al. (2025). Evaluating the impact of FSDOI physical design strategies on automotive-grade semiconductor chip reliability and performance. *Journal of Computational Methods in Sciences and Engineering*. https://www.semanticscholar.org/paper/91edf111641ccf09903bedaef3f1308636390c73 (accessed on 27th September 2025).

CHAPTER 2

SEMICONDUCTOR MATERIAL SCIENCE

Contents

Introduction 47
2.1. Crystal Structures of Semiconductors 51
2.2. Doping in Semiconductors 56
2.3. Defects in Semiconductor Materials 61
2.4. Bandgap Engineering 66
2.5. Wide-Bandgap Materials 71
2.6. Two-Dimensional Materials 76
Summary 81
References 83

INTRODUCTION

The science of semiconductor materials is the pillar of today's electronic and optoelectronic innovations, which are undergoing significant phase transitions defined by advancements in materials and engineering processes. The magic behind this discipline lies in the crystal structure, doping processes, and defect engineering, whose functions in governing the electrical characteristics of semiconductors play a central role. Through skillful control of these factors, semiconductors can be engineered to meet specific performance levels required for certain applications, ranging from transistors to lasers. This characteristic is extremely robust in wide-bandgap semiconductors such as GaN, SiC, and diamond, which, due to high thermal stability and breakdown voltage, are most appropriate for high-frequency and high-power devices. Moreover, the recent discovery of new two-dimensional materials like graphene and MoS_2 further expanded the application scope of semiconductors, proving all the more relevant to the existing process of technological development. Being a period where performance with respect to energy and efficiency is given maximum importance, bandgap engineering emerges as an indispensable method applied in semiconductor engineering. By adjusting the energy gap, researchers can engineer materials for specific applications, such as lasers with specific wavelengths or technology that must have better electronic properties. Creating highly mismatched semiconductor alloys, such as GaNAs and GaBiAs, is just one of the fascinating possibilities in this field. These alloys possess unique electronic characteristics with large energy gap reduction within a brief period and greater spin-orbit splitting energy with better composition, all of which directly affect their functioning as laser material. GaInNAs telecommunication lasers, in accordance with recent texts, with recent progress now demonstrate performance levels equivalent to traditional devices and hence the importance of the study of new materials and their engineering methods for possible applications (Broderick et al., 2012). Finally, the growing interest in self-organization phenomena in thin film growth highlights a fundamental area of research that brings together basic science with technological promise. Advances in the field are transforming the synthesis of nanostructures such that both one-dimensional and zero-dimensional structures are now achievable that were previously inaccessible using conventional lithographic methods. These self-assembled systems not only contribute to the fundamentals of low-dimensional physics but also promise new applications in semiconductor technology. Integration of these nanostructured materials into device structures is anticipated to enhance performance and expand application fields, as illustrated in the recent special issue on self-organization in surface materials (Arakawa et al., 2005). The combination of these new methods in their interconnection makes semiconductor material science a dynamic and cohesive discipline that is poised to meet the challenges of tomorrow's technology.

Definition of Semiconductor Materials

The sophistication of semiconductor materials extends beyond their definitions and encompasses the intense involvement of crystal structures, doping, and defects. Semiconductor materials are characterized by their unique bandgap properties, which can be engineered in a way to design various electronic devices. The intrinsic crystal structure of a semiconductor determines its electrical as well as optical properties. One of the most significant methods for modifying the electrical conductivity of semiconductors is doping, which involves adding deliberate impurities. Doping silicon with phosphorus or boron, for instance, can create n-type or p-type material, respectively. These modifications are responsible for creating fundamental devices such as diodes and transistors, which form the heart of modern technology. The presence of defects can also be impacting the performance of semiconductor materials in a manner that intentional and unintentional changes are central to the technology of semiconductor materials. Bandgap engineering is another feature of semiconductor materials, which provides the capability to design the electronic and optical characteristics of materials for specific applications. Bandgap refers to the energy difference between a semiconductor's valence and conduction bands that dictates its conductivity and the wavelengths of light it will absorb or emit. By varying the composition and structure of semiconductor materials, engineers can create tailored bandgaps for nearly any variety of applications, from photovoltaic cells to light-emitting diodes (LEDs). For example, the combination of gallium and arsenic leads to the creation of GaAs, a direct bandgap semiconductor very appropriate for laser diodes. Bandgap engineering not only enhances the efficiency of already available technology but also provides new avenues of innovative usage in electronic and optoelectronic devices, hence expanding the scope of semiconductor material usage. Wide-bandgap semiconductors such as Gallium Nitride (GaN), Silicon Carbide (SiC), and diamond have gained a lot of attention for their high-power and high-frequency device potential. Wide-bandgap semiconductors are superior to traditional semiconductors such as silicon with narrower bandgaps in that they can function efficiently at high temperatures and under extreme conditions and are thus well-suited for use in fields such as power electronics and RF devices. Additionally, the discovery of two-dimensional materials, for instance, graphene and transition metal dichalcogenides like MoS_2, reflects the ongoing development of semiconductor material science. These materials have unique electronic and mechanical characteristics, which can be utilized in the design of the next generation of sensors and transistors. Hence, learning about these varied semiconductor materials and their properties is necessary to enhance technology and introduce innovation in the science of semiconductor material because it finds significance in material properties in its applications today.

Importance in Modern Technology

Semiconductor material science development plays a vital part in driving innovation in modern technology. At the center of this evolution is the understanding of crystal structure, doping, and defects in semiconductor materials, which affects their electronic properties quite profoundly. Doping is the incorporation of impurities into the crystal structure, thereby changing the electrical conductivity and enabling the production of p-type and n-type semiconductors that are central in producing electronic devices. Knowledge of defects is also important because although they can weaken material performance, they can strengthen it depending on their type and quantity. This interaction is crucial in devices ranging from integrated circuits to solar cells, where precision of these parameters determines device efficiency. With technology demanding compactness and enhanced performance, the roles of these parameters become more important in creating breakthroughs that can keep pace with the rapid rate of advancement in the semiconductor field. One of the central research fields for modern semiconductor research is bandgap engineering, which allows material scientists to tailor the energy band gap of the semiconductor materials according to specific applications. This capability has made it possible to produce wide-bandgap materials such as GaN and SiC, which are important for high-power and high-frequency applications. These materials not only increase thermal stability and minimize energy loss in devices but also widen the list of materials available for use in optoelectronic devices. Engineered with an optimal bandgap for performance, GaN and SiC are fast becoming indispensable in telecommunication and electric car technology (Figure 2.1).

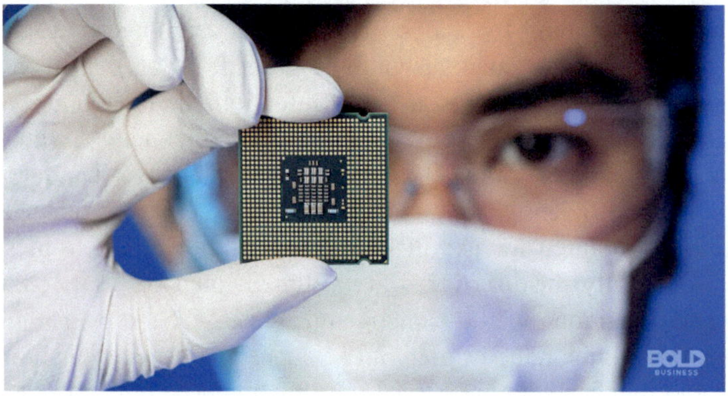

Figure 2.1. *The specification of semiconductor material science.*

Source:https://www.boldbusiness.com/wp-content/uploads/2019/05/Semiconductor_in-article.jpg.

The emergence of two-dimensional materials, including graphene and MoS_2, also constitutes an additional aspect to bandgap engineering because of their unique electronic properties, making possible new device geometries and highly efficient applications, being of vital importance for future microelectronics. Cutting-edge research demonstrates that the combination of different semiconductor compounds can lead to the formation of new compounds with even better properties, demonstrating the ongoing validity of semiconductor material science in modern technology. For instance, quaternary material synthesis such as SiBCN with the structural analogue of graphene shows immense potential using the wide direct bandgap of approximately 2.63 eV, between the bandgaps of traditional semiconductors such as SiC and GaN (Deng et al., 2017). Moreover, it is endowed with excellent carrier mobility, which is key to the realization of high performance in electronic devices. The ability to maintain such a structure at any temperature extremes marks the materials ultimate utility in varying high-performance conditions. With researchers discovering these new materials, their work underscores the necessity of semiconductor material science in the development of solutions that can blend into existing silicon-based technology, thus ushering future longevity in the world of modern technology (Singha et al., 2004).

2.1. CRYSTAL STRUCTURES OF SEMICONDUCTORS

The intricate interdependence between crystal structure and the performance of semiconductors is essential within semiconductor material science. Doping techniques are applied to enhance electronic properties by introducing impurities into the crystal lattice and thus altering electrical conductivity. Since the crystal structure of a semiconductor defines its bandgap and charge transport mechanisms, strict control of such structures is of paramount importance for high-end applications. For instance, successful dopant integration can enhance carrier concentrations and mobility, leading to improved device performance. Defects will make this situation more difficult, however. Lattice flaws have a tendency to disorder charge carrier channels, undermining the semiconductors efficiency. Material scientists can tailor semiconductor properties for specific applications by understanding how different crystal structures influence defect formation and doping efficiency. This will ultimately improve microelectronic and optoelectronic device performance. Engineering of the bandgap is another critical parameter influencing the performance of semiconductor materials. The bandgap defines the energy required for electrons to travel from the valence band to the conduction band, which directly influences a semiconductors ability to emit and absorb light. For instance, wide-bandgap semiconductors such as GaN, SiC, and diamond enable high-temperature and high-power device applications due to their good structural integrity and increased thermal stability (Figure 2.2).

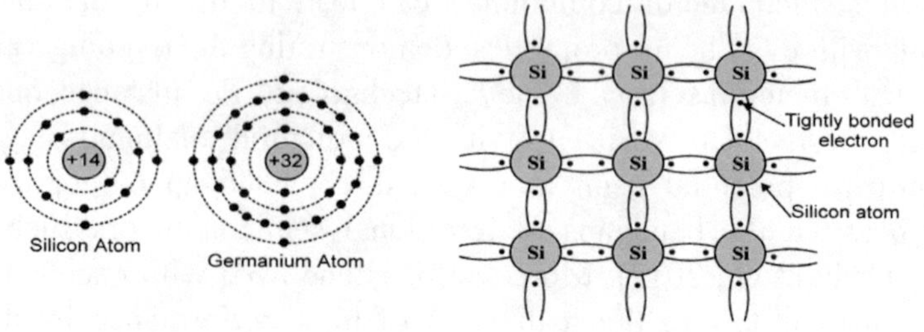

Figure 2.2. *Crystal structure and behavior of semiconductor.*

Source: https://vsa.edu.in/wp-content/uploads/2023/11/image.png.

Moreover, advanced two-dimensional materials, e.g., monolayer MoS_2, have also opened avenues for new applications due to their direct bandgap nature. As described by Berkelbach et al. (2013), a deeper study of the microstructure of such materials leads to greater crystallinity, which advances electronic and optical properties, therefore benefiting nanoscale electronics and optoelectronic devices to move forward. New architectures, such as photonic crystal structures, are found to be used to improve performance in semiconductors by developing newer designs. Planar photonic crystal fabrication, especially with honeycomb lattices, is of interest due to the ability to fabricate devices with complete in-plane photonic bandgaps for specific applications, as observed in Bahriz et al. (2007). This not only alleviates fabrication steps but also alleviates any potential semiconductor material damage, hence enhancing overall device reliability. With the advancement of semiconductor technology, examination of the integration of other crystal structures with advanced doping techniques will be the driving force for achieving new heights in the functionality of semiconductors. Lastly, a solid understanding of crystal structures, defects, and how they interact with external parameters is still the key to advancing the semiconductor industry to high-performance, cutting-edge solutions.

2.1.1. Common Crystal Structures (Diamond and Zinc Blende)

The relationship between semiconductor material properties and crystal structures is central to semiconductor material science. Diamond and zinc blende are two of the most universal structures, which are both distinguished by their exclusive atomic arrangements. The diamond crystal structure possesses a tetrahedral carbon atom coordination, which forms an extremely stable face-centered cubic lattice with excellent electronic characteristics. This structure not only contributes to diamonds renowned hardness but also constitutes a decent backdrop for semiconductor applications. In particular, diamond's broad bandgap makes it an excellent option for high-voltage and high-temperature electronic devices, consequently widening its range of application across varying technological fields. Proceeding to the zinc blende structure, arguably one of the most documented examples is indium phosphide (InP), which owes enormous significance in semiconductors' realm. This crystal lattice has alternating indium and phosphorus atoms in face-centered cubic structure with characteristic electronic properties required for optoelectronic devices. Unlike diamond, there is efficient electron mobility via zinc blende structure, making materials like InP suitable for high-performance photonic devices and high-frequency electronics. The internal characteristics of the zinc blende structure can also be even further improved by the process of doping, an activity that enhances the semiconductors functionality by modifying its electrical properties. Consequently, the versatility of zinc blende structures accommodates a wide range of applications, particularly for telecommunications and imaging technologies. Research on the electronic properties of diamond and zinc blende structures indicates encouraging topological properties that contribute to their functionality. Recent studies have proved that such semiconductor structures possess non-trivial topological character, particularly of their valence band topology. Such peculiar nature of the electronic band structure can lead to the formation of topological surface states, which are of immense significance for future electronics. For example, in GaAs and InSb materials, spin-orbit coupling action gives rise to strange phenomena that can potentially enhance the efficiency of semiconductor devices. These findings not only reinforce the relevance of crystal structure knowledge in semiconductor material science but also offer scope for novel engineering strategies that exploit the unique features of diamond and zinc blende systems to leapfrog current technology (Rauch et al., 2019; Salman et al., 2024).

2.1.2. Role of Crystal Structure in Electronic Properties

There is a complicated interplay between the crystal structure of semiconductor materials and their electronic characteristics, defining how such materials are utilized in applications. Crystal structure in semiconductor physics governs lattice arrangement of the atoms, influencing directly carrier mobility, band structure, and overall conductivity. Particularly, defects and doping can either facilitate or hinder these electronic attributes. For instance, new advancements in the growth of highly crystalline monolayer molybdenum disulfide revealed a close correlation between the orientation of the microstructures and the electronic behavior of the material. In even greater detail, triangular islands of this semiconductor become single crystals that show better optical and electric properties, showing that crystal boundaries play an important role in such phenomena as photoluminescence and conductivity, thus underpinning the importance of structural integrity towards enhancing electronic performance (Berkelbach et al., 2013). The electronic property engineering through bandgap engineering has emerged as a point of interest for semiconductor material science. Structural modification-based bandgap tailoring is possible, leading to electronic transition control and allowing materials to be suitable for some purposes, such as photovoltaics and light-emitting diodes. For example, the encapsulation of quantum dots in artificial ferromagnetic crystals demonstrates that unusual crystal structures can produce flat band features which lead to remarkable electronic phenomena like ferromagnetism at half-filling without incorporating magnetic ingredients. This demonstrates how nanoparticles are able to make use of unusual electronic behavior, by being prompted by crystalline structure. These advances increase the material stability for next-generation electronic device development in enabling tailored functionality based on complexities of crystal structures (Kenji et al., 2001). Wide-bandgap semiconductors such as Silicon Carbide (SiC) and Gallium Nitride (GaN) also demonstrate the significance of crystal structure in the characterization of electronic properties. Their lattice structure stability offers excellent electrical and thermal performance, making them ideal for high-temperature and high-power devices. The potential for engineering these wide bandgaps expands their usage across several fields, from power electronics to optoelectronics. Likewise, two-dimensional material such as graphene and molybdenum disulfide also have interesting properties closely tied to the nature of their crystal structure. Electronic effects manipulation by these materials through structural adjustments highlights the key role that crystal order has to the overall efficiency of semiconductors. Thus, understanding and leveraging these structural nuances is a continuous crucial undertaking in semiconductor material science, with the promise of innovation in many fields.

2.1.3. Techniques for Analyzing Crystal Structures

Crystal structure analysis is critical in semiconductor material science, particularly with regard to doping and defects that heavily influence material properties. Techniques such as X-ray diffraction (XRD) and transmission electron microscopy (TEM) are critical for establishing the atomic structure within a crystal lattice. XRD provides accurate information about the orientation of crystals, phase analysis, and the quality of the material. By diffraction pattern measurement, researchers can determine lattice parameters and observe defects, such as dislocations and vacancies, that can affect electronic properties (because defects can be centers for charge carrier scattering). TEM also enables atomic-level imaging so microstructural defects can be observed directly and their effects on charge transport measured quantitatively. These techniques are not only critical in revealing fundamental properties but also in developing materials based on some functional requirements in semiconductor devices. The evaluation of bandgap engineering techniques needs comprehensive knowledge of crystal structures. Bandgap is the energy difference between the valence band and the conduction band, which decides a semiconductors electrical and optical properties. Techniques such as photoluminescence spectroscopy (PL) and angle-resolved photoemission spectroscopy (ARPES) are significant in the assessment of energy levels and band structure in semiconductors. PL allows researchers to study the emission of light from a material upon excitation and can provide information regarding the energy levels for introduced defects or impurities via doping processes (Figure 2.3).

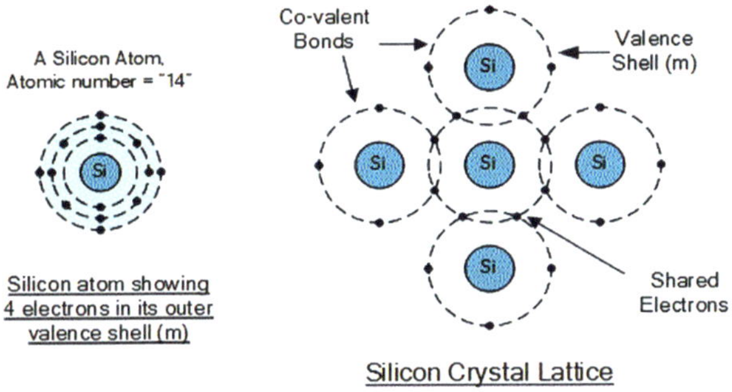

Figure 2.3. *The analysis of crystal structures in semiconductor material.*

Source:https://static.wixstatic.com/media/9d9fe8_79b46651aa3243ccbd32e00255 0025fe~mv2.gif.

Besides, ARPES can supply valuable information regarding the electronic structure and dispersion relations of semiconductors, which reflect the effects of dimensionality as well as any potential quantum effects. These techniques enable bandgap optimization to be conducted effectively, as it is vital for novel applications in optoelectronic devices and photonic materials. Keeping in mind the growth and characterization of wide-bandgap materials, crystal structure versus functional performance becomes increasingly significant. Materials like GaN and SiC are in the spotlight with their unique electronic and thermal properties making them the candidates for high-power and high-frequency devices. Thin film growth has also evolved with methods of molecular beam epitaxy (MBE) and chemical vapor deposition (CVD), allowing precision control of the layer thickness and composition. Techniques mentioned in newer literature stress the importance of self-organization processes that contribute to the formation of nanostructures with properties worthwhile for their functional characteristics (Arakawa et al., 2005). In addition, studying two-dimensional materials such as graphene and MoS_2 demands new characterization methods to reach the excellent electrical and mechanical characteristics. This interaction between multiple crystal characterization methods and the development of semiconductor materials provides a base for technological advances in the future (Boschker et al., 2016).

2.2. DOPING IN SEMICONDUCTORS

Doping of semiconductors is the most critical process in contemporary electronics, enabling the manipulation of electrical, thermal, and optical properties needed in device operation through the variation of the semiconductor's conductivity. Through the intentional introduction of impurities into a semiconductor crystal, the material's conductivity can be altered to produce an n-type or p-type semiconductor. It plays a critical role in the device manufacturing of diodes, transistors, and solar cells.

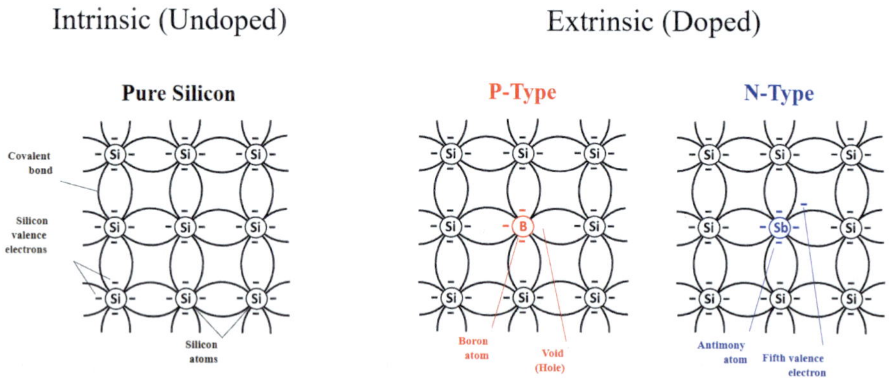

Figure 2.4. *The manipulation of doping in semiconductor properties.*

Source: https://upload.wikimedia.org/wikipedia/commons/5/58/Silicon_doping_-_Type_P_and_N.svg.

Accurate control of doping concentration and distribution is required to achieve favorable electrical properties without the occurrence of defects that are detrimental to carrier mobility. Surprisingly, advances in the technologies of doping have recently improved tremendously using ion implantation and molecular beam epitaxy technologies, significantly increasing the yield of high-quality semiconductor layers, hence making electronic devices' overall performance much better (Gao et al., 2018). Doping within bandgap engineering is behind the use of semiconductors for applications far beyond their traditional scope. Through the inclusion of dopants appropriate to a given electronic property, researchers are able to engineer materials into specific band structures. For example, doping wide-bandgap semiconductors like Gallium Nitride (GaN) or Silicon Carbide (SiC) with n-type and p-type dopants allows them to be controlled for their electronic characteristics under different optical and thermal conditions (Figure 2.4).

It is especially crucial in power electronic applications where performance and efficiency under high-temperature conditions are the main concern. It has been recently demonstrated through studies that the ability to control the bandgap through intentional doping techniques can be utilized for the development of next-generation optoelectronic materials, which can have the ability to transform photonic devices and solar cells of high efficiency (Boukhvalov et al., 2015). Though research goes on in two-dimensional materials like graphene and transition metal dichalcogenides (MoS_2), doping is a dominant method of functionalization of the materials for their desired applications. The intrinsic nature of 2D materials will naturally give rise to unique electron transport properties, and these can be tuned or even engineered using dopant inclusion. Not only does this affect carrier density but also has implications on light-matter interactions, such that this process is a critical element in the development of future photonic and electronic devices. Study has suggested doped forms of these materials could improve carrier mobility and integrate them into existing semiconductor technology and expand their field of applications in ultrafast optoelectronics and flexible electronics. Further study and adjustment of doping techniques will likely play an important role in realizing the ultimate capability of semiconductor materials on several technological fronts (Gao et al., 2018).

2.2.1. Purpose and Process of Doping

The properties of semiconductors are controlled to a great extent by intentional impurity introduction through doping, an inherent process in the science of semiconductor materials. By this process, it is possible to vary electrical conductivity to enable materials to shift between intrinsic and extrinsic levels of conductivity. By incorporating dopants—elements from either Group III or Group V of the periodic table—the semiconductors can be designed to have either n-type or p-type conductivity. N-type doping, in which donor atoms are introduced, creates a surplus of electrons, while p-type doping introduces acceptor atoms that create holes, resulting in a predominance of positive charge carriers. This control of concentration to such a high degree of precision is critical to the optimizing of devices ranging from transistors to diodes and impacts their efficiency and performance in use throughout power devices and photovoltaic devices. With an increasing need for new materials within electronic equipment, the process of doping becomes ever more crucial to understand. Techniques in doping are at the forefront of device enhancement in high-temperature and high-voltage applications within wide-bandgap semiconductors like Silicon Carbide (SiC) and Gallium Nitride (GaN). For instance, while 3C-SiC is wanted for its excellent semiconductor quality, there are challenges to activating p-type dopants as a result of low-temperature annealing procedures, which hamper full activation of dopants in ion implantation (Figure 2.5).

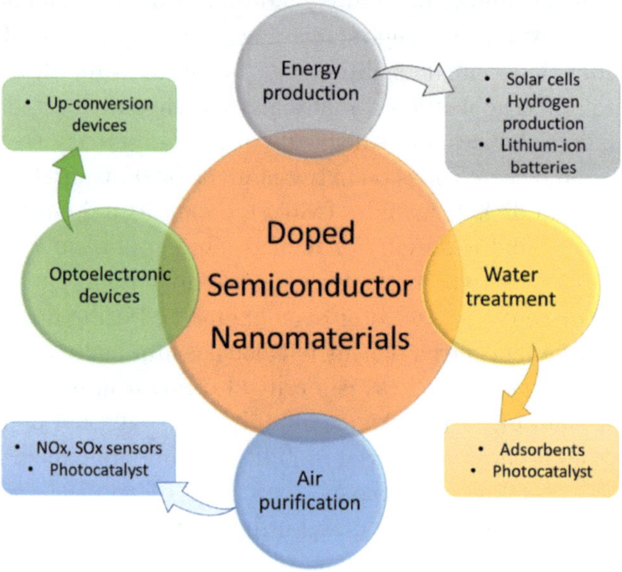

Figure 2.5. *Fundamental doped process in semiconductor material science.*

Source: https://media.springernature.com/lw685/springer-static/image/chp%3A10.1007%2F978-3-030-36268-3_138/MediaObjects/477767_1_En_138_Fig2_HTML.png.

The design and simulation of future structures, e.g., the new vertical 3C-SiC-on-Si MOSFETs, illustrate the activity to overcome these limitations, promising to revolutionize power electronics into more efficient and retaining the critical blocking device performance. These advancements highlight the critical role played by successful doping methods in the evolution of semiconductor technologies. Also, recent developments in solar cell technology highlight the quintessential role doping plays to the efficiency and performance of photovoltaic systems. In dye-sensitized solar cells (DSSCs), application of nanostructured material at the cathode and anode enables enhanced charge transport and separation, significantly increasing energy conversion efficiency. Both n-type and p-type semiconductors applied in DSSCs illustrate the diverse applications of doping beyond conventional electronics and show its importance in renewable energy technology. They are less efficient than silicon-based solar cells but possess simpler processes of manufacture and lower costs. This is in accordance with a broader trend in semiconductor material science, where doping is not only a means to optimize electrical properties but also an integral factor in facilitating sustainable technologies. Research on various types of semiconductors confirms the pivotal role of doping in the relentless drive to innovate.

2.2.2. Types of Dopants (N-type and P-type)

Semiconductor control through doping is a fundamental aspect of semiconductor material science. Specifically, the introduction of dopants enables the modulation of electrical conductivity as well as the creation of materials for a broad spectrum of applications. Doping refers to a deliberate introduction of impurities within an intrinsic semiconductor crystal structure that results in n-type or p-type material. N-type semiconductors are formed when those elements with more valence electrons than the semiconductor, such as phosphorus or arsenic within silicon, are introduced. These free electrons can migrate freely, thereby making the material electrically conductive p-type semiconductors are, however, created by doping with atoms with fewer valence electrons, e.g., boron. Doping forms holes in the crystal lattice, leading to positive carriers that also contribute to conduction. The precise control over dopant concentration and type is required to create desired electrical characteristics in semiconductor devices. The significance of p-type and n-type semiconductors is shown through their complementary use in the formation of p-n junctions, which are vital in various electronic devices. The two dopant types interact to produce areas of varying electrical properties, creating a junction through which charge can flow when an external voltage is applied. For instance, in photovoltaics, an n-type layer adjacent to a p-type layer enables charge separation, ultimately leading to improved efficiency in solar cells. Improvements in dopant selection and incorporation techniques enhance performance factors such as junction depth and resistivity and expand the utility of the materials. The recent advancements in the wide-bandgap materials GaN and SiC show the function of doping strategies to enhance thermal stability and breakdown voltage significantly to make them viable for high-power electronics and optoelectronics, a clear reflection of the used relevance of doping in semiconductor material science. The study of non-metal dopants within the semiconductor matrix shatters the boundaries of traditional doping strategies. A classic case is carbon doping in zinc oxide (ZnO) to induce fascinating phenomena like intrinsic ferromagnetism in the system. The research indicates that carbon replacing oxygen in ZnO not only gives it a magnetic moment but also boosts its conductivity to render it an n-type semiconductor. These developments open up novel opportunities for applications in spintronics and magnetoelectronics, in which the interplay between charge transport and magnetism is the focus. The carbon-doped ZnO research supports the potential of alternative dopants to create material that transgresses conventional limits of semiconductor performance, highlighting the ongoing need for new studies on semiconductor material science in the pursuit of next-generation electronic and optoelectronic devices (Alexander et al., 2025; Pan et al., 2006).

2.2.3. Effects of Doping on Conductivity and Performance

The intricate relationship between doping and conductivity in semiconductor materials is central to enhancing them for various applications. Doping introduces impurities that modify the electric characteristics of semiconductors by the generation of charge carriers, and this has a revolutionary impact on their conductivity. One such application is the incorporation of beryllium (Be) as a dopant acceptor in GaAs, which leads to noticeable enhancement of nanowire transistors' performance. These devices may exhibit remarkable characteristics like greater on-state conductivity and an enormously large on-off ratio, based on doping density employed. But extremely high doping can provide poor device performance with consequences of freezing of conduction at low temperatures and challenges in fabrication of a visible off-state under gating. Consequently, doping level optimization is an important task in order to balance conductivity and performance and therefore an important research area in semiconductor material science (Gluschke et al., 2017). The advances of recent years in bandgap engineering have highlighted the role of doping in governing the electronic properties of materials. By the deliberate tailoring of the bandgap through doping, the electronic properties of semiconductors can be regulated, increasing their application in devices like transistors and photodetectors. For two-dimensional materials such as MoS_2, for instance, studies have established that the conductive nature can be significantly tuned with doping. With the introduction of strain in addition, specific strains possess the ability to induce semiconductor-metal transitions, which in the end contribute to enhancing thermoelectric efficiency. This highlights the significance of tailored doping techniques to achieve optimal performance of semiconductor devices in various conditions, thus pushing semiconductor material science to new uses and higher device functionality (Bhattacharyya et al., 2014). While the quest for high-performance semiconductor materials continues, a knowledge of the impact of doping persists at the core. The host-lattice-doping-agent interaction can lead to the formation of defects, which influence the electronic states' transparency and material stability. Comprehensive research on wide-bandgap semiconductors including gallium nitride (GaN) and silicon carbide (SiC) has found doping to be useful in enhancing electronic and thermal conductivity, making the materials suitable for high-frequency and high-power applications. Furthermore, the development of new materials like diamond and other 2D systems reveals that doping not only modulates conductivity but is also found to influence mechanical properties through defect engineering. Therefore, extensive studies examining the multi-dimensioned effects of doping are necessary for further promoting the functionality and reliability of future semiconductor devices in a world increasingly digitalized (Bhattacharyya et al., 2014; Gluschke et al., 2017).

2.3. DEFECTS IN SEMICONDUCTOR MATERIALS

The intricate dependence of crystal structures and defects on semiconductor materials defines their electronic characteristics and overall functionality at a fundamental level. As a method to enhance semiconductor functionality, doping procedures are used to incorporate impurities that modify carrier densities and thereby maximize conductivity.

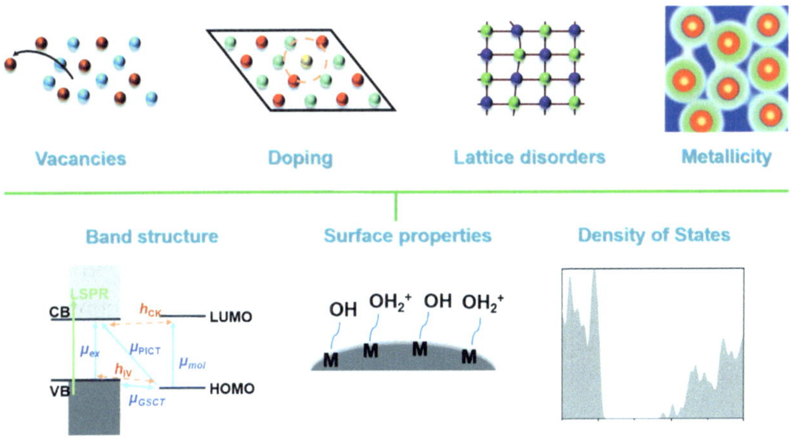

Figure 2.6. *The crystal structures and defects in semiconductor materials.*

Source:https://pubs.rsc.org/image/article/2022/SC/d1sc05940h/d1sc05940h-s1_ hi-res.gif.

Sadly, dopant insertion also generates a variety of native defects, such as vacancies and interstitials, that can detrimentally affect electronic characteristics. For example, in semiconductors like GaN and SiC, defect control through precise doping has been essential in achieving required bandgap properties. The combined effect of engineered doping with such defects typically decides the performance of devices like high-power transistors and light-emitting diodes. It is necessary to be aware of this balance for the evolution of semiconductor technology since limitation by defects can severely constrain the performance and reliability of semiconductor devices. Following advancements in semiconductor materials, bandgap engineering has become a vital concept for enhancing the performance of devices in a variety of applications (Figure 2.6).

By tailoring the electronic band structure, researchers can create materials with targeted optical and electronic properties. Semiconductors such as diamond and GaN with wide-bandgaps are increasingly used in high-temperature and high-voltage applications, effectively overcoming limitations in traditional silicon-based technologies. Moreover, two-dimensional materials such as graphene and MoS_2 have further enriched the bandgap tuning landscape. These materials possess novel properties, allowing for heterostructure formation that can possess novel electronic states and enhanced functionalities. Understanding the role played by defects in these engineered systems is important as extraneous imperfections could influence charge transport properties with the consequence of hindering device fabrication and properties of resulting semiconductors. Improvements in semiconductor materials continue to explore the character of defects in relation to their applications, particularly in emerging technologies like photovoltaics. Recent findings have detailed the unique phenomena of amphoteric defects in halide perovskite solar cells, which are a technology that has rapidly evolved to popularity for its unparalleled efficiency and versatility. The synergetic interaction between acceptor-like and donor-like configurations of the defects facilitates the transport of charge carriers, which plays a crucial role in the optimization of photovoltaic performance (Asta et al., 2018). Tuning these defect dynamics is at the core of excising stability issues conventionally associated with perovskite materials. This is to illustrate a broader trend in semiconductor material science, where defects are not just nuisances but can be utilized as functional constituents of new materials, providing new areas of scientific study and application in quantum-matter heterostructures and beyond (Boschker et al., 2016).

2.3.1. Types of Defects (Point, Line, and Surface)

The sophisticated interplay between crystal structures, doping, and defects contributes significantly to semiconductor material electrical and optical properties. Defects in semiconductors are usually classified into three broad groups: point, line, and surface defects. Point defects, including vacancies, interstitials, and substitutional impurities, are localized distortions in the crystal lattice. These very small distortions can have substantial impacts on carrier concentrations and the electrical conductivity and optical properties of semiconductors. For instance, in dichalcogenides of the transition metals like MoS_2, doping with manganese creates mid-gap states that enhance magnetic behavior and thereby enable possible application in electronic devices and spintronics (Zerger et al., 2021). Here, point defect knowledge becomes central to purposeful material property engineering and therefore semiconductor application innovation. Line defects, or dislocations, are generated by atomic misfit in a crystal lattice and can have significant influence on the performance of semiconductors. They can enable strain accommodation during the growth of semiconductor films and can serve as electrical charge carrier channels, thereby modifying their mobility. Also, dislocation occurrence tends to be for the use of semiconductors in wide-bandgap material, such as GaN and SiC, where control of dislocations enhances device reliability. Control of line defects is imperative to maintain energy efficiency in power electronics and photonic devices where thermal stability must be high. A complete understanding of line defect formation and evolution can thus be employed to find maximum device performance and lifetime under real-world situations. Surface defects, formed by termination of the crystal lattice on the materials surface, entwine with vast implications for charge transport and surface reactivity. Such defects present opportunities for exciting electronic properties because their local shape can dictate band bending and trap charge carriers. In two-dimensional material innovation like graphene and MoS_2, defects on the surface have the capability to drastically alter carrier dynamics, consequently impacting photonic and electronic applications. Surface defect types and energetics are determined with very high accuracy by techniques such as scanning tunneling microscopy and spectroscopy. Such properties can be controlled by researchers through the application of appropriate surface treatments and defect engineering methods to achieve higher device efficiency and novel functionalities (Schroder, 1990). Therefore, understanding surface defects opens up future directions for the development of semiconductor material applications in different fields of technology.

2.3.2. Impact of Defects on Semiconductor Behavior

The complexity of semiconductor materials stems from their complicated crystal structures, which have a determining influence on their electrical properties. In these structures, defects can have an important impact on semiconductor behavior through modifications in carrier concentration and mobility. For instance, point defects such as vacancies and interstitials upset the periodicity of the atoms and cause localized states within the bandgap. These may trap charge carriers or allow recombination processes that can worsen the device performance. Moreover, the addition of dopants—done to make the semiconductors conductivity higher—can create additional defects, rendering the connection between defects and electronic response complicated. By intelligently managing these defects during fabrication, bandgap engineering can be a powerful method to produce custom semiconductor materials like the quantum-matter heterostructures with new electronic states and emergent behaviors (Boschker et al., 2016). Temperature is also key to the determination of how defects will affect semiconductors' behavior. As temperature changes, so does energy landscape of defects, particularly in high dielectric constant and volatile oxide layers (Figure 2.7).

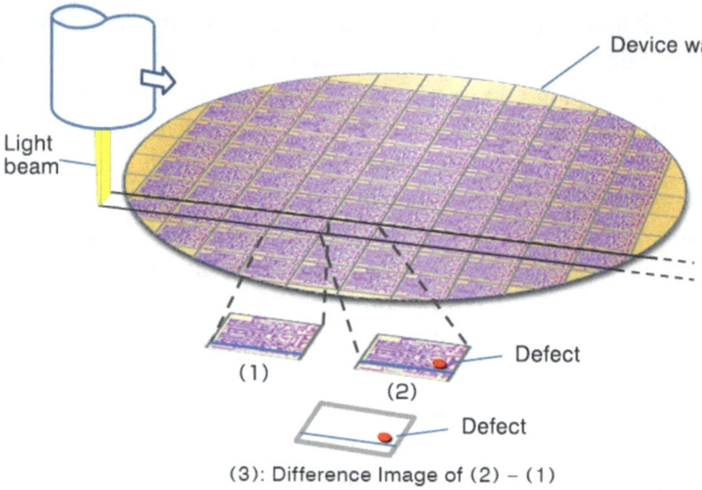

Figure 2.7. *The semiconductor technical approach to defect pattern analysis.*

Source: https://www.chetanpatil.in/wp-content/uploads/2025/02/image.png.

Experiments have shown that at a given temperature, some defects in interface areas can dominate capacitance behavior, as seen in MOS stacks fabricated with high-k dielectrics. For instance, for the ambient temperature, charge carrier behavior is controlled by traps in the deeper parts of the oxide layer and not on the semiconductor interface itself. Temperatures that are lower than ambient, however, enhance more localized effects due to defects and can impact nano-electronics performance. This diversity underscores the significance of studying spatial defect distributions in an effort to predict and control capacitance frequency dispersion more accurately, thus enhancing semiconductor technology and informing design principles for successful device manufacture (Aguirre et al., 2018). Another feature of defect study is material selection impact. Wide-bandgap semiconductors, like gallium nitride (GaN) and silicon carbide (SiC), show how defects can be employed to achieve sophisticated properties for specific applications, like high-power electronics and optoelectronic devices. Intrinsic resistance of these materials to large electric fields often leads to characteristic defect behavior, which may be advantageous and detrimental simultaneously depending on the application. For example, while defects can mitigate thermal effects in some applications, they can create unwanted trap sites that decrease device efficiency. This neutrality compels scientists and manufacturers to consider not only the impacts of defects during material synthesis at the start but also throughout the device's life. Therefore, ongoing advancements in semiconductor material science aim at the identification and control of defect states to attain optimal performance, bridging the theory-applications gap (Boschker et al., 2016).

2.3.3. Methods for Defect Characterization and Control

The intricacy of semiconductor materials necessitates robust techniques for defect characterization and control for optimal performance in applications. These defects, which are typically seen in growth processes or through extrinsic stresses, can significantly influence the electronic and optoelectronic properties of materials. Advanced techniques like atomic force microscopy (AFM) and transmission electron microscopy (TEM) allow high-resolution imaging of the positioning and study of defects at the atomic level. Through these methods, not only do researchers determine the nature of existing defects but also their spatial locations. Such data are of relevance for tailoring fabrication protocols and material quality optimization, and ultimately for device performance enhancement. The addition of quantum materials, as discussed in Boschker et al. (2016), introduces yet other challenges where engineered interfaces may be locations for defects that must be controlled if the unique electronic states they provide are to be maintained. The role played by doping in semiconductor technology cannot be overstated because it has direct relevance to the electrical conductivity and overall performance of such materials. Uncontrolled doping, however, will bring in extraneous defects, and therefore careful doping techniques have to be employed to preserve the integrity of the crystal lattice. Among such techniques, the employment of liquid-phase epitaxy (LPE) techniques for doping is notable as a means of being able to exercise better control over the incorporation of dopants into a crystal lattice. Templated liquid-phase (TLP) growth process, included in Amani et al. (2020), exhibits a significant advancement in enabling low-temperature growth of high-quality III-V semiconductors on non-traditional substrates. Not only does this process eliminate thermal stress, but it also enables the achievement of tailored electronic properties by controlling defects through precise manipulation of growth conditions and material interactions. Since more semiconductor research is shifting towards two-dimensional and wide-bandgap materials, defect control presents new challenges. The demand for such materials as SiC, GaN, and graphene requires new approaches to defect sensing and control to unleash their full potential for application in optoelectronic and high-power devices. Techniques such as photoluminescence spectroscopy and metalorganic vapor phase epitaxy prove to be invaluable in evaluating the electronic properties of such metal oxides and defect states that could affect their functionality. Furthermore, research on methods to engineer 2D material bandgaps can reveal approaches for creating materials with desired electronic properties free of harmful defects. As researchers make inroads deeper into this field of research, the interplay between defect characterization and control will more and more set the course of future semiconductor material science, driving advancement in various areas of technology.

2.4. BANDGAP ENGINEERING

Electronic property manipulation in semiconductors has gained significant traction in recent years, primarily through processes in the field of bandgap engineering. Controlled modulation of semiconductor material composition is used by researchers to create alloys with modifiable bandgaps, an important characteristic that determines the materials optical and electronic function. Most importantly, the development of nanoscale semiconductor materials like traditional II-VI and III-V semiconductors, and recent 2D materials, signifies their extensive scope and promise for application. For instance, their nanoscale nature demonstrates incredible resilience to misfits in the lattice, permitting almost arbitrary control over alloy composition without being subjected to the harmful defects of bulk material. These advancements bring forth a plethora of applications in photonics, ranging from solar cells to lasers with tunable wavelengths, demonstrating the revolutionary impact of bandgap engineering in modern semiconductor material science (Dou et al., 2017). The establishment of wide-bandgap materials such as silicon carbide (SiC), gallium nitride (GaN), and synthetic diamond has expanded the scope of semiconductor applications. These materials possess improved thermal conductivity, large breakdown voltages, and low power losses and hence are suitable for high-temperature and high-frequency applications (Figure 2.8).

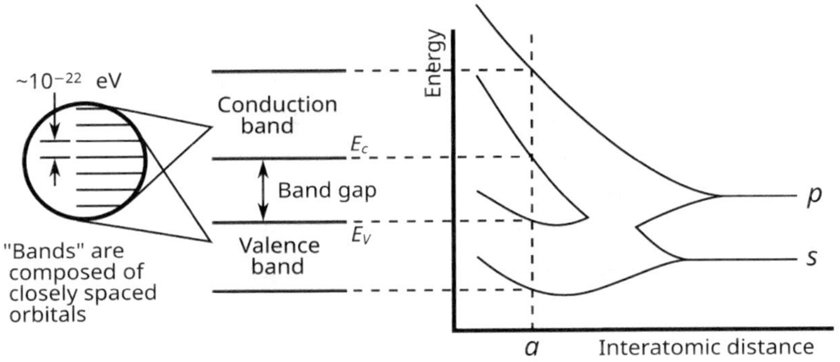

Figure 2.8. *The analysis bandgap energy.*

Source:https://upload.wikimedia.org/wikipedia/commons/e/ef/Solid_state_electronic_band_structure.svg.

The advent of 2D materials, including graphene and transition metal dichalcogenides MoS_2, offers additional prospects for bandgap manipulation according to their unique electronic and optical characteristics. These materials not only allow for higher device efficiency but also enable one to study new physical phenomena. By leveraging the advantages of wide-bandgap materials in combination with emerging 2D materials, scientists can create more efficient devices and make them capable of operating at extreme conditions, representing a significant leap in semiconductor technology (Dou et al., 2017). Generally, the combination of graphene with traditional semiconductor materials like HgCdTe for infrared detection is an indication of bandgap engineering feasibility. The use of graphene overcomes key disadvantages of HgCdTe detectors by reducing photogenerated carrier recombination and enhancing overall detection efficiency. The hybrid idea is also helpful for carrier mobility and lifetime enhancements, key parameters for ideal device functioning. The potential for use extends beyond the raw performance enhancement; it unravels the potential for tiny, low-cost infrared detectors that find applications ranging from satellite earth observation to the military. As researchers optimize the hybrid devices, the synergy of materials reveals the vast potential behind bandgap engineering and how it is expanding the limits of semiconductor material science.

2.4.1. Definition and Significance of Bandgap

The unique properties of semiconductor materials are primarily a consequence of their bandgap, the energy difference between the valence band and the conduction band. The bandgap determines significantly the electrical and optical properties of a semiconductor, influencing its conductivity, thermal stability, and sensitivity to external stimulations. Especially, the size of the bandgap determines which wavelengths of light a semiconductor absorbs or emits. Understanding the nature of the bandgap allows scientists and engineers to engineer semiconductors for specific applications, whether electronics, photonics, or other advanced technologies. The ability to manipulate the bandgap through mechanisms like doping and defect engineering then becomes an essential aspect of semiconductor materials science. Apart from its intrinsic importance in determining material properties, the bandgap is also important in facilitating next-generation applications, particularly optoelectronics and power devices. Wide-bandgap materials such as gallium nitride (GaN) and silicon carbide (SiC) are becoming more sought after due to their high-temperature, voltage, and power operation, which are the most critical features for next-generation energy solutions. As an illustration, GaN's large bandgap facilitates the creation of high-efficiency deep ultraviolet photodetectors that suit environmental monitoring and military reconnaissance. Specialists note that while traditional silicon compounds are cost-effective and integratable, wide-bandgap semiconductors enhance the quality figures of devices significantly, illustrating the key function of the bandgap in modern technology applications (Meng, 2024). Moreover, the discovery of new material in the realm of two-dimensional (2D) compounds further widens bandgap engineering. Graphene and transition metal dichalcogenides (e.g., MoS_2) are such materials, which possess superior bandgap properties that are easily modulated by outer parameters such as strain or the application of an electric field. Such control places 2D materials as key players in future electronic and optoelectronic devices, where efficiency and flexibility are paramount. Aside from independent applications, bandgap engineering's integration into broader structures of semiconductor design underscores the significance of this concept in subsequent technology progress. By way of example, h-BN's ultrawide bandgap properties not only allow the development of new detectors but also set unprecedented performance benchmarks for materials in aggressive environmental conditions (Li et al., 2025).

2.4.2. Techniques for Bandgap Modification

There are numerous ways in which the bandgap of semiconductor materials can be successfully modified, all with the purpose of enhancing their performance and use in various sectors. One such commonly employed procedure involves doping, where specific impurities are added to a semiconductor crystal lattice. Under this process, controlled adjustment of electronic features can be made by adjusting the carrier concentration, thereby widening or narrowing the bandgap.

For instance, in wide-bandgap materials such as gallium nitride (GaN) and silicon carbide (SiC), precise doping methods can shift the energy levels in order to enable bandgap modulation for the purposes of high-temperature and high-power operation. Furthermore, inducing defects in the crystal structure may also be utilized in bandgap engineering, generating localized states that can effectively influence electron behavior. Such approaches illustrate the vital trade-off between electronic alteration and structural integrity during the optimization of semiconductor performance. Another significant approach to bandgap alteration is novel synthesis techniques used to create new materials with target electronic characteristics. The development of two-dimensional materials, for example, transition metal dichalcogenides such as MoS_2 and graphene, has presented new avenues in semiconductor research. These materials possess unique band structures due to their dimensionality, which allow for tunability of their electronic properties with accuracy through methods such as modulation of the number of layers or strain engineering. Scientists have demonstrated that favorable bandgap modification can be achieved by varying the stacking sequence and introducing intercalants (Figure 2.9).

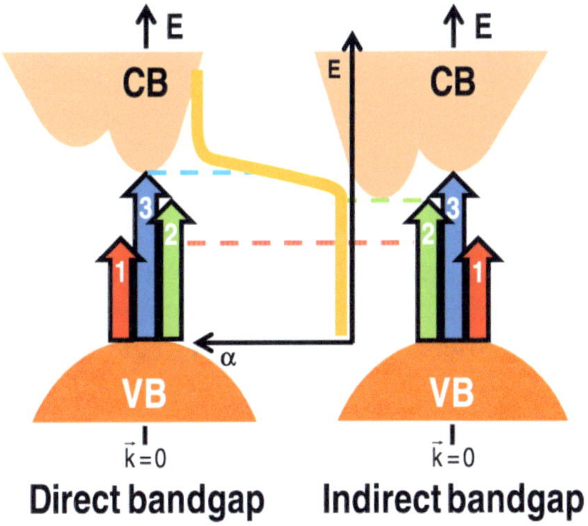

Figure 2.9. *Revisiting the optical bandgap of semiconductors.*

Source: https://media.springernature.com/lw685/springer-static/image/art%3A10.1038%2Fs41598-019-47670-y/MediaObjects/41598_2019_47670_Fig3_HTML.png.

This diversity also renders 2D materials very promising for application in optoelectronics and flexible electronic devices, where conventional semiconductor materials may not be sufficient. With these new methods, the potential for discovering new functionality in bandgap-engineered materials also increases. Further study of bandgap modification methods also encompasses the use of hybridization and composite structures to achieve better performance characteristics. For instance, composite semiconductor synthesis combines the beneficial properties of different materials to create enhanced charge carrier kinetics and photocatalytic performance.

Such operations like the deposition of a noble metal on semiconductor surfaces, as shown in the titanium dioxide (TiO_2) photocatalyst example, operationally optimizes the efficiency of light absorption and reduces electron-hole recombination rates, effectively modifying the bandgap. The successful alteration of TiO_2 by employing methods like transition metal and rare earth metal doping presents additional ways to optimize its photocatalytic behavior under visible light-based processes. With the integration of these various forms of modification, semiconductor material science continues to evolve, providing new opportunities for energy conversion and storage applications based on strategically designed bandgap profiles.

2.4.3. Applications of Bandgap Engineering in Devices

The recent advancements in semiconductor materials science have revolutionized the capability of devices in performance, particularly through the novel concept of bandgap engineering. By manipulating the electronic attributes of materials such as II-VI and III-V semiconductors, researchers have achieved the formation of alloy materials with variable bandgaps that greatly enhance their performance across different applications. The nanoscale quality of the materials has superior mismatch tolerance in the lattice, which is vital while merging several different materials into a single device structure. This is one important feature that makes possible applications like tunable lasers and efficient solar cells.

Although controllability of light absorption and emission bandwidth enhances not only device performance characteristics but also the potential for new applications in solid-state lighting and artificial photosynthesis significantly (Dou et al., 2017), integration of strain engineering into the family of semiconductors has seemed to be a vital strategy in optimizing capabilities of conventional materials like germanium (Ge). Ge, as a primarily indirect-bandgap semiconductor, is challenging to use in light-emitting applications on silicon substrates. However, its introduction of tensile strain can easily convert its nature into a direct-bandgap type, making the material's efficiency at optoelectronic applications optimal. Recent studies involving mechanically stressed Ge nanomembranes demonstrate the high promise of this method, depicting improved light emission and population inversion under optical pumping. These developments pave the way for integrating such nanomembranes into silicon photonic circuits, thus allowing the integration of electronic and photonic functionalities without sacrificing mechanical flexibility (Lagally et al., 2018). In addition to the traditional semiconductor materials, studies in wide-bandgap and two-dimensional materials further expanded the avenue of bandgap engineering. These substances, such as gallium nitride (GaN) and silicon carbide (SiC), are already used extensively in high-frequency and high-power devices, demonstrating the robustness of their wide bandgaps. Contrarily, 2D substances such as graphene and molybdenum disulfide (MoS_2) have introduced new paradigms in device structure that enable one to realize lightweight and conformable structures. The capacity to engineer the electronic characteristics of such materials by adjusting them using bandgap engineering not only enhances the performance of devices but also presents new functionalities, such as the realization of efficient and compact sensors, photodetectors, and transistors. Generally, these advancements underscore the revolutionary aspect of bandgap engineering in the future of semiconductor devices to facilitate technological innovation in the study of material science.

2.5. WIDE-BANDGAP MATERIALS

Development of semiconductor material science has witnessed tremendous shift towards the study of wide-bandgap materials, which offer unique properties useful for various applications. Gallium Nitride (GaN), Silicon Carbide (SiC), and diamond are a few such materials, which have bandgaps in excess of 2 eV and hence can operate efficiently at high temperatures, voltages, and powers than normal semiconductors. The quest for wide-bandgap materials has led to increased interest in novel structures like heterostructures and quantum wells in enhancing the ability to design electronic and optoelectronic devices. The material's crystal structure also determines its desirable performance characteristics, including enhanced electron mobility and thermal conductivity. All these characteristics make wide-bandgap materials a natural building block of future electronic devices, including high-power transistors, blue and ultraviolet light-emitting diodes, and laser diodes. The process of bandgap engineering is necessary in order to modify the electronic properties of wide-bandgap materials as per specific technological needs. By controlling composition via alloying or by doping element introduction, researchers can tailor the bandgap and acquire specific electronic and optical properties. This capability is particularly evident in the case of nanoscale semiconductor materials, whose alloy composition is adjusted to enhance their tolerance for lattice mismatch in monolithic structures (Figure 2.10).

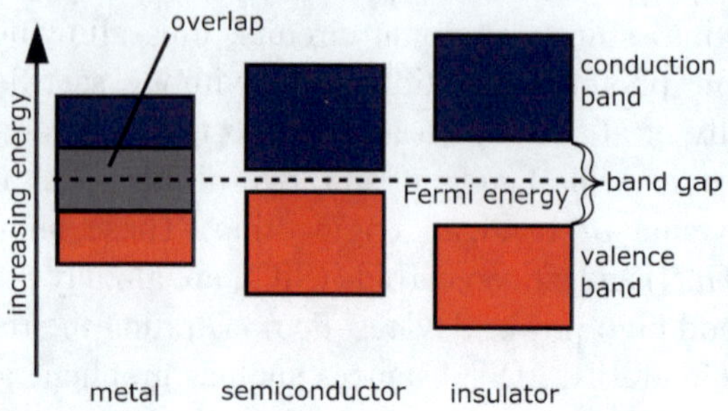

Figure 2.10. *Wide bandgap semiconductors.*

Source: https://www.mouser.com.gt/images/microsites/article-wide-band-gap-fig01.jpg.

I-VI and III-V alloys serve as examples of how minor structural modification gives rise to high bandgap difference disparities, enabling a wide range of applications. Most significantly, this tunability is critical in photonic device engineering, such as lasers and solid-state lighting devices, since optimization of bandgaps has direct implications on device efficiency and performance (Dou et al., 2017). Emerging research on intersubband transitions reports the increasing range of ultrawide bandgap material use, such as Aluminum Gallium Oxide and Gallium Oxide, in intersubband device applications. Classic semiconductors, while useful in mid- and long-wave infrared devices, are limited by their highest conduction band offsets. In contrast, the new structure of $(AlxGa_1\text{-}x)_2O_3/Ga_2O_3$ quantum well systems offers a potentially potent route for short and near-infrared device development necessary for optical communications and biomedical imaging. Theoretical analysis reveals that this family of compounds has the potential to cover a broad range of transition wavelengths, which is huge versatility. Impressively, the projections are predicting excellent figures of performance, including superior room temperature detectivity, further increasing the prospects for such devices in real-world applications. Therefore, the development of wide-bandgap materials continues to unlock potential for groundbreaking innovation in semiconductor technology (Krishnamoorthy et al., 2020).

2.5.1. Overview of Wide-Bandgap Semiconductors (Gallium Nitride (GaN), Silicon Carbide (SiC), and Diamond)

Exploration of the nature of wide-bandgap semiconductors, such as GaN, SiC, and diamond, reveals unique properties that distinguish them from traditional silicon-based technologies. The crystal structures are pivotal in dictating their electronic and thermal characteristics and thus their potential for lucrative applications in power electronics and high-frequency devices. Their potential is further enhanced through the ability to tailor the bandgap through doping and composition. GaN, for instance, possesses higher electric field strength and electron mobility, thereby being of immense application in high-frequency communication and cutting-edge photonic devices. Similarly, SiC is also cited for its improved thermal conductivity, enabling the devices to operate under extreme conditions. Thus, understanding these properties is vital to enhance the devices' performance that rely on these semiconductors and drive future innovations in the sector. The use of wide-bandgap semiconductors in modern technology relies significantly on device modeling and simulation development. Traditional TCAD tools, which were primarily developed for silicon, have difficulty working with GaN, SiC, and diamond devices. The material physics and device parameter differences necessitate special methodology for valid modeling. Specific focus on convergence problem solving and defect/trap state simulation is critical since these effects have a significant influence on semiconductor performance (Antoniou et al., 2018) (Figure 2.11).

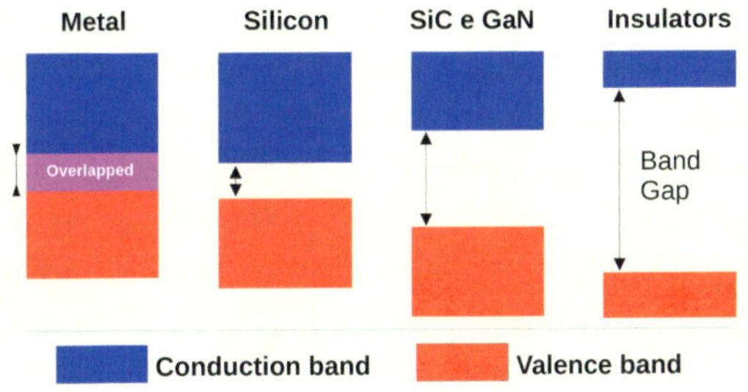

Figure 2.11. *The exploration of wide-bandgap semiconductors.*

Source: https://www.powerelectronicsnews.com/?attachment_id=67693.

With the further development and refinement of modeling approaches with ongoing research, reliability and accuracy in prediction of performance can be enhanced, thus driving research and development in wide-bandgap technology. This greater understanding not only enhances performance measures but also addresses the requirements of expanding markets, creating a complete package for future semiconductor developments. The market situation is poised for revolution, spurred by the use of wide-bandgap semiconductors. GaN has proven itself as a top contender, with extreme improvements in efficiency and speed over its silicon-based counterparts. Its high-performance character has propelled its commercial significance, particularly for power conversion, more and more a necessity for modern electronic systems (Amano et al., 2018). Meanwhile, SiC continues to serve automotive and industrial applications in which reliability under extreme conditions is paramount. Meanwhile, diamond semiconductors, in as yet nascent stages of commercialization, promise exceptional thermal and electronic properties. Combined, these materials are shaping the semiconductor future of industries worldwide, with massive investment being required in research and development. The surge initiated by these wide-bandgap semiconductors holds promise for an era marked with increased device efficiency, faster switching speeds, and new circuit topologies tailored to new applications.

2.5.2. Advantages of Wide-Bandgap Materials in High-Power Applications

Semiconductor material research has fueled critical advance in high-power usage, and that is primarily because of the application of wide-bandgap analogs such as Gallium Nitride (GaN) and Silicon Carbide (SiC). The former materials possess enhanced thermal conductivity and higher breakdown voltage compared to traditional silicon-based semiconductors. The unique properties of wide-bandgap materials enable efficient operation at high temperature and are thus appropriate for applications where a requirement for high-power density exists. In addition, the ability to engineer the bandgap through doping and defects allows for better performance potential for such materials, which is then carried forward into device innovations such as power amplifiers and inverters. As demand continues to rise for higher efficiency power conversion in renewable energy systems, the application of wide-bandgap semiconductors becomes an essential aspect of controlling energy challenges while maintaining overall system efficiency. In applications involving high frequencies, wide-bandgap materials reveal their capability to operate efficiently at microwave and millimeter-wave frequencies. Benefits of higher electron mobility in materials like GaN yield switches capable of switching quickly with reduced energy lost. This usefulness reveals itself most prominently in the designing of high-efficiency RF amplifiers, which are essential for telecommunications and radar systems. Further, their use in modern technologies is in line with future needs for small, efficient, and strong electronic devices. The advancing knowledge that wide-bandgap semiconductors are capable of withstanding high levels of power without rapid thermal degradation makes them highly valued within the technology society. Improvements in these fields not only result in enhancing existing applications but also drive the development of future-generation devices necessary for industrial applications. Their use also extends into new fields such as photovoltaic devices in which their application can be meaningfully equated to enhanced energy conversion efficiencies. Thin-film solar cells have registered advancements where the material structural properties such as $CuInSe_2$ allow them to work optimally under high-radiation conditions, essential for space application (Bailey et al., 2025). By utilizing the unique properties of these semiconductors, researchers can study multibandgap systems that take advantage of the efficiencies of various materials and demonstrate a viable path toward enhanced performance in solar arrays used for powering space missions. The introduction of several bandgap energies in cascaded cell configurations demonstrates the potential for efficiency records being shattered due to such systems, which have a broad range of solar energy capture. Lastly, the advantages offered by wide-bandgap materials not only allow near-term performance in power electronics but also establish innovations that can transform energy systems for numerous industries.

2.5.3. Challenges in the Fabrication and Use of Wide-Bandgap Materials

The semiconductor material growth dynamics have been strongly influenced by the search for an enhancement of the electronic properties and expanding the possibilities for use with wide-bandgap materials. Gallium nitride (GaN), silicon carbide (SiC), and diamond have better performance in high-power and high-frequency devices. However, the fabrication problems inherent in them weigh significantly against their widespread use. The issues related to lattice mismatches and defect densities are particularly severe for wide-bandgap materials and directly impact the overall reliability and performance of the device. Single-crystal growth of high quality is a technology challenge in that conventional growth methods introduce defects that can destroy electrical performance. Avoiding these crystal defects while remaining economically viable is critical to advancing the integration of wide-bandgap materials into commercial applications. Adding to the complexity of wide-bandgap materials is the complexity of doping effectively, which is necessary to modify the electrical behavior of semiconductors. The intentional addition of dopants into such materials can dramatically increase their conductivity, but p-type doping, especially in III-nitride materials like AlGaN/GaN, is found to be extremely difficult to achieve. Recent breakthroughs have unveiled potential solutions, as indicated through the development of p-channel field-effect transistors from AlGaN/GaN superlattices employing magnesium as the p-type dopant (Hatui et al., 2019). With the integration of different concentrations of doping, the issue remains to maintain optimal electrical performance without detrimental effects on mobility and sheet resistance. Thus, ongoing research continues to play a key role to address these doping complexities while maintaining stability in changing operational environments. The integration between bandgap engineering and crystallography in the area of wide-bandgap materials introduces new challenges in semiconductor fabrication. The need for precise control of the material composition and the resulting electronic structure demands advanced techniques in material synthesis and crystal growth. Wide-gap materials can offer bandgap and properties tunability but typically at the expense of applying novel techniques such as alloying other semiconductors and introducing new 2D materials such as graphene and MoS_2 (Dou et al., 2017). These processes have to address the challenge of maintaining homogeneity while limiting defects to a low level, which can be detrimental to performance in real applications. Therefore, the intricate interdependence of manufacturing processes and material properties remains an area of special interest for researchers who aim to unlock the full potential of wide-bandgap materials in future semiconductor technologies.

2.6. TWO-DIMENSIONAL MATERIALS

Researches on two-dimensional materials have begun a new era in semiconductor material science, particularly due to their unique electrical, optical, and mechanical properties. Characterized by their atomic thinning without compromising major structural strength, materials such as graphene and transition metal dichalcogenides (TMDs) have opened up novel possibilities for applications ranging from flexible electronics to advanced sensors. Maybe the most pivotal challenge in the application of such materials, however, is bandgap engineering. For instance, while graphene boasts superior electron mobility, the fact that it has a zero bandgap acts as a major limitation to its use in semiconductor devices. On the other hand, TMDs like MoS_2 possess a tunable bandgap, which means that they can be very functional in transistors and photodetectors (Avsar et al., 2017). Being capable of engineering wide-bandgap semiconductors through the addition of two-dimensional materials raises their chances of contributing to electronic and photonic devices of the future. Fabrication of two-dimensional semiconductor devices is a gigantic step towards discovering effective electronic devices. The inclusion of such a material as black phosphorus (bP) can be most beneficial with its favorable spin transport properties and the ability for bandgap modulation. Recent studies have proved the fabrication of spin valves with ultra-thin bP, where both room-temperature spin detection and injection have been proven. Utilization of boron nitride layers between magnetic electrodes and bP minimizes issues related to conductivity mismatch, hence optimizing spin injection efficiency (Avsar et al., 2017) (Figure 2.12).

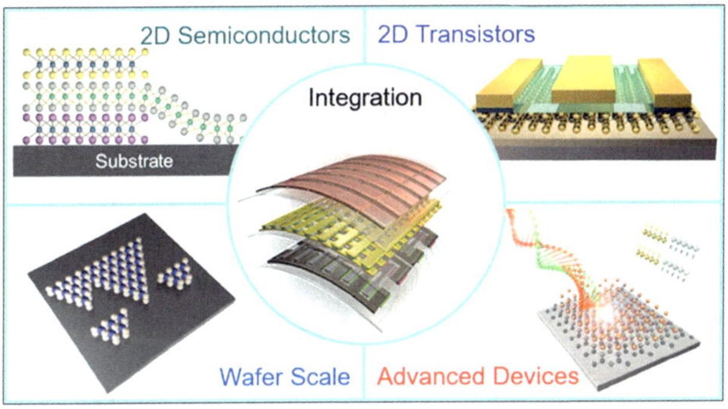

Figure 2.12. *Two-dimensional semiconductors and transistors for future integrated circuits.*

Source:https://pubs.acs.org/cms/10.1021/acsnano.3c10900/asset/images/medium/nn3c10900_0013.gif.

These strategies not only emphasize the potential of two-dimensional semiconductors for spintronics but also highlight a turning point in the utilization of layered material configurations to side-step inherent limitations of traditional semiconductor technology. In addition to their intrinsic character, the possibility of making heterostructures from numerous two-dimensional materials opens up paradigm-changing opportunities for semiconductor science. By assembling layers in cascades of different quantum materials, it is possible to design interfaces to exhibit collective and emergent properties not found in isolated systems. Heterostructural assembly in this manner allows the emergence of electronic states, therefore opening up opportunities for scientific discovery and technological innovation. The biocompatibility and flexibility of quantum-matter heterostructures show promise in manipulating electronic band structures, offering potential for new quantum computing and photonic-based technologies. Despite the promising potential, challenges in ascertaining interactions at such interfaces and optimizing production techniques to produce optimal functional efficiency remain to be addressed, which would take precedence as the technology develops towards useful applications.

2.6.1. Introduction to Two-Dimensional (2D) Materials (Graphene and Molybdenum Disulfide (MoS_2))

The invention of 2D materials has changed the landscape of semiconductor material science by discovering an interesting interaction between atomic-scale structures and macroscopic physical properties. Among all of these materials, graphene stands out due to the fact that its atomic structure comprises carbon atoms in a monoatomic layer of a hexagonal lattice. Its structure provides it with very high electrical, thermal, and mechanical properties. Further, MoS_2 being a prototype TMD has great promise with the semiconducting nature. Unlike graphene possessing a zero bandgap, MoS_2 has a direct bandgap for the monolayer structure, and thus it is highly sought after to be utilized in optoelectronic devices and sensing. Understanding how the crystal structure influences these properties in unleashing their full potential in future semiconductor technology is crucial. Doping and defect engineering are of great importance that determines the properties of 2D materials. For example, doping in MoS_2 with impurities has the ability to drastically alter its electronic properties such that the device's efficiency can be tailored to suit desired purposes. Emergence in the field of photodoping, where carriers are remotely induced using optical means, has recently enabled new directions in controlling transport properties. For example, in such devices where MoS_2 is partly coated with hexagonal boron nitride (hBN) and interfaced with graphene for electronic contact, quantum confinement effects become apparent at low temperature. This is a strategy that demonstrates the importance of heterostructures since they combine various 2D materials to optimize electronic and optical performance and yet enhance their applicability in semiconductor technology (Banszerus et al., 2018). 2D materials bandgap engineering provides an avenue towards the fabrication of new devices with higher functionalities. The integration of materials like MoS_2 into heterostructures with discrete electronic character will lead to the formation of type II band structures. In these geometries, the conduction band maximum and valence band minimum are within distinct layers, allowing spatially separated electron-hole pairs, which is preferable for photodetector and photonic device applications. The strong interlayer coupling that these geometries allow also highlights the potential for band structure properties to be controlled by careful material selection and layer stacking techniques. The continuous exploration of 2D materials, their relationship with wide-bandgap materials like silicon carbide (SiC), gallium nitride (GaN), and diamond, is expected to form the foundation for significant progress in semiconductor material science (Li et al., 2018).

2.6.2. Unique Properties and Potential Applications

Continuing evolution of semiconductor material science has uncovered special properties of a range of materials, particularly 2D materials like graphene and transition metal dichalcogenides (MoS_2). Graphene, for example, has excellent electrical conductivity and mechanical rigidity because of its intrinsic two-dimensionality made up of an atomic monolayer of hexagonal lattice carbon atoms. The possibility of bandgap manipulation through chemical doping or structural distortion presents prospects for application in numerous areas, including optoelectronics, sensors, and flexible electronics. In addition, graphene nanoribbon (GNR) research has been optimistic to enhance spintronic performance by their edge states, thus realizing the multifunctional benefits of low-dimensional material utilization in future electronics (Bostwick et al., 2010). In addition to 2D materials, wide-bandgap semiconductors like gallium nitride (GaN) and silicon carbide (SiC) have also become critical building blocks in high-power and high-frequency applications. These materials are known for very good thermal stability, appropriate for applications in harsh environments, e.g., aerospace and automotive. The potential of bandgap engineering through the precise doping makes it possible to achieve tailored electronic properties, guaranteeing efficiency and performance enhancement. For instance, GaN's big bandgap facilitates more effective light-emitting diodes (LEDs) and power amplifiers, while SiC excels in power electronics through its lower on-resistance and higher breakdown voltage. Generally, these characteristics position wide-bandgap materials at the forefront of the transition toward more effective energy devices and clean technologies in the semiconductor sector (Davidson et al., 2009). Another major area in semiconductor material science is the fabrication of electrochemical sensors from carbon materials, such as carbon nanotubes (CNTs) and synthetic diamond. CNTs have received a lot of interest due to their high electrochemical reactivity that can noticeably enhance the sensitivity of sensors for biomolecules. For instance, CNT-modified electrodes can be coated with biomolecules and minimize the surface fouling effect, allowing real-time detection of analytes at ultralow concentrations. Similarly, synthetic diamonds exhibit better electrochemical performance, chemical inertness, and biocompatibility and are therefore preferred to be used for biosensing applications. The use of diamond nanowires in sensing platforms has also facilitated additional development in label-free detection methods, rendering carbon-based materials vital components in the next wave of electrochemical biosensors (Davidson et al., 2009).

2.6.3. Current Research Trends and Future Prospects

Significant advances in semiconductor material science are currently placing priority on the synthesis and characterization of wide-bandgap materials such as gallium nitride (GaN) and silicon carbide (SiC). These materials are increasingly important since they can operate effectively at high voltages and temperatures and are, thus, viable for application in power electronics. They are focused on by research with a view to enhancing their characteristics through bandgap engineering and defect control in their crystalline structures. Through precise control of the bandgap, researchers aim at making devices more efficient and effective, particularly high-power devices. As such innovation is attempted to be commercialized by makers, the interaction between crystal growth, doping techniques, and resulting device performance remains a focus in the field. It not only contributes to the overall body of knowledge about semiconductor behavior but also allows for the development of next-generation electronic components. Two-dimensional (2D) materials have been one such area of focus in recent times with their unique electronic characteristics and potential applications. Graphene and transition metal dichalcogenides such as molybdenum disulfide (MoS_2) are quintessential examples of this new and promising area of investigation, whose reduced dimensionality leads to emergent properties not observed in their bulk counterparts. It is currently investigated in the self-organization of such materials onto substrate surfaces to enable the incremental formation of nanostructures with controlled functionalities specifically, this self-assembly is now also employed to develop novel and advanced electronic devices that exploit the intrinsic characteristics of 2D materials. The development of flexible, light, and energy-saving electronic devices displays a front-runner direction in semiconductor material science that converges materials science, nanotechnology, and engineering disciplines to facilitate cross-disciplinary innovations in device fabrication and application (Figure 2.13).

Figure 2.13. *Important advancements in semiconductor material science.*

Source: https://asia.kyocera.com/products/uploads/thumb_pd_all.png.

Besides, the most encouraging of the newer trends in recent research indicate rising interest in the integration of 3C-SiC with silicon (Si) substrates, both increasing innovative prospects and concerns. Cubic silicon carbide provides excellent isotropic properties, making it highly appropriate for power metal oxide semiconductor field effect transistors (MOSFETs). Despite its desirable characteristics, the integration process shows material constraints, such as the difficulty of achieving effective p-type doping after ion implantation due to the annealing temperature requirements. Investigation is ongoing to develop methods that improve activation efficacy and device performance, as shown in recent vertical 3C-SiC-on-Si MOSFET process flows (Antoniou et al., 2019). These developments mark a turning point in the relief of current obstacles and optimization of manufacturing techniques that could ultimately lead to marketable solutions within power electronics, ultimately paving the way for semiconductor technology.

SUMMARY

When one synthesizes the various motifs addressed in semiconductor material science, it is evident that the field is marked by intricate and dynamic interaction between crystal structures, doping, defects, and innovative engineering tactics. The manipulation of these constituents renders material properties for modern electronic devices tunable. The invention of doping has significantly increased semiconductors' electrical conductivity and enabled charge carrier concentrations to be manipulated. Secondly, defects cannot be exaggerated; they are both a problem and a chance to optimize materials and impact photovoltaic cell and transistor performance to an equal extent. In the end, understanding such constituents proves to be essential in the future of establishing reliable semiconductor technologies that meet the demands of contemporary electronic devices. As mentioned throughout this study, bandgap engineering is an essential method in semiconductor design and application. With the modification of bandgap in materials, scientists can establish semiconductors based on specific purposes, whether in optoelectronics, solar cells, or power electronics. The work on the wide-bandgap materials like GaN and SiC stands particularly out, as these materials are more thermally stable and efficient than the traditional silicon-based semiconductors. Further, the unique properties exhibited by 2D materials like graphene and MoS_2 have opened new avenues for new applications because they are flexible, highly conductive, and miniaturizable. Taken together, these developments highlight the intense influence of bandgap engineering on the future state of semiconductor material science. In reflecting on the future direction of semiconductor research, it is critical to recognize challenges and opportunities that are facing us. Industrialization of prospective materials, as seen in initiatives such as the low-cost silicon solar array (LSSA), calls for a close scrutiny of manufacturing capabilities and scalability (Gates et al., 2025). In addition, the interesting findings on 2D transition metal dichalcogenides, where unique properties at reduced dimensions are noted, suggest that progression in these will increasingly rely on advances in nanoscale materials

(Chakrabarty et al., 2014). Overcoming the production and commercialization of such advanced materials will be crucial to their applications across various fields. Sustained collaboration of theoretical understanding and experimental use has the capability to propel the semiconductor industry, producing the innovations that can transform technology and society as a whole.

In semiconductor materials science, the interplay of crystal structures and doping processes is central in the determination of the electronic properties of materials. The atomic makeup of a semiconductor not only establishes its conductivity but also the facility with which it might be doped, thereby allowing for optimal electrical characteristics. For instance, the introduction of specific dopants will yield the required charge carriers, optimizing performance from transistor to solar panels. Moreover, the existence of defects, intrinsic or extrinsic, can change the conductivity and trap the charges, impacting the efficiency of semiconductor devices as a whole. Such underlying phenomena are important for developing technology and optimizing material properties in electronic devices. Bandgap engineering is found to be a unifying concept in semiconductor physics, particularly in device fabrication for selective optical and electronic properties. By adjusting the bandgap, researchers can prepare materials that respond suitably to various wavelengths of light, a property crucial in optoelectronics and photovoltaic applications. The ability to design semiconductors with specific bandgaps allows devices to be operated at varying levels of energy. For instance, GaN and SiC are broad-bandgap materials that have better efficiency and thermal stability and are perfectly suited for high-power and high-frequency applications. This deliberate bandgap behavior design is critical to ensure performance optimization across a broad range of semiconductor-based technologies. The investigation of new materials, specifically two-dimensional (2D) compounds such as graphene and MoS_2, has presented new opportunities to push semiconductor technology ahead. These materials exhibit unusual electronic characteristics due to reduced dimensionality and quantum confinement effects. For instance, MoS_2 has been of interest due to its ability to hold valley coherence at room temperature, as documented by research citing preservation of valley coherence through linear polarization resolved photoluminescence techniques (Kallatt et al., 2016). In addition, the defect engineering capability and the degree of control over the electronic environment in these 2D materials provide researchers the platform to tailor them to suit a particular application, e.g., sensors or transistors. It is, therefore, important to understand the nature and possibilities of wide-bandgap materials and 2D compounds for semiconductor technologies to continue to evolve.

REFERENCES

1. Aguirre, F. L., Eizenberg, M., Krylov, I., et al. (2018). Influence of the spatial distribution of border traps in the capacitance frequency dispersion of Al_2O_3/InGaAs. https://core.ac.uk/download/287883706.pdf (accessed on 27th September 2025).
2. Ajoy, A., Jana, R. K., Jena, D., et al. (2015). Transistor switches using active piezoelectric gate barriers. http://arxiv.org/abs/1505.00378 (accessed on 27th September 2025).
3. Alexander, P., Burger, D., & Gallagher, B. (2025). Flat-plate solar array project. Volume 5: Process development. https://core.ac.uk/download/pdf/42837487.pdf (accessed on 27th September 2025).
4. Alonso-Álvarez, D., Ekins-Daukes, N., Farrell, D., et al. (2018). Solcore: A multi-scale, Python-based library for modelling solar cells and semiconductor materials. http://arxiv.org/abs/1709.06741 (accessed on 27th September 2025).
5. Amani, M., Chrzan, D. C., Chueh, Y. L., et al. (2020). Shape-controlled single-crystal growth of InP at low temperatures down to 220 °C. https://core.ac.uk/download/287624057.pdf (accessed on 27th September 2025).
6. Amano, H., Baines, Y., Borga, M., et al. (2018). The 2018 GaN power electronics roadmap. https://core.ac.uk/download/154431388.pdf (accessed on 27th September 2025).
7. Antoniou, M., Arvanitopoulos, A., Gyftakis, K. N., et al. (2019). Viable 3C-SiC-on-Si MOSFET design disrupting current material technology limitations. https://core.ac.uk/download/228157973.pdf (accessed on 27th September 2025).
8. Arakawa, H., Asada, M., Bardotti, L., Bauer, M., Berger, J., Brune, H., Brune, C., et al. (2005). Self-organization on surfaces: Foreword. https://core.ac.uk/download/580035061.pdf (accessed on 27th September 2025).
9. Asta, M. D., Broberg, D., Han, G., Jaquez, R., et al. (2018). Bistable amphoteric native defect model of perovskite photovoltaics. http://arxiv.org/abs/1805.04977 (accessed on 27th September 2025).
10. Avsar, A., Avsar, A., Brown, A., Avsar, A., Radisavljevic, B., Zimmermann, B., Özyilmaz, B., et al. (2017). Gate-tunable black phosphorus spin valve with nanosecond spin lifetimes. http://arxiv.org/abs/1706.02076 (accessed on 27th September 2025).
11. Babu, S. R., Dhar, N. K., Ganguly, S., Ghosh, S., et al. (2025). Development of high-performance graphene-HgCdTe detector technology for mid-

wave infrared applications. https://core.ac.uk/download/pdf/288485151.pdf (accessed on 27th September 2025).

12. Bahriz, M., Colombelli, R., Crisafulli, O., Moreau, A., et al. (2007). Design of mid-IR and THz quantum cascade laser cavities with complete TM photonic bandgap. https://core.ac.uk/download/216115678.pdf (accessed on 27th September 2025).
13. Bailey, S. G., Flood, D. J., & Landis, G. A. (2025). Advances in thin-film solar cells for lightweight space photovoltaic power. https://core.ac.uk/download/pdf/42827406.pdf (accessed on 27th September 2025).
14. Banszerus, L., Beschoten, B., Epping, A., Güttinger, J., et al. (2018). Quantum transport through MoS_2 constrictions defined by photodoping. http://arxiv.org/abs/1612.01118 (accessed on 27th September 2025).
15. Berkelbach, T. C., Chenet, D. A., Heinz, T. F., Hone, J., et al. (2013). Grains and grain boundaries in highly crystalline monolayer molybdenum disulfide. http://arxiv.org/abs/1301.1985 (accessed on 27th September 2025).
16. Bhattacharyya, S., Pandey, T., & Singh, A. K. (2014). Effect of strain on electronic and thermoelectric properties of few layers to bulk MoS_2. http://arxiv.org/abs/1407.7522 (accessed on 27th September 2025).
17. Bonomo, M., Cavallo, C., Di Pascasio, F., Dini, D., et al. (2017). Nanostructured semiconductor materials for dye-sensitized solar cells. https://core.ac.uk/download/54538567.pdf (accessed on 27th September 2025).
18. Boschker, H., & Mannhart, J. (2016). Quantum-matter heterostructures. http://arxiv.org/abs/1607.07239 (accessed on 27th September 2025).
19. Bostwick, A., Cresti, A., Dato, A., Castro Neto, A. H., Giesbers, A. J. M., Javey, A., Geim, A. K., et al. (2010). Towards graphene nanoribbon-based electronics. http://arxiv.org/abs/1002.4461 (accessed on 27th September 2025).
20. Boukhvalov, D. W., Han, S. H., Kim, E. K., Lee, H., et al. (2015). Charge transfer induced molecular hole doping into thin film of metal-organic-frameworks. http://arxiv.org/abs/1508.02139 (accessed on 27th September 2025).
21. Broderick, C. A., O'Reilly, E. P., et al. (2012). Band engineering in dilute nitride and bismide semiconductor lasers. http://arxiv.org/abs/1208.6441 (accessed on 27th September 2025).
22. Chakrabarty, S., Das, G. P., & Wasey, A. H. M. A. (2014). Quantum size effects in layered VX_2 (X = S, Se, Te) materials: Manifestation of metal

to semimetal or semiconductor transition. http://arxiv.org/abs/1408.1777 (accessed on 27th September 2025).

23. Davidson, J. L., Gurbuz, Y., Gürbüz, Y., Kang, J., et al. (2009). Review on carbon-derived, solid-state, micro and nano sensors for electrochemical sensing applications. https://core.ac.uk/download/pdf/11741398.pdf (accessed on 27th September 2025).
24. Deng, K., Kan, E., Qian, Y., Wu, X., et al. (2017). Graphene-like quaternary compound *SiBCN*: A new wide direct band gap semiconductor predicted by a first-principles study. http://arxiv.org/abs/1703.03893 (accessed on 27th September 2025).
25. Du, Y., Liu, E. K., Liu, Z. Y., Wang, W., et al. (2013). Crossover of magnetoresistance in the zerogap half-metallic Heusler alloy Fe_2CoSi. http://arxiv.org/abs/1304.3517 (accessed on 27th September 2025).
26. Gao, W., & Kono, J. (2018). Science and applications of wafer-scale crystalline carbon nanotube films prepared through controlled vacuum filtration. http://arxiv.org/abs/1810.02928 (accessed on 27th September 2025).
27. Gates, W. R., Lee, T., & Smith, J. L. (2025). *Historical Evidence of Importance to the Industrialization of Flat-Plate Silicon Photovoltaic Systems, Volume 2.* https://core.ac.uk/download/pdf/42874274.pdf (accessed on 27th September 2025).
28. Gluschke, J. G., Krogstrup, P., Micolich, A. P., Nygård, J., et al. (2017). Towards low-dimensional hole systems in Be-doped *GaAs* nanowires. http://arxiv.org/abs/1704.03957 (accessed on 27th September 2025).
29. Hatui, N., Jang, R., Keller, S., Koksaldi, O., et al. (2019). *AlGaN/GaN* superlattice based p-channel field effect transistor (pFET) with TMAH treatment. http://arxiv.org/abs/1902.02022 (accessed on 27th September 2025).
30. Kallatt, S., Majumdar, K., & Umesh, G. (2016). Valley coherent hot carriers and thermal relaxation in monolayer transition metal dichalcogenides. http://arxiv.org/abs/1605.06623 (accessed on 27th September 2025).
31. Krishnamoorthy, S., & Lyman, J. E. (2020). Theoretical investigation of optical intersubband transitions and infrared photodetection in $\beta\text{-}(Al_xGa_{1-x})_2O_3/Ga_2O_3$ quantum well structures. http://arxiv.org/abs/2001.07299 (accessed on 27th September 2025).
32. Lagally, M. G., & Paiella, R. (2018). Optical properties of tensilely strained *Ge* nanomembranes. https://core.ac.uk/download/286139374.pdf (accessed on 27th September 2025).

33. Li, H., Wang, H., Li, Y., Lu, X., Li, L., Yan, Y., & Guo, W. (2025). Micro-nanoscale laser subsurface vertical modification of *4H-SiC* semiconductor materials: Mechanisms, processes, and challenges. *Discover Nano, 20.* https://www.semanticscholar.org/paper/e03bca56af9f2b033cbdbbce8cd377a45791ef5f (accessed on 27th September 2025).

34. Li, J., Lu, H., Ni, G., Peng, X., et al. (2018). Electronic, optical and transport properties of van der Waals transition-metal dichalcogenides heterostructures: A first-principal study. https://core.ac.uk/download/223021738.pdf (accessed on 27th September 2025).

35. Li, J., Tingsuwatit, A., Alemoush, Z., Lin, J. Y., & Jiang, H. (2025). Ultrawide bandgap semiconductor *h-BN* for direct detection of fast neutrons. *APL Materials.* https://www.semanticscholar.org/paper/d76bf92576178b56c48716afeff88b7475600558 (accessed on 27th September 2025).

36. Meng, L. (2024). Material selection and performance optimization of deep ultraviolet photodetectors: A comparative study of silicon-based and wide-bandgap semiconductors. *Highlights in Science, Engineering and Technology.* https://www.semanticscholar.org/paper/811de91d5818ac9998e23cddd58f6848c793fd33 (accessed on 27th September 2025).

37. Muhammad, S., Ullah, A., Zaman, S., Mahmoud, E. E., & Belay, M. (2024). 3D molecular structural modeling and characterization of indium phosphide via irregularity topological indices. *BMC Chemistry, 18.* https://www.semanticscholar.org/paper/6c92965868f65212e41872a14dc8c1fbc3ce90da (accessed on 27th September 2025).

38. Pan, H., Yi, J. B., Ding, J., Yang, J. H., Yin, J. H., Lin, J. Y., Van, L. H., et al. (2006). Carbon-doped *ZnO*: A new class of room temperature dilute magnetic semiconductor. http://arxiv.org/abs/cond-mat/0610870 (accessed on 27th September 2025).

39. Rauch, T., Rogalev, V., Bauernfeind, M., Maklar, J., Reis, F., Adler, F., Moser, S., et al. (2019). Nontrivial topological valence bands of common diamond and zinc-blende semiconductors. *Physical Review Materials.* https://www.semanticscholar.org/paper/d6af5ae9674207cde311e8cd73ebb8e1a42867ff (accessed on 27th September 2025).

40. Schroder, D. (1990). *Semiconductor Material and Device Characterization.* https://www.semanticscholar.org/paper/032a594d5edf02fd92b14b0fb3e853ee8067871c (accessed on 27th September 2025).

41. Shiraishi, K., Takayanagi, H., Tamura, H., & Shiraishi, K. (2001). Design of a semiconductor ferromagnet in a quantum dot artificial crystal. https://core.ac.uk/download/56642719.pdf (accessed on 27th September 2025).
42. Singha, A., Roy, A., Grethlein, C. E., Lamb, H., Dhar, P., & Rytov, S. M. (2004). A non-destructive analytic tool for nanostructured materials: Raman and photoluminescence spectroscopy. http://arxiv.org/abs/cond-mat/0406096 (accessed on 27th September 2025).
43. Tan, M., Zha, J., Hu, Y., Chen, Q., Zhang, J., & Luo, X. (2024). Research progress of TiO_2 modification and photodegradation of organic pollutants. *Inorganics.* https://www.semanticscholar.org/paper/bde1961887e7e83c121aed7f67fbc09b97706a2d (accessed on 27th September 2025).
44. Triplett, G. E. (2013). Extending device performance in photonic devices using piezoelectric properties. https://core.ac.uk/download/215485112.pdf (accessed on 27th September 2025).
45. Zerger, C., Contryman, A. W., Lee, C., Patankar, S., Orenstein, J., Layden, T. J., Kastner, M., et al. (2021). Quantum imaging of single-atom spin-splitting in a monolayer semiconductor. https://www.semanticscholar.org/paper/7cb63bb84b9989c6dcd73982c68106b874e1e87d (accessed on 27th September 2025).

CHAPTER 3

CARRIER TRANSPORT MECHANISMS

Contents

Introduction 89
3.1. Drift Mechanism 93
3.2. Diffusion Mechanism 98
3.3. Recombination and Generation 103
3.4. High-Field Effects 108
3.5. Ballistic Transport 113
3.6. Quantum Confinement 118
Summary 124
References 127

INTRODUCTION

The mechanistic principles behind carrier transport phenomena are the cornerstones of modern electronics and optoelectronics, spreading their reach from semiconductor design to nanotechnology. Understanding how charge carriers behave in varying circumstances is essential in the development of devices that offer improved performance and reliability. In semiconductor physics, drift, diffusion, recombination, and generation processes are all very important to understand how carriers respond to electric fields, temperature gradients, and defects in the material. As devices are scaled down to the nanoscale, these transport processes become ever more complex and interconnected, necessitating a deeper understanding of them to maximize the device's functionality. The dynamism of such mechanisms underlines the necessity to develop theoretical models and experimental approaches further to adequately define carrier behaviors in materials and architectures of interest. Moreover, recent insight into high-field effects and ballistic transport has also demonstrated the necessity to re-evaluate common paradigms of electron flow in solid-state systems (He et al., 2022). The study of quantum confinement in nanoscale devices has changed the character of carrier transport, unveiling phenomena beyond everyday experience. Once the size of materials is brought to dimensions approaching the de Broglie wavelength of charge carriers, quantum effects are significant, and the transport characteristics are drastically altered. In such circumstances, the quantized energy levels and increased scattering results highlight the need for novel descriptions of physics to accurately describe electronic phenomena. This environment also raises questions on how drift and diffusion mechanisms interact in quantum-confined systems because carriers possess both classical as well as quantum mechanical properties. Furthermore, more than pure science, these findings dictate useful applications in photovoltaic cells, lasers, and transistors. The strategic engineering of nanoscale devices more relies on awareness of quantum mechanics in order to successfully harness these unusual transport phenomena (Xiao et al., 2021). With the continuous advances in device engineering and material science, studies on carrier transport mechanisms are becoming ever more significant. New materials such as graphdiyne have been piquing interest for their outstanding properties, which provide potential for innovation in energy storage, conversion, and catalysis. The unique properties of these materials enable carrier mobility manipulation, which is important in enhancing the efficiency of electronic devices and delivering maximum performance in various applications. Furthermore, the development of vehicle for manipulating recombination and generation rates holds promises for new possibilities in system design with high performance. Understanding carrier transport mechanisms in such sophisticated materials in detail is necessary to reach their full potential and drive the fields of sustainable energy and efficient electronic technologies. As much as ongoing research continues to uncover the complexities of carrier dynamics, it remains pertinent to bring theory and experiments together to generate new solutions that address real technological challenges (He et al., 2022).

Definition of Carrier Transport Mechanisms

In condensed matter physics, numerous processes control the transport of charge carriers and, therefore, the efficiency of electronic and optoelectronic devices. Drift is the transport of carriers due to an electric field, by which electric currents are conducted through conductive materials. On the other hand, diffusion is the movement of carriers from high to low concentration regions, a thermally driven process of inherent randomness. The two processes can be interactive, since diffusion will create a steady state in carrier concentration if not offset by drift. Besides, generation and recombination are significant processes by which carriers may annihilate one another or be created, respectively, resulting in changes to the material's total carrier concentration. It is crucial to comprehend these basic transport processes in order to achieve the maximum device performance and design their functionalities for utilization. With technology advancing towards nanodevices, new phenomena occur due to carrier transport physics itself in high electric fields. High-field effects result in dramatic enhancements of carrier mobility and can lead to breakdown of the ohmic conduction model, necessitating more finer details to be known regarding transport. Here, carriers can experience higher acceleration, which results in effects known as impact ionization and avalanche breakdown.

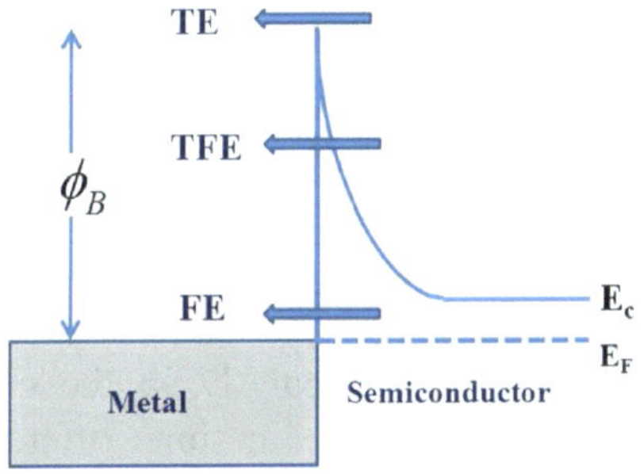

Figure 3.1. *Different types of current transport mechanism in metal/semiconductor Schottky diodes.*

Source:https://www.researchgate.net/publication/282331580/figure/fig2/AS:1143 1281208274153@1701382174463/Different-types-of-current-transport-mechanism-in-metal-semiconductor-Schottky-diodes.jpg.

Also, the ballistic transport is the major in these systems, whereby carriers move across the material with minimal scattering to maintain their kinetic energy over larger distances than their mean free path. This is the high-speed transport regime crucial in the production of high-frequency electronic devices to enable improved speed and efficiency in communication networks. The integration of such advanced mechanisms allows for the development of faster, more efficient devices that exploit the new properties of nanoscale materials. Quantum confinement is one of the dominant factors in controlling carrier transport mechanisms in nanoscale devices, such as semiconductor quantum dots and nanowires. In these structures, spatial confinement of carriers leads to quantized energy levels, drastically altering their transport behavior compared to bulk materials. The confinement effects enhance electron and hole interaction, so recombination and generation rates are greater due to smaller sizes and higher surface-to-volume ratios. Charge carriers' effective mass thereby changes, impacting their mobility and diffusion characteristics. Such an effect is particularly significant in performance-critical applications such as thermoelectric materials and photonic devices where carrier transport needs to be optimized. Advancements in their new quantum effects significantly improve technological innovation, enabling the development of materials to enhance carrier mobility and efficiency, as well as lowering loss (He et al., 2022; Ozawa et al., 2019) (Figure 3.1).

Importance in Semiconductor Physics

Dynamics of charge carrier action in semiconductor materials are especially significant in device operation, where roles played by carrier transport mechanisms become extremely important. Drift and diffusion are fundamental processes that account for motion of charge carriers under electric field and concentration gradients, respectively. When electrons and holes are subjected to an application of electric field, they exhibit drift, with the current flow enhanced. In contrast, diffusion plays a critical role in ensuring charge carrier equilibrium in a semiconductor, particularly in the case of heterogeneous semiconductors. The coupling of these motions offers a fertile ground for carrier generation and recombination, processes that play a crucial role in devices like photovoltaics and light-emitting devices. Understanding these processes enables one to optimize the efficiency of semiconductors by realizing optimal conditions that will enable desired carrier dynamics. This is particularly pertinent when addressing problems like bimolecular recombination that can prevent the device from working, as exemplified in recent work on new organic semiconductor mixtures (Gasparini et al., 2017). Advances in semiconductor physics have also included sophisticated behavior, like high-field effects and ballistic transport, which are growing more relevant to modern devices. High-field effects are seen when the applied electric field is sufficiently strong to affect carrier mobility, and mechanisms such as impact ionization and velocity saturation are formed. Such effects could introduce extra power dissipation and limits of operation in traditional devices. Ballistic transport, however, is concerned with carrier transport without scattering over a short distance, which is typical for nanoscale devices in which quantum confinement is extremely strong. Understanding these modes of transport is made simpler with which to design more efficient semiconductor devices through minimizing energy loss and ensuring carrier mobility. The breathtaking advances of two-dimensional materials have established a paradigm for such rules by leveraging unique transport properties that are unique to low-dimensional systems, initiating a new era of electronic applications (Novoselov et al., 2016). In addition, the interaction between carrier transport and materials structure has widespread significance for next-generation semiconductor technology, particularly that based on two-dimensional and organic materials.

Organic semiconductor research, for instance, highlights the importance of material property optimization in order to maximize carrier mobility and suppress bimolecular recombination losses, an issue that is compatible with high-efficiency device goals. Experiments have demonstrated that the crystalline structure of members in organic blends significantly influences carrier transport phenomena through the reduction of undesirable interactions responsible for recombination losses (Gasparini et al., 2017). Similarly, studies of heterostructures based on two-dimensional materials exhibit remarkable transport properties that find applications in innovative electronic and optoelectronic devices. These advances underscore the pivotal role of materials science to semiconductor physics, as a clear understanding of carrier transport mechanisms has a direct impact on device fabrication with historically unmatched performance parameters (Novoselov et al., 2016).

3.1. DRIFT MECHANISM

In semiconductor physics, the intricate behavior of charge carriers is primarily controlled by an underlining drift mechanism, under the influence of external electric fields. This is a crucial phenomenon that makes us understand carrier transport because it determines the manner in which charge carriers, such as electrons and holes, mobilize in semiconductor materials. With an applied electric field, carriers gain kinetic energy and relocate in directions that are opposite to electrodes, enhancing the efficiency of devices such as solar cells. The drift process needs to be examined diligently along with other phenomena such as diffusion and recombination to understand material transport dynamics fully. Interestingly, innovations in modeling methods, especially the drift-diffusion model, reveal how these interactions can influence device performance and result in efficient designs in new photovoltaic technologies (Choy et al., 2017). Beyond the basic drift mechanism, high-field effects also become a dominant aspect of carrier transport, especially for devices exposed to extreme conditions. Under high electric fields, carriers possess enough energy to accelerate significantly, so much so that they can be subject to ballistic transport where their motion is less burdened by scattering events. This regime is quite different from the normal drift and diffusion behavior and involves complex modeling approaches to simulate transport phenomena accurately (Figure 3.2).

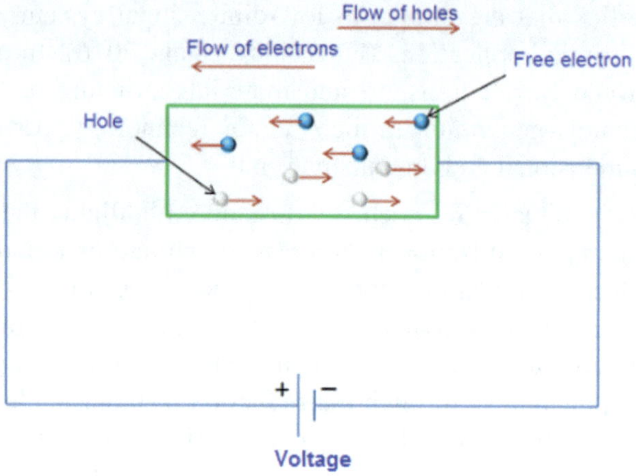

Figure 3.2. *The interplay between drift and diffusion processes.*

Source:https://www.physics-and-radio-electronics.com/electronic-devices-and-circuits/semiconductor/images/drift-current.png.

The interaction between drift and diffusion processes is quite intricate in such high-field conditions, which may have a remarkable impact on the performance of nanoscale devices. Additionally, such high-field effects are also important in the new field of spintronics where electron spin control combined with their charge transport enables new functions and devices (Abalmassov et al., 2010). Drift is also influenced by quantum confinement effects, particularly in nanoscale devices where the length scales are comparable to carrier wavelengths. In such devices, simple modifications in carrier statistics and energy levels necessitate reassessment of traditional transport models. Because carriers are under confinement, carrier energy states become quantized, thus altering carrier drift dynamics. Understanding these quantum effects is required to maximize device performance in devices like quantum dots and nanowires, where carrier transport mechanisms are significantly different from those in bulk materials. By integrating a profound understanding of both drift and quantum confinement, researchers will make possible new strategies for achieving the optimum performance and efficiency for next-generation semiconductor devices, across a broad spectrum of applications from photovoltaics to high-end electronic components (Choy et al., 2017).

3.1.1. Definition and Explanation of Drift

Carrier transport mechanisms are complemented by an extensive examination of drift, a process inherent in charge carrier motion in solid-state materials. Drift refers to the movement of charge carriers, say electrons and holes, due to an electric field. This is essentially related to the utilization of external electric fields that exert a directional force, propelling carriers to travel more quickly and inducing currents to flow. Drift can be quantitatively explained by equations derived on the basis of semiconductor physics theory, i.e., the relationship between drift velocity and electric field intensity and carrier mobility.

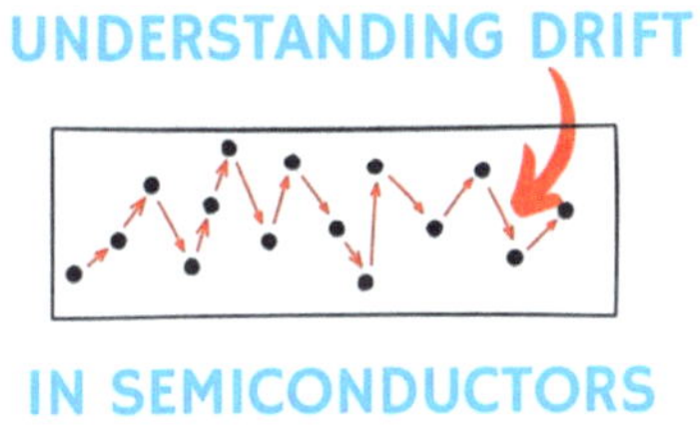

Figure 3.3. *A comprehensive examination of drift in semiconductor.*

Source: https://substackcdn.com/image/fetch/$s_!7N4G!,w_520,h_272,c_fill,f_auto,q_auto:good,fl_progressive:steep,g_auto/https%3A%2F%2Fsubstack-post-media.s3.amazonaws.com%2Fpublic%2Fimages%2F37398dec-d305-482c-9940-f933574812b2_1456x1048.png.

Surprisingly, the effect can be controlled by material impurities and lattice vibrations, which can significantly influence carrier mobility. Lastly, understanding drift is essential to optimize the performance of electronic devices that need precise control of carrier dynamics, particularly in advanced applications like spintronics, where charge and spin are manipulated (Aurélien Manchon et al., 2019). As part of a general investigation of carrier transport, how drift plays with other processes like diffusion and recombination demonstrates the complexity of semiconductor action. While drift is highly dependent on the external electric field, diffusion is mediated by concentration gradients, driving carriers from high concentration to low concentration regions. This makes it clear that carrier transport has a dual nature, where drift is what can increase overall material conductivity under the influence of an applied field, and diffusion is what acts as a key mechanism for reaching equilibrium in charge distributions. Interestingly, drift and diffusion interaction can generate responses like the generation of charge accumulation within junctions, which influences the functioning of devices across various applications, such as the case for medical technology and security systems (Yanagida, 2018) (Figure 3.3).

Knowledge of these interactions is the basis for innovations in carrier transport mechanism and device structures. Apart from being influenced by external fields, drift is also significantly influenced by high-field effects as well as quantum confinement effects for nanoscale devices. With the increase in strength of the electric field, carriers can be subjected to more intense scattering events due to vibrations in the lattice, which makes the current-voltage characteristics non-linear—a property that is highly crucial in high-speed electronic applications. Further, in nanomaterials, quantum confinement effects can change the energy levels of charge carriers and thus cause drift and modify mobility. In such systems, drift properties become more complex due to reduced dimensions, and it takes advanced theoretical models to accurately predict behavior. This richness is also to be found in future fields like spintronics, where the spin-orbit coupling can enable an enhancement of drift through new mechanisms that are related to quantum characteristics (Aurélien Manchon et al., 2019). The understanding of these high-level drift phenomena is essential for the advent of next-generation electronic devices that operate effectively at the nanoscale.

3.1.2. Factors Affecting Drift Velocity

Understanding drift velocity involves a thorough exploration of many factors influencing carrier transport in semiconductor materials. The electric field imposed across a material can significantly alter the drift velocity of charge carriers like electrons and holes. When the field is stronger, the carriers are accelerated more and hence the drift velocity is higher. This relationship is, however, not linear at very high strengths of field, as effects like velocity saturation arise due to the effects of scattering. At high energies, the carriers are scattered by lattice defects or phonons, which ultimately slow them down and put a limit on the drift velocity. The higher breakdown voltage materials are also able to withstand higher-strength electric fields without loss, and it is a measure of their efficiency directly when in use.

Therefore, one should be aware of how field strength interferes with carrier dynamics to design and improve devices. Temperature is another factor that controls drift velocity through the influence of carrier concentration and mobility in a semiconductor. Thermal vibration of the lattice structure becomes higher with an increase in temperature, which means more instances of scattering occur. Such scattering might reduce the mobility of charge carriers and thereby reduce their drift velocity. Carrier mobility is more temperature dependent in silicon than in wide bandgap semiconductors such as silicon carbide. The superior thermal conductivity and the ability to handle higher operating temperatures at effective drift velocities are provided by the latter, as exemplified by the widespread use of SiC MOSFETs in power devices. Higher drift velocities in such hard materials may result in higher power conversion efficiency, and temperature control and material selection are paramount in optimizing carrier transport mechanisms. Quantum confinement effects also have a major role to play in drift velocity in nanoscale devices, and the carrier transport dynamics are changed accordingly. As dimensions shrink to the nanoscale, the behavior of electrons varies from classical physics and gives rise to quantized energy and the altered mobility characteristics. For material such as inorganic perovskite quantum dots, the ability to manipulate their electronic characteristics allows for additional optimization of carrier transport. Quantum confinement enhances the quantum yield of photoluminescence and allows one to control precisely the recombination rate and therefore influence the effective drift velocity of charge carriers. There is also a need for advanced synthesis methods because exactly controlling growth conditions will enable materials to capitalize on these quantum effects. Understanding such principles is increasingly crucial in designing the next generation of electronic devices that are capable of operating with high efficiency under multiple conditions, as demonstrated in recent research into inorganic perovskite QDs (Chien-Huang et al., 2022) and SiC power devices (Langpoklakpam et al., 2022).

3.1.3. Applications of Drift in Electronic Devices

In the field of modern electronics, drift phenomena become critical in optimizing the functionality of electronic devices, particularly in organic electrochemical transistors (OECTs). The combined transport of ionic and electronic charge carriers in OECTs due to their different architecture leads to high transconductance and huge gate-to-channel capacitance. Unlike conventional field-effect transistors, the added complexity due to ion-electron coupling provides an opportunity for awesome breakthroughs in device performance. Since sophisticated mixed charge transport is involved, advanced theoretical models like one- and two-dimensional drift-diffusion formalisms have been employed to describe the OECT operating mechanism. These models are pivotal in deciding key parameters influencing device performance, which ultimately paves the way for innovative applications in digital circuits and biosensors, where response and precision are of the highest concern in delivering effective electronic solutions (Tu et al., 2020). In addition, perovskite compounds, especially methylammonium lead halides, have demonstrated massive advancements in optoelectronic applications with tremendous improvement arising mostly due to their distinctive carrier transport behavior under the influence of ionic drift phenomena. The migration of ions in such materials can dramatically alter electronic behavior, normally causing problems of the form of hysteresis and degradation in device reliability. New studies show that doping with other mobile I^- ions can enhance ion migration, hence altering the electrical properties of perovskite films. By using ordered devices, i.e., vertical metal oxide semiconductors and lateral metal-semiconductor-metal diodes, researchers can effectively decouple the role of ion migration from electronic transport. Differences in measurement circumstances, such as electric field history and scan rate, also underline the sophisticated role of drift behaviors on device operation, pointing to high-efficiency applications ranging from memristors to solid electrolyte batteries (Pang et al., 2018). The effects of drift in electronic devices extend beyond the operation efficiencies themselves to encompass even the rudiments of transport phenomena at lower dimensions. In nanodevices such as quantum dots and nanowires, drift behavior becomes coupled with quantum confinement effects to produce fascinating transport behavior. As device dimensions become smaller, classical transport models must change to encompass quantum mechanical effects as well as drift and diffusion. Interplay between these variables can result in variable conductivity and current modulation based on the device structure and materials of choice. More importantly, onset of high-field phenomena in ultrathin devices increases the importance of drift dynamics, necessitating advanced simulation techniques for predictive modeling across a wide range of operating conditions. Thus, the merging of drift applications not only enhances current technologies but also facilitates further research on new structures and materials, which is essential to further develop the electronic world.

3.2. DIFFUSION MECHANISM

The diffusion theory is fundamental in describing the mechanisms by which carriers transfer themselves from one semiconductor material to another. The process is directly related to the concentration gradients in the material such that carriers will flow from regions of higher concentration to regions of lower concentration. The implications of such actions are substantial in device operational performance such as in perovskite solar cells, where the dynamics of ion migration play a major role in determining the measures of performance. Understanding the coupling of mobile ions in the absorber layer is vital in such systems so that band alignment and transport of charge may be maximized. Recent studies with techniques like X-ray photoelectron spectroscopy (XPS) provide valuable insights in these procedures, particularly concerning the influence of light on carrier generation and charge accumulation modulation at interfaces, thus emphasizing the role of diffusion mechanisms in device function optimization (López-Varo et al., 2024) (Figure 3.4).

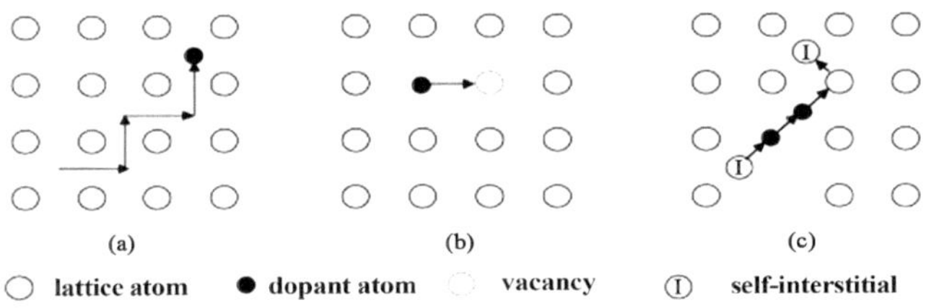

Figure 3.4. *The mechanism of diffusion.*

Source:https://ars.els-cdn.com/content/image/3-s2.0-B0080431526015242-gr1.jpg.

Because carrier transport in semiconductor devices is controlled not just by diffusion but also by other basic processes, the interaction between these mechanisms is worthy of extensive investigation. Drift, for instance, goes hand in hand with the application of external electric fields, affecting carrier mobility and potentially enhancing conduction as a whole. Diffusion itself is not influenced by such fields and is tasked with achieving thermal equilibrium. Furthermore, generation and recombination processes can vie with diffusion and affect the net carrier number available for conduction. In photoconductive devices such as GaAs switches, the transient electric fields that are generated can lead to complex phenomena, like nucleation of charged domains when the photogenerated carriers interact under specific biases. These interactions indicate the profound effect of diffusion on carrier dynamics and device performance of state-of-the-art devices (Li-Qiang et al., 2023).

In nanoscale devices, quantum confinement introduces additional complexities to the diffusion process. As carriers are constrained in very small dimensions, their mobility can significantly alter relative to bulk materials, affecting the effectiveness of processes such as ballistic transport and leading to non-classical transport phenomena. The carrier transport rules need to be altered in order to consider changes in energy levels and localized states, which are seen as a consequence of quantum effects. Information on these dynamics is highly significant in designing new semiconductor devices, particularly in optoelectronics and high-speed electronics. The diffusion mechanisms must be discussed in conjunction with other transport phenomena to fully appreciate their importance for device performance, marking milestones in material science and engineering technology for the next generation of application in this rapidly emerging field (Li-Qiang et al., 2023; López-Varo et al., 2024).

3.2.1. Definition and Explanation of Diffusion

Understanding of the intricate processes that govern carrier transport is crucial to comprehension of diffusion, a significant process that underlies particle motion in various media. It is a process that takes place as a result of thermal random motion of particles, leading them to move from regions of high concentration to regions of low concentration, hence reaching equilibrium. Diffusion is not a passive process that occurs; it is regulated by particle size, temperature, and viscosity of the media. The mathematical treatment which describes diffusion usually is condensed in Ficks laws, which define the relationship between the particle flow and concentration gradient. This model is important in the understanding of the transport of energy in semiconductor materials and living organisms, with diffusion playing a key role in nutrient assimilation and the elimination of waste products within living cells. Diffusion is also involved in active transport processes, where particles, in this instance ions, move through membranes against their concentration gradient through the use of energy in the form of ATP (Figure 3.5).

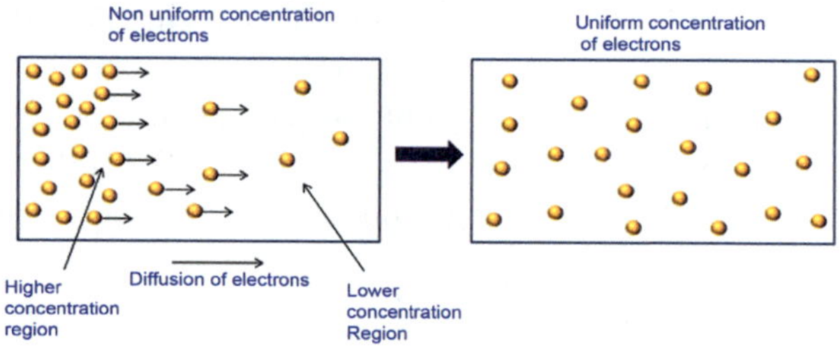

Figure 3.5. *The diffusion mechanisms in semiconductor materials.*

Source:https://www.physics-and-radio-electronics.com/electronic-devices-and-circuits/semiconductor/images/diffusion-current2.png.

This equilibrium between passive diffusion and active transport is of immense significance within biological organisms, where cells actively regulate their internal environments to attain homeostasis. For instance, gas exchange between oxygen and carbon dioxide through the alveolar membrane of the lungs is concentration gradient based, providing an example of species exchange for respiration. In contrast, glucose entry into cells would generally need to involve active transport processes, different from the complexity of diffusion in biology. These differences are crucial in differentiating various transport processes under other conditions, e.g., high-field effects, which can facilitate ionic conduction through non-linear behaviors in electric fields. In addition, diffusion is applied in advanced materials science, specifically in the case of nanoscale devices where quantum confinement will change diffusion behavior. When particle sizes are at the nanoscale, typical models of diffusion will not be able to predict correctly due to altered surface-to-volume ratios and the greater relevance of quantum effects. In this context, researchers are exploring artificial autonomous particles, inspired by nature, that exhibit unique diffusion patterns in concentrated systems (Bechinger et al., 2016). The artificially crafted particles can mimic biologically inspired behaviors, and the door opens to applications in medicine, sustainability, and defense. Understanding diffusion in these cases not only enhances our understanding of transport processes but also brings to the forefront the intricate interdependencies between microscopic actions and macroscopic effects, revealing the nuances involved in efficiency and energy innovation systems (Sorrell, 2015).

3.2.2. Fick's Laws of Diffusion

The intricacies of carrier transport processes are essentially influenced by fundamental principles that control the particle behavior in most materials. Ficks Laws of Diffusion play a central role in the explanation of migration of carriers such as electrons and holes in concentration gradients.

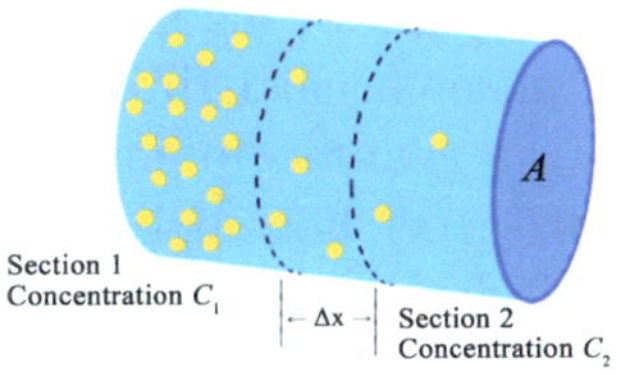

Figure 3.6. *Fick's first law of diffusion.*

Source: https://study.com/cimages/multimages/16/mydiffusion3845222956040873962.png.

Fick's first law specifically addresses steady-state diffusion, with the assertion that the rate of flux of particles is directly proportional to the negative concentration gradient. Fick's law acquires a special role in the arena of materials science, especially in the context of thermoelectric devices, when there has to be charge carrier transport with high efficiency for optimum performance. As advancements in thermoelectric materials depend both on intrinsic properties as well as on interface interactions, understanding Ficks laws is central to explaining how these materials can be designed towards better thermoelectric utilization for energy conversion and waste heat recovery, as demonstrated in the literature drawing attention to the need for a greater device-level understanding (Liu et al., 2019). Systematic study of diffusion effects is enhanced by incorporation of Fick's second law, which commences exploration of transient diffusion through the explanation of how the concentration varies with time. This evolving insight is of central importance not only for conventional semiconductors but also for mixed ionic-electronic conducting materials in which the coupling of the ions and electrons dictates transport behavior (Figure 3.6).

In ambipolar transport, for instance, the synergistic movement of positive and negative charge carriers may significantly alter diffusion characteristics, which is difficult to analyze under varied conditions (Kerner et al., 2017). Through Ficks second law, scientists can model these complex phenomena, thereby enhancing the advancement of thermoelectric materials and interfaces with maximum energy conversion efficiencies and maximum device lifetimes. As such, application of these principles to study of carrier transport mechanism is inevitable for the purpose of pushing material performance frontiers. Furthermore, implications of Ficks Laws extend as far as nanoscale devices, where quantum confinement is more apparent. With the reduction in dimensions, the behavior of charge carriers shifts from bulk to quantum mechanics, and therefore diffusion processes become more challenging. In such situations, ballistic transport effects and high-field phenomena can dominate, and one requires a proper understanding of how the laws of Fick need to be extended to include these emerging phenomena. Drift-diffusion and quantum effects interacting with each other can give rise to novel behaviors in charge carrier dynamics, particularly in next-generation applications based on thermoelectric materials and hybrid electronic devices. Placing an understanding of Ficks Laws into the context of such complex transport phenomena, therefore, is integral to the design of future materials and devices that leverage the full power of classical and quantum transport effects for meeting global energy needs.

3.2.3. Role of Diffusion in Carrier Concentration

Understanding the subtleties of carrier concentration is critical in semiconductor science and materials research. The dynamic carrier mechanisms such as drift and diffusion are essential in determining the overall electrical and optical properties of a material. Diffusion, which is the movement of charge carriers from regions of high density to regions of low density, is the principle underlying carrier mobility. This flow is not a simple passive movement in accordance with concentration gradients but gives rise to a very strong relationship between local carrier density and material conductivity. Perhaps most significantly, the relationship between diffusion and carrier concentration is of very deep significance in the optimization of electronic devices where effective charge transport is implicated. The overall consideration of diffusion mechanisms is therefore essential for the identification of the effective carrier concentration that sustains device operation. The connection between diffusion rates and carrier concentration is also explained by the nature of solid-state interfaces. The mechanism by which lithium-ion transport in lithium metal batteries works, for instance, is a key demonstration of how interfaces affect ionic conductivity. The low ionic conductivity that is felt in bulk inorganic solid electrolyte interphases (SEIs), contributed primarily by low carrier concentration, is a sign of the requirement for expanded diffusion channels to facilitate efficient ionic transport (Xia-Ma et al., 2022). In such materials such as Li 2 O/LiF, there is an increase in carrier concentration with enhanced ionic conductivity even when there is a high diffusion barrier. This demonstrates the importance of material design with optimized diffusion characteristics, which not only enhances ionic transport but also supports overall performance of energy storage devices where carrier concentration is a key aspect. Investigation of carrier transport phenomena in metal-organic frameworks (MOFs) reveals yet another facet of diffusions role in carrier concentration. The crystalline character and tunable properties of MOFs allow for extensive chemical changes affecting porosity and stability, which, in turn, alter diffusion paths. While some of these MOFs are very porous to make them useful for adsorption and separation of gases, charge carrier diffusion within the framework also relies on both their structural stability and functionalization (Furukawa et al., 2013). Such features enable carriers to freely travel through the framework, which may be applied for catalysis and energy storage. The cooperation of carrier concentration and diffusion in MOFs shows how material science can apply the underlying physical laws to engineer surfaces and interfaces to create new technological breakthroughs.

3.3. RECOMBINATION AND GENERATION

Charge carrier dynamic interaction in semiconductor materials is one of the keys to the photovoltaic device efficiency, particularly for the processes of recombination and generation. When a semiconductor absorbs light, the electrons are bumped to higher energy levels such that electron-hole pairs are formed. The fate of these formed carriers decides the quality of the device. When the carriers recombine before they can be collected for current generation, efficiency is significantly lost. It is important to learn the radiative and non-radiative recombination processes as it determines how to optimize energy conversion processes most optimally. Better materials, such as fullerenes, have been shown to enhance charge separation and transport and hence reduce recombination losses. Control of these interactions provides an opportunity to further enhance solar cell design and other electronic systems (Waqas et al., 2025). Charge carrier transport efficiency requires careful study of varying mechanisms like drift and diffusion, which are critical intervening processes between recombination and generation (Figure 3.7).

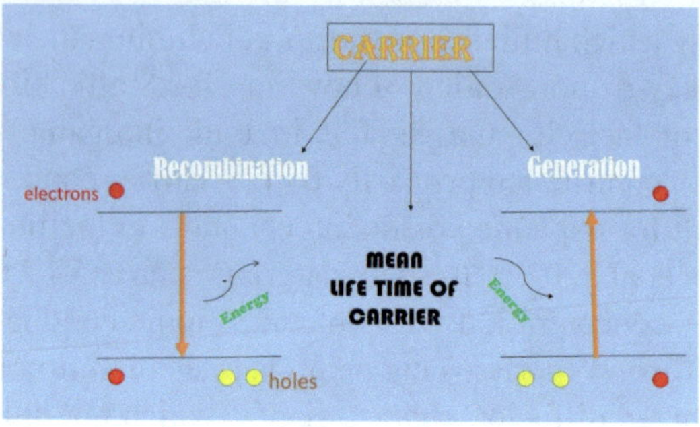

Figure 3.7. *Generation, recombination of charge carriers.*

Source:https://i.ytimg.com/vi/4y1PePVUmgA/maxresdefault.jpg.

Drift occurs when the movement of the carrier is caused by the influence of an electric field, while diffusion defines the movement due to the concentration difference. Both effects are central to determining how well charge carriers created can be collected at the electrodes to ensure high output. For materials like organic-inorganic halide perovskites, high carrier mobility is a key property, allowing for higher drift velocities that can increase the likelihood of collection before recombination occurs. Also, defects and impurities in the material can create trap states, which increase the rates of recombination. Covering these issues is critical for maximizing photovoltaic efficiency and realizing greater conversion efficiencies in new solar technologies (Zeng, 2025). Quantum confinement effects in nanoscale devices add a further level of complication in carrier generation and recombination. Nanoscale structures tend to have distinctive electronic properties because of their lower dimensionality, substantially changing carrier dynamics. In these kinds of systems, the energy levels get quantized and thus are able to provide a boost to the photonic absorption and induced transitions. But it also enhances the possibility of recombination because of localized states that can be trapping the carriers. Thus, the generation-recombination balance in these confinement systems must be finely tuned in order to realize their potential. By researching new material compositions and engineering methods which can resist recombination loss—e.g., defect passivation methods and incorporating tailored electron-accepting agents—engineers aim to develop carrier transport mechanisms definitively in nanoscale solar cell technologies (Waqas et al., 2025; Zeng, 2025).

3.3.1. Definition of Recombination and Generation

Understanding the major processes of recombination and generation in semiconductor physics is required to fully describe carrier transport mechanisms. In semiconductor materials, recombination is the mechanism whereby free charge carriers—i.e., holes and electrons—annihilate each other and therefore decrease the number of free carriers. In semiconductors, recombination can significantly affect electrical and optical characteristics since the operation of devices such as solar cells and light-emitting diodes depends to a great extent on the presence of an advantageous balance between generation and recombination. Generation, however, typically follows thermal excitation or photon absorption, resulting in the generation of electron-hole pairs. Temperature, material purity, and external fields control these processes and, consequently, exemplify the complex interplay accountable for controlling the electronic device's performance and highlighting the importance of achieving maximum carrier lifetimes and mobilities for efficient application in organic electronics (Scaccabarozzi et al., 2021) (Figure 3.8).

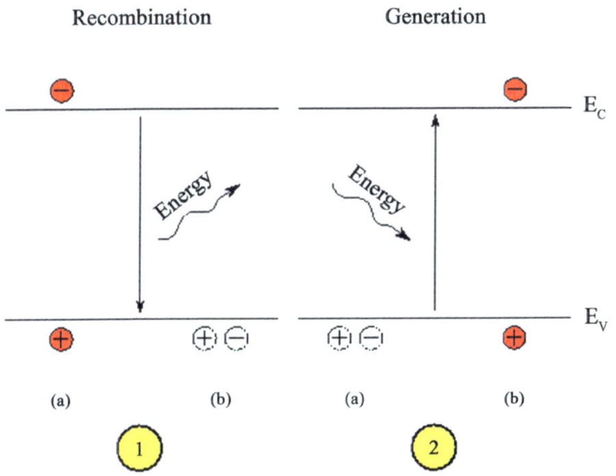

Figure 3.8: *Direct generation/recombination process.*

Source: https://www.iue.tuwien.ac.at/phd/entner/img158.png.

With increasing efforts by scientists to advance semiconductor technology, the recombination-generation interaction becomes clearly apparent in advanced materials. Sufficient doping methods, for instance, are invoked to reverse the effect of recombination and enhance charge carrier generation in organic semiconductors. The recent research interest in dopants has opened opportunities for increased conductivity of organic material, and hence they can be able to have properties which can make high-performance electronic devices possible. Awareness of the associated physical processes governing these mechanisms facilitates savvy creations of organic electronics that continue to advance device functions such as efficiency and reliability. In particular, analysis of popular mechanisms along with novel theories that have been put forward allows for a complete overview of how generation and recombination respond under various operating conditions, especially under operating conditions where maximum mobility and minimum recombination losses are best for the maximization of device performance (Scaccabarozzi et al., 2021). The newly emerging technologies that rely on quantum confinement effects in nanoscale devices also form a significant background upon which recombination and generation concepts can be elaborated. At the nanometer scales, the generation dynamics of the charge carriers are radically altered by the spatial confinement, which introduces discrete energy levels. Such a confinement can enhance the recombination rate but may also offer new routes for improved generation dynamics upon illumination. For instance, with scientists exploring the field of localized surface plasmon resonance and photothermal effects, these effects are capable of influencing generation paths, thus total transport efficiency. These advances result in a further exploration of how much the engineered nanostructures may alter the recombination-generations balance, ultimately resulting in application-oriented outcomes in photo-catalysis as well as chemical reaction enhancement. These observations could have a deep influence on the design methods of materials for the creation of high-performance materials used in energy conversion and storage applications.

3.3.2. Types of Recombination Processes

Recombination, a key process in semiconductor physics, is a fundamental player in determining carrier transport mechanism efficiencies. The interaction between electrons and holes forms the basis for several types of recombination processes, each with different properties and consequences. Perhaps the most ubiquitous are radiative and non-radiative recombination. Radiative recombination occurs with the emission of a photon when an electron recombines with a hole, and this is particularly important in optoelectronic devices such as light-emitting diodes and solar cells. Non-radiative recombination occurs via defect states and trap states, ejecting energy in the form of heat and not light. Comprehending these processes becomes essential, as they have direct consequence on the performance and efficiency of devices that are charge carrier dynamics-based and hence highlight the demand for the enhancement of carrier lifetimes for maximizing overall device performance. The architecture of heterojunctions plays a significant role in recombination mechanisms within semiconductor devices. By engineering at the nanoscale and utilizing two-dimensional (2D) materials, scientists have observed enhanced carrier separation and reduced rates of recombination. Heterojunction photocatalysts are a prime example, in which the spatial segregation of electron-hole pairs avoids the speedy recombination otherwise accountable for constraining photocatalytic efficiency. As described in Low et al. (2017), these advances facilitate improved light utilization and aids in elevated photocatalytic activity, which is essential in renewable energy and environmental remediation. Moreover, effective management of recombination processes with heterojunction engineering can allow for new insights into carrier transport physics and unlock new device designs that exploit the unique properties of 2D materials and their heterostructures. Analysis of various recombination mechanisms also extends into the quantum confinement regime in nanoscale devices. In these kinds of systems, charge carrier dynamics are significantly altered by confinement effects, leading to novel recombination dynamics. For instance, reduced dimensionality can enhance the interaction between light and matter, leading to the formation of 2D excitons, which are crucial in efficient energy transport. The properties of the new 2D crystals and their hybrids provide the capability to fabricate new devices, such as tunneling transistors and resonant tunneling diodes, as indicated in Novoselov et al. (2016). The phenomena reflect the inseparable connection between recombination processes, carrier transport mechanisms, and next-generation electronic and optoelectronic devices' functional capability, further confirming the need for greater research and innovation within this field.

3.3.3. Impact on Carrier Lifetime and Device Performance

The intricate interplay between carrier lifetime and device performance of semiconductor materials is pivotal in the construction of a broad span of optoelectronic applications. Carrier lifetime controls the duration that charge carriers can survive before recombining and, therefore, the efficiency of the photovoltaic device. In particular, recent research on halide double perovskites such as $Cs_2AgBiBr_6$ reveals that while such materials exhibit long lifetimes of charge-carriers that are greater than 1 μs, their performance in thin-film solar cells is still far from the theoretical limit. The findings indicate that grain boundaries are a significant issue because they are the prevailing sites for nonradiative recombination, hence restricting carrier mobility and device performance in general (Li et al., 2021). Therefore, grain size and structure optimization is needed to bring the performance of such thin films into register with the intrinsic single-crystal properties, and hence increasing carrier lifetime can have drastic effects on device performance improvement. Barring that, further work needs to be done in the study of carrier dynamics and how altering carrier density is related to recombination processes and how this can have a direct correlation effect on total device efficiency. Carrier generation and recombination kinetics are regulated through processes such as excitation intensity and the inherent properties of the material. In perovskites halides, Auger, bimolecular, and monomolecular recombination processes interact to offer critical insight into the optoelectronic properties of such materials. Exactly, at low carrier densities, carrier lifetime can appear to be extended because of lower recombination rates, but at higher densities, bimolecular recombination takes the upper hand, thereby reducing the carrier lifetime. The connection emphasizes the importance of achieving the right balance between carrier generation and the ability to sustain an adequate carrier lifetime to achieve desirable device performance in applications ranging from photovoltaics to light-emitting diodes and laser technology (Qin et al., 2020). To achieve even more insight into the carrier dynamics implications, high-field effects and ballistic transport phenomena introduce additional complexity to device operation. High electric fields may lead to quick carrier drift with enhanced transport efficiency but at the same time may introduce nonuniformities to cause increased carrier scattering and recombination losses. In nanoscale devices, where quantum confinement is a factor, the character of charge carrier behavior can be fundamentally different; for instance, confined carriers may enjoy greatly extended lifetimes as a result of reduced recombination rates over their bulk counterparts. This feature highlights the crucial importance of maximizing devices to utilize efficiently the new characteristics of nanoscale materials while mitigating recombination losses. Thus, carrier transport mechanisms at the nano and macro level must be tuned to devise effective optoelectronic devices that incorporate better carrier lifetime in their design (Li et al., 2021; Qin et al., 2020).

3.4. HIGH-FIELD EFFECTS

The behavior of charge carriers in semiconductor materials is fundamentally influenced by various transport mechanisms, particularly under the influence of high electric fields. As the applied field strength increases, carriers can gain significant energy, leading to conditions that deviate from normal drift and diffusion processes. High-field effects encompass a range of phenomena, including increased drift velocities and saturation due to scattering mechanisms that undergo rapid changes under these conditions. In many traditional materials, this can result in non-linear current-voltage characteristics that can be critical for device performance, particularly in high-frequency applications. Understanding these dynamics is essential for optimizing materials and structures in advanced electronics and can provide insights into the limits of carrier mobility under extreme conditions, an important factor in developing high-efficiency transistors and other semiconductor devices. Analyzing the interplay between high-field effects and other transport mechanisms enhances our understanding of carrier behavior in semiconductors. For instance, when carriers are subjected to intense electric fields, they can experience a transition from diffusive transport to ballistic transport, where they traverse distances without experiencing substantial scattering events. This transition is particularly pivotal in nanoscale devices, where quantum confinement effects begin to dominate carrier dynamics (Figure 3.9).

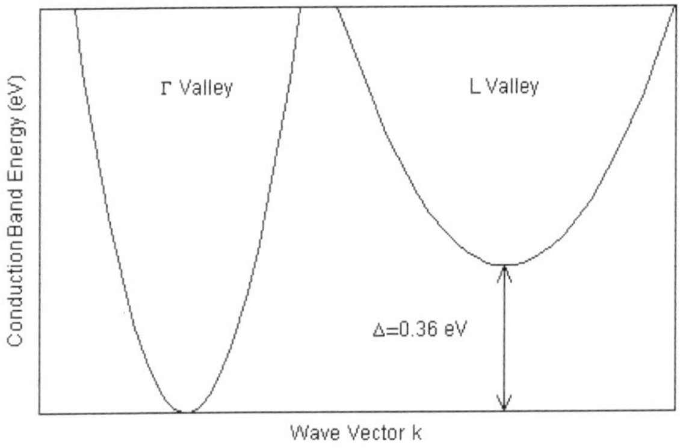

Figure 3.9. *The high-field effects of semiconductor physics.*

Source: https://britneyspears.ac/physics/highfields/images/Image79.gif.

In such settings, the design of the device geometry becomes crucial, as it can dictate the effective transport regime and the overall performance characteristics. As highlighted in Song et al. (2024), the need to address the charge capture and defect-induced scattering becomes even more pressing, affecting the reliability and efficiency of semiconductor devices, especially in applications requiring high operational stability. Recent advancements in material science have further illuminated the significance of structural properties on carrier transport mechanisms in high-field scenarios. The exploration of materials like copper iodide (CuI) demonstrates how manipulating anion vacancy states through thermal annealing can enhance carrier concentration and mobility under high electric fields. The findings in Lee et al. (2022) accentuate the role of chemical treatments and passivation layers in improving device performance, revealing that structural alterations can directly influence the electrical characteristics of semiconductors. As a result, understanding how high-field effects interact with material properties not only offers pathways for improving current technologies but also opens avenues for developing next-generation devices that are more efficient and reliable, thereby advancing the field of semiconductor physics and engineering.

3.4.1. Explanation of High-Field Effects

An understanding of carrier transport mechanisms is essential for optimizing electronic devices, particularly as they operate under varying field strength conditions. In high-field environments, traditional drift-diffusion models begin to falter, as carriers experience altered mobility due to increased energy levels imparted by electric fields. This phenomenon leads to high-field effects where carriers are subjected to significant accelerations, resulting in the possibility of ballistic transport. Ballistic transport occurs when carriers traverse the material without scattering, thus providing an avenue for enhanced current flows. In devices with nanoscale dimensions, such effects are particularly pronounced due to quantum confinement, altering both the energy distribution and the carrier dynamics. Consequently, researchers are urgently exploring these high-field effects, as they can significantly impact device performance parameters such as speed, efficiency, and overall functionality. The intricate relationship between material composition and high-field phenomena is exemplified by the research into organic semiconductors. Innovations such as halogen substitution in bay-annulated indigo-based polymers have been demonstrated to yield significant improvements in spin transport performance, as evidenced by the increase in carrier mobility and prolonged spin lifetime (Figure 3.10).

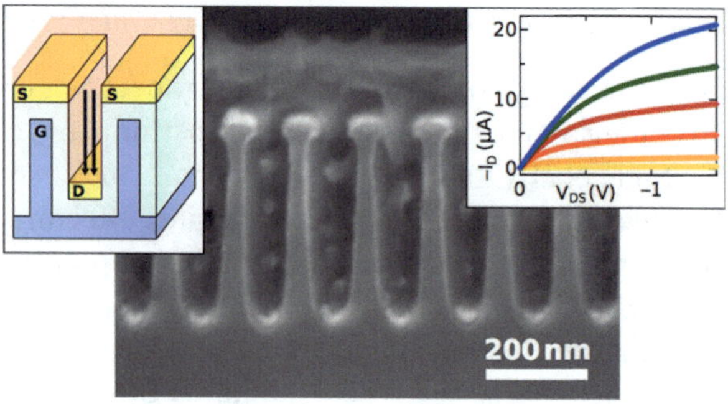

Figure 3.10. *Understanding of carrier transport mechanisms in high-field environments.*

Source: https://scx2.b-cdn.net/gfx/news/hires/2013/onevoltopera.jpg.

These attributes are crucial when operating within high-field conditions, as improved mobility facilitates faster carrier movement, essential for high-frequency applications. Furthermore, the results suggest that modifications in molecular structure can lead to more effective spin diffusion lengths, critical for enhancing the operational efficiency of devices relying on spintronics. Notably, the understanding of how spin-orbit coupling and hyperfine interaction affect these high-field conditions holds promise for advancing the design and functionality of organic electronic materials, paving the way for improved application in next-generation technologies (Yang et al., 2024). The exploration of high-field effects in p-type semiconductors, such as copper iodide, reveals additional complexities in carrier transport mechanisms. Recent investigations highlight the sensitivities of CuIs electrical characteristics to the generation of anion vacancies under varying thermal annealing conditions. High-field applications necessitate robust materials capable of maintaining carrier density and mobility despite intrinsic defects. In this case, the introduction of an Al_2O_3 passivation layer significantly enhances the electrical performance, allowing for superior field-effect mobility and stability. These enhancements not only underline the importance of maintaining structural integrity but also emphasize the need for innovative approaches to manage carrier dynamics in high-field settings. The findings suggest that addressing issues related to anion vacancy states can be pivotal in enabling p-type semiconductors to perform efficiently under high-field conditions, thus enriching the field of semiconductor technology (Lee et al., 2022).

3.4.2. Breakdown Mechanisms in Semiconductors

The multifaceted nature of carrier transport mechanisms in semiconductors manifests prominently during breakdown events, where critical electric fields lead to various effects on device performance. High-field effects such as avalanche and Zener breakdown mechanisms are particularly pivotal in understanding the limitations imposed on semiconductor devices. As electric fields exceed a threshold level, the drift velocity of carriers increases significantly, which can culminate in collisional ionization, thereby generating additional electron-hole pairs. This process can lead to an avalanche effect, where the cascade of charge carriers escalates rapidly, resulting in device failure if not managed appropriately. By examining these high-field effects, researchers can better optimize semiconductor structures, particularly in the context of applications requiring robust operation under elevated electric fields and stringent high-frequency conditions. In doping-dependent semiconductors, the role of deep traps further complicates breakdown mechanisms through Poole-Frenkel emission, observed in devices operating at elevated temperatures. This phenomenon becomes essential in understanding why semiconductors often experience increased leakage currents at high bias voltages. Detailed studies of materials, such as Al_{0.6}Ga_{0.4}N, indicate that under thermal excitation, negatively charged traps can significantly alter the barrier height, which subsequently affects the injection of carriers during breakdown events. Specifically, the interaction of photogenerated carriers with deep traps can act to reduce the barrier height upon illumination, destabilizing the conducting path and eventually leading to catastrophic failure under excessive voltage conditions. Therefore, the identification and characterization of these deep traps are crucial for advancing semiconductor design and enhancing device stability. Photoconductive gain is another aspect intricately linked to the mechanisms associated with breakdown phenomena. In certain configurations, the presence of deep traps can influence the time-dependent behavior of carriers, thereby augmenting photoconductive response. The resultant trapping and slow response times can lead to challenges in device performance, notably in applications where rapid switching is essential. Devices such as the β-Ga₂O₃-based photodetectors demonstrate that both thermionic and Poole-Frenkel mechanisms are dominant at varying electric fields, complicating the dynamics of carrier transport. These interactions not only inform the design of the devices but also necessitate comprehensive assessments of the effects of temperature and bias on efficiency. As researchers continue to explore these complexities, they pave the way for innovations in semiconductor technology that can withstand the rigors of modern electronic applications, particularly where high-temperature and high-voltage tolerances are mandated.

3.4.3. Implications for Device Design and Operation

The intricacies of carrier transport mechanisms significantly dictate the design and functionality of electronic devices. Understanding how drift, diffusion, recombination, and generation operate at the nanoscale is crucial for optimizing device performance. As these processes influence the charge carrier mobility and overall efficiency, engineers must account for variations in material properties, such as the unique characteristics of nanoparticles. For instance, due to their high surface area, nanoparticles demonstrate altered recombination rates and diffusion paths, which are pivotal in applications ranging from sensors to energy harvesting devices. As noted in the literature, metal nanoparticles exhibit distinct optical properties that are size-dependent, underscoring the need to tailor device materials that maximize these benefits while mitigating potential drawbacks, such as degradation or environmental toxicity associated with rigid nanoparticles like lead and mercury (Khan et al., 2017). The implementation of advanced materials, such as metal-organic frameworks (MOFs), opens new paradigms for device design and operation within the field of solid-state microelectronics. When optimized for integration,

MOFs can enhance gas storage and separations alongside conventional electronic applications. However, their successful incorporation relies heavily on the compatibility of their synthesis with existing manufacturing protocols. Transitioning from traditional microcrystalline powder synthesis to scalable film deposition techniques is essential for expanding their role in electronic devices. As highlighted in current research, addressing fabrication-related challenges and focusing on the fundamental properties of MOFs can yield significant advancements in device performance, particularly in areas requiring improved carrier transport—such as in next-generation transistors and sensors (Stassen et al., 2017). The exploration of high-field effects and ballistic transport is imperative for refining the electrical characteristics of nanoscale devices. As device dimensions shrink, the behavior of charge carriers under strong electric fields becomes increasingly relevant, affecting current flow and response times. Quantum confinement phenomena introduce additional complexities that must be navigated during device design. For instance, the dimensional constraints of materials result in altered energy levels, thus influencing how carriers are transported. Understanding these dynamics facilitates the development of devices with enhanced speed and efficiency. Designers must also remain cognizant of the trade-offs between performance and material stability, as seen with nanoparticles. Balancing these factors will be crucial for achieving reliable operation in advanced electronic applications, ensuring that innovations in material science translate into practical advancements in technology.

3.5. BALLISTIC TRANSPORT

In the exploration of carrier transport mechanisms, the role of ballistic transport emerges as a significant phenomenon that influences the efficiency and performance of nanoscale devices. Unlike conventional diffusion mechanisms that rely on scattering events, ballistic transport describes the motion of charge carriers, such as electrons, that travel through a medium without experiencing significant collisions (Figure 3.11).

This highly efficient transport occurs over short distances and is paramount in devices where reduced dimensions play a critical role, such as in quantum wells and nanowires. The ballistic transport mechanism helps to maintain the high-speed operation of devices, as carriers traverse through materials rapidly. This efficiency can be particularly beneficial in advancing technology for applications like field-effect transistors (FETs) and other electronic components that require swift charge transport and reduced power loss. Examining ballistic transport through the lens of quantum confinement reveals fascinating implications for the design of modern electronic devices. When materials are confined to nanoscopic dimensions, quantum effects become pronounced, leading to alterations in the energy levels and the dynamics of charge carriers. In such cases, the mean free path of electrons can exceed the dimensions of the device itself, facilitating a regime where ballistic transport predominates. This is particularly relevant in applications where high speeds and efficiencies are desired. For instance, in solid-state lithium-ion batteries, understanding the dynamics at the nanoscale can inform strategies for enhancing interfacial charge transfer, potentially improving overall battery performance (Deng et al., 2024). As researchers continue to delve into ballistic transport, the insights generated could inform the development of innovative materials and structures that capitalize on these quantum effects. The consequences of ballistic transport extend beyond mere theoretical constructs; they have practical implications across various fields, including materials science and biomedical applications. Investigations into charge carrier dynamics reveal that factors such as device architecture, material purity, and interfacial properties significantly affect transport mechanisms (Veshkini et al., 2024). The findings from recent studies provide a deeper understanding of how alterations in gene expression related to epithelial barriers and transport systems can influence the function of biological membranes. By drawing parallels between electronic transport in solid-state devices and carrier dynamics in biological systems, insights into how to enhance carrier mobility and transport efficiency become increasingly pertinent. Ultimately, this multidisciplinary approach underscores the necessity of comprehending ballistic transport as a crucial element in the evolution of both technological and biological systems.

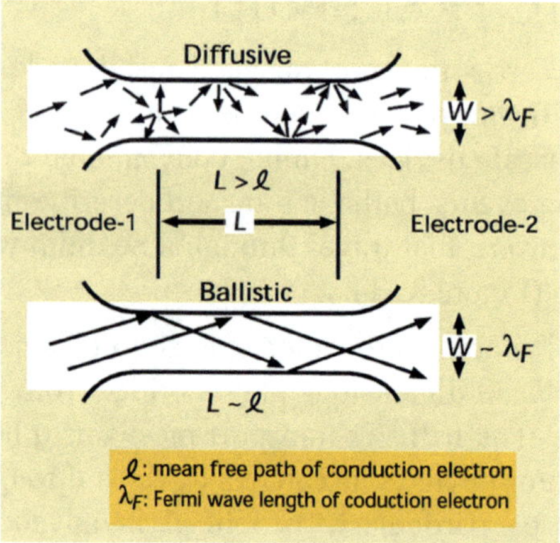

Figure 3.11. *Ballistic transport in action.*

Source:https://dare3v3.wordpress.com/wp-content/uploads/2014/03/diffusive-vs-ballistic.png.

3.5.1. Definition and Characteristics of Ballistic Transport

The intricacies of carrier transport mechanisms play a critical role in the efficiency and functionality of electronic devices, particularly those operating at the nanoscale. Among these mechanisms, ballistic transport stands out as a phenomenon wherein charge carriers traverse a medium without scattering. This behavior is most pronounced in systems where the mean free path of the carriers exceeds the dimensions of the device, which commonly occurs in nanoscale materials. In this regime, carriers move in straight lines, governed by their initial velocity and the influence of external electric fields, resulting in a highly efficient transfer of charge. The absence of scattering within ballistic transport allows for minimal energy loss, a characteristic particularly advantageous for enhancing performance in advanced electronic applications. As devices shrink in size, the role of ballistic transport becomes increasingly significant, especially in terms of high-field effects that are prevalent in nanoscale structures. The efficiency realized through ballistic transport can lead to remarkable improvements in the speed and energy efficiency of devices. Moreover, this transport mechanism is closely intertwined with quantum confinement effects, where the spatial dimensions of semiconductor structures influence the electronic properties due to discrete energy levels. These effects can generate new pathways for charge transport and enhance device functionality (Figure 3.12).

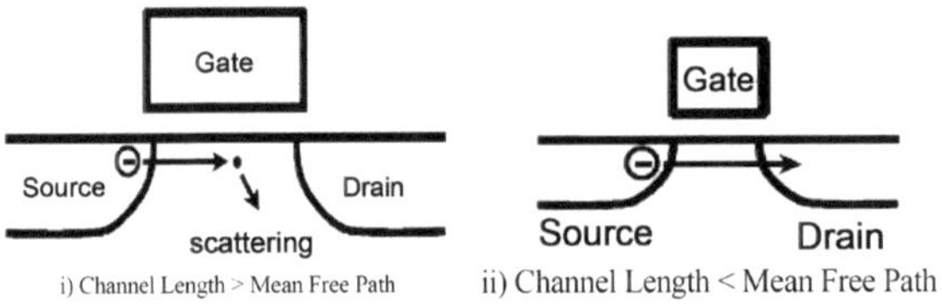

Figure 3.12. *Ballistic transport in transistor channel.*

Source: https://www.researchgate.net/profile/Rahmat-Sanudin/publication/48268079/figure/fig6/AS:307391976493064@1450299366769/Ballistic-transport-in-transistor-channel.png.

In view of these intricacies, engineers and researchers are exploring the potential of ballistic transport in spintronic devices, as mentioned in Hirohata et al. (2020). The integration of such mechanisms is believed to pave the way for next-generation nanoelectronics that leverage both charge and spin for superior performance. Additionally, advancements in artificial and self-propelled systems, as explored in recent studies, highlight the relevance of ballistic transport in non-equilibrium scenarios, particularly within biological contexts. The concept of active particles, as described in Bechinger et al.

(2016), illustrates how directed motion and energy conversion can parallel the behaviors seen in ballistic transport. Such insights reveal a framework where charge carriers mimic these active particles, leading to advances in the manipulation and application of charge transport phenomena. As researchers continue to uncover the complex interactions among carriers in crowded nanostructures, the principles of ballistic transport will undoubtedly continue to bridge theoretical physics and practical applications, thereby enriching our understanding of carrier dynamics and leading to innovative solutions in electronic device design.

3.5.2. Conditions Favoring Ballistic Transport

In the study of carrier transport mechanisms, the conditions that promote ballistic transport are essential for harnessing the efficient functionality of electronic devices. Ballistic transport refers to the unhindered movement of charge carriers, such as electrons, over distances significantly greater than their scattering lengths. This phenomenon is influenced by various factors, including device dimensions and the purity of semiconductor materials. In nanoscale devices, where quantum confinement becomes increasingly significant, characteristics such as reduced dimensionality and enhanced surface-to-volume ratios can facilitate ballistic transport. Moreover, minimizing impurities and defects within the material is critical, as these scattering centers impede the migratory capacity of charge carriers, thus favoring conditions conducive to ballistic transport in advanced electronics. One notable example highlighting the conditions favorable for ballistic transport can be observed in the performance of graphene field-effect transistors (GFETs). The unique properties of graphene, distinguished by its two-dimensional structure and exceptional carrier mobility, make it a prime candidate for nanoscale applications. By utilizing the Boltzmann transport equation (BTE) in modeling, researchers have uncovered that under various operating conditions, GFETs exhibit ballistic behavior characterized by minimal scattering events. The optimization of parameters such as channel length and impurity concentration plays a pivotal role in achieving uniformity in electron distribution, thereby enhancing the overall performance of the device. The findings from these studies underscore the importance of carefully engineered conditions that empower ballistic transport, providing insights for further advancements in nanotechnology. Additionally, the interplay between dimensionality and carrier dynamics significantly shapes the potential for ballistic transport within semiconductor materials. In systems experiencing tensile strain, for instance, research has shown that a softening of chemical bonds and alterations in phonon interactions encourage a two-dimensional character of heat dissipation while enhancing carrier mobility. Under such conditions, the distinct profiles of electronic band structures, marked by degeneracy and anisotropic dispersions, enable carriers to traverse the material with minimal resistance. Conversely, compressive strains have been identified to enhance three-dimensional electronic transport, further demonstrating the critical role of strain engineering in manipulating carrier paths. These findings not only facilitate the exploration of novel thermoelectric materials but also illustrate the broader implications of strain-induced modifications, reinforcing the significance of exploring diverse conditions to promote ballistic transport in advanced carrier systems (Ai et al., 2025; Herrera-González et al., 2024).

3.5.3. Applications in Nanoscale Devices

The physics of nanoscale devices involves sensitive consideration of carrier transport mechanisms, especially the behavior of electrons and holes under various conditions. As devices are miniaturized, the traditional models cannot provide a proper prediction of performance due to the effects such as quantum confinement in which carriers possess discrete energies. The confinement significantly enhances material properties so that they become competent for advanced applications. Above all, carrier drift and diffusion processes in such dimensions must be optimized since they are carefully balanced to result in optimum electronic performance. Synthesis breakthroughs like those that yield control over morphology and dimensions are fundamental to tailor-made materials for better efficiency and performance in use at the nanoscale. Such variable control has immediate implications, influencing as wide a range of things as communication technology to storage of energy. Interfacial thermal resistance then becomes an important issue in nanoscale device devices, as it affects thermal management system performance. Studies show that with decreasing sizes of devices, effective heat transport across materials is constrained by thermal boundary conductance, an interaction that roots at the atomic level. It is important to know these thermal transport processes so as to determine device reliability and efficiency. Recent advances in experimental and computational techniques have provided a helpful glimpse into how the phonon gas models must be re-reviewed to include localized interfacial modes that play an important role in energy transport. Such results are significant because they highlight potential ways to enhance thermal management methods, which are critical to applications ranging from microelectronics to optoelectronics where heat dissipation is important. Apart from this, nanoparticles (NPs) have also garnered immense attention in the area of nanoscale devices due to their unique physical properties and their very diversified applications. Between 1 nm and 100 nm in size, metal, ceramic, and polymeric NPs exhibit unique features that are very much different from their bulk counterparts. Their extensive surface area and quantum-scale effects increase reactivity with them as prime candidates for a host of applications from catalysis to energy storage technology. The material's tunable optical properties through dimensional control enable novel applications in the fields of imaging and sensing devices. For instance, employment of heavy metal NPs like mercury and lead creates concerns over environmental toxicity and demands proper regulation in their application and synthesis processes. This is the equilibrium between utilization of the nanoscale phenomenon and avoidance of the possible risk, which will determine the future direction of nanoscale device applications.

3.6. QUANTUM CONFINEMENT

Quantum confinement is realized when the dimensions of a semiconductor are reduced to the nanoscale, typically below the exciton Bohr radius, leading to quantized energy levels. As carriers are confined to a reduced dimensional space, they undergo dramatic alterations in their quantum behavior that influences both their transport phenomena as well as material electronic properties.

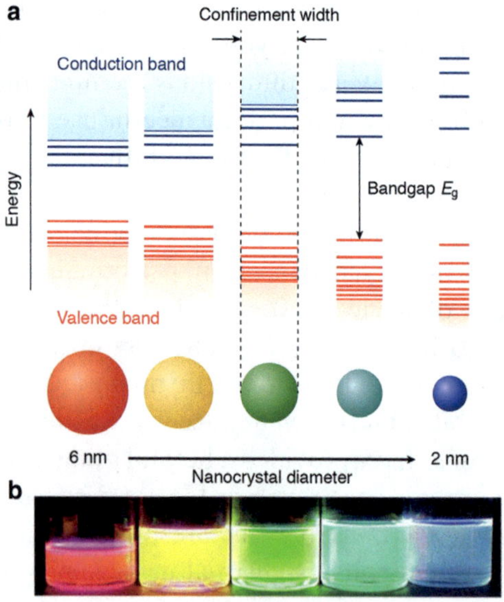

Figure 3.13. *Schematic representation of the quantum confinement effects.*

Source: https://www.researchgate.net/publication/306001639/figure/fig4/AS:393164155310108@1470749048148/Schematic-representation-of-the-quantum-confinement-effects-the-bandgap-or-HOMO-LUMO.png.

This confinement results in a shift in the energy levels, thereby leading to an increase in the energy bandgap, hence affecting charge injection and transport properties. For instance, in compounds like antimony (Sb), as depicted in recent studies, thin film geometries show enormous electronic band structure alterations due to quantum confinement, leading to a topological phase transition. These transitions influence carrier mobility and the overall performance of various optoelectronic devices, thereby highlighting the dominant interplay between size, quantum effects, and electronic properties in nanoscale systems (Wijesinghe et al., 2025). In the case of colloidal quantum dot (CQD) systems, the quantum confinement effect is particularly significant on carrier transport mechanisms. The discrete energy levels result in unique charge transport properties that are governed to a great extent by hopping mechanisms, which differ strongly from the bulk systems. Charge injection efficiency and mobility in such systems depend crucially on interplay of variables such as interdot coupling, ligand chemistry, and defect state presence. Experimental evidence indicates that size and composition fluctuations in quantum dots can induce dramatic performance variations due to the subtle effect of quantum confinement on carrier dynamics. Changing the size of the quantum dots influences the hopping rates of the carriers with the potential barriers between them arising as being important (Figure 3.13).

This intricacy presents challenges which must be surmounted in order to optimize the performance of CQD-based optoelectronic devices (Bisri, 2025). Future technologies in quantum confinement seek to exploit such unique transport characteristics for new applications, particularly for spintronics and quantum computing. The incorporation of strong spin-orbit coupling materials, such as Sb, into nanoscale devices uses quantum confinement to achieve unexpected phenomena, such as the emergence of topological states. When these states couple with carrier transport mechanisms, phenomena such as weak antilocalization can be achieved, providing insights into electronic state coherence in these confined systems. It is then necessary to study the connection between phase decoherence processes and quantum confinement since the knowledge of this area will enable the design of devices with enhanced performance features. Lastly, understanding the intricacies of how quantum confinement influences carrier transport mechanisms will be important to the formulation of nanoscale technologies and optimizing their functionalities in actual applications (Bisri, 2025; Wijesinghe et al., 2025).

3.6.1. Explanation of Quantum Confinement Effects

Quantum confinement is becoming increasingly important as a factor for explaining carrier transport mechanisms in nanoscale devices. With dimensions at the nanoscale level, electronic material properties undergo tremendous alterations due to spatial confinement. Under this regime, the energy levels of charge carriers become quantized, leading to discrete states rather than continuous bulk bands.

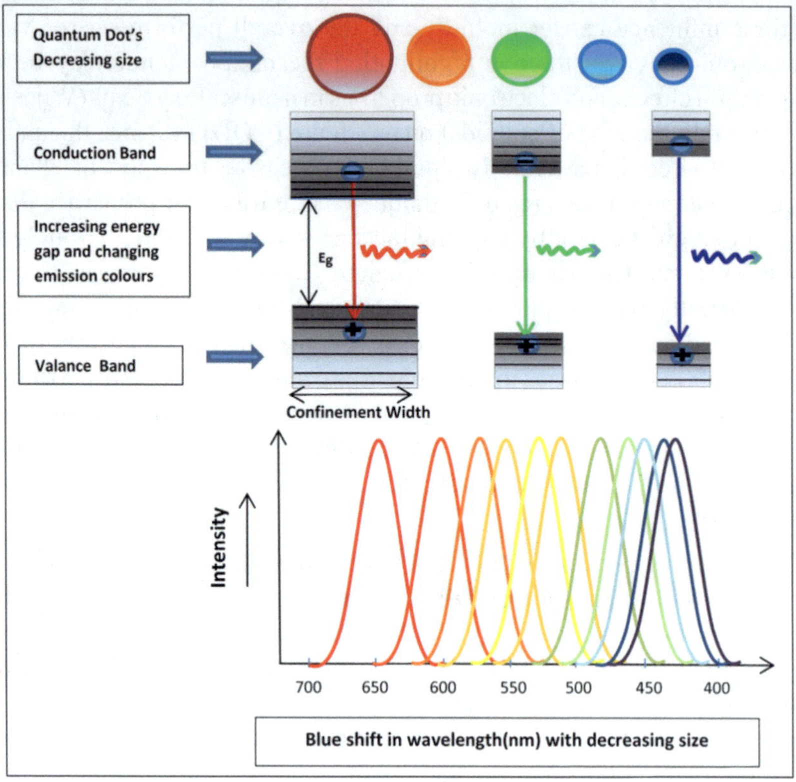

Figure 3.14. *Illustration of quantum confinement effect.*

Source: https://www.researchgate.net/profile/Anuradha-Bhalerao-Pawar/publication/331868642/figure/fig3/AS:741320318533633@1553755945105/llustration-of-quantum-confinement-effect-125.jpg.

Quantization further affects the dynamics such as drift and diffusion, whereby carrier mobility is governed by their energy landscape. Carriers subsequently exhibit a distinctive behavior such as higher scattering rates and changed transport properties upon traversing nano-scale structures. The intricate relationship between quantum confinement and carrier transport phenomena emphasizes the need to incorporate the influence of these phenomena into theoretical models to make realistic predictions of material behavior on smaller dimensions. Interestingly, the impact of quantum confinement extends to the high-field effects and ballistic transport phenomena, which are a deviation from conventional theories. Upon application of higher electric fields to charge carriers, they travel along the confined directions with enhanced momentum, at times overcoming the forces exerted by scattering. Here, transport ballistic can be the dominant one with carriers having high speeds for extended distances in contrast to bulk materials. Consequently, this phenomenon holds the promise of enhanced performance in electronic devices, especially in an age where miniaturization is the catchword (Figure 3.14).

Quantum effects along with high-field conditions provide new operating regimes and designs that are able to leverage these properties effectively. The understanding of these interactions plays a critical role in optimizing the device architectures that exploit quantum confinement advantages, offering insights into more functional and efficient devices. In addition to this, the application of quantum confinement effects is demonstrated through developments in materials such as metal-organic frameworks (MOFs), which possess unique transport characteristics despite being focused on the separation and storage of gas. Existing research indicates the potential of MOFs in microelectronics, with their porous nature being integrated into mechanisms of carrier transport (Stassen et al., 2017). Applications of this nature call for parallel concern with the fundamental characteristics of MOFs and with the methods necessary to make MOFs in cleanroom environments. As the technology evolves, challenges from integrating these materials into existing microelectronic platforms become paramount. A general understanding of quantum confinement effects will not only promote the innovation of device functionality but will also open doors to revolutionary approaches in the synthesis and use of future electronic materials.

3.6.2. Impact on Carrier Transport in Nanoscale Structures

The underlying physics of charge carriers in nanoscale structures goes against common sense, primarily due to the consequences of quantum confinement and reduced dimensionality. When semiconductor material dimensions approach the nanometer regime, carrier transport mechanisms undergo significant changes. Drift and diffusion dynamics that are typically dominated by classical physics are found to have special properties in the context of quantum confinement. Specifically, these confined geometries can lead to higher carrier mobility for efficient transport with characteristics much different from bulk materials. The mutual influence of external fields on the quantum effects also generates high-field effects that can become the dominant factor in transport behavior, especially in devices in which high voltage is common. An understanding of these alterations is at the heart of optimizing the operation of nanoscale devices, e.g., transistors and photodetectors, in which conventional theories underestimates what is measured, necessitating a more comprehensive theory to account for these special properties. As recent studies reveal, not only does quantum confinement affect drift and diffusion, but it also complexly affects recombination and generation processes. In nanoscale structures, the surface-to-volume ratio is greater, significantly enhancing the role of the surface states that can trap carriers and subsequently affect recombination rates. This increased recombination will diminish the efficiency of devices, an important parameter in devices like photovoltaics and light-emitting diodes. Furthermore, exciton creation in quantum dots, wires, and wells is unlike that of bulk due to the altered energy band configurations.

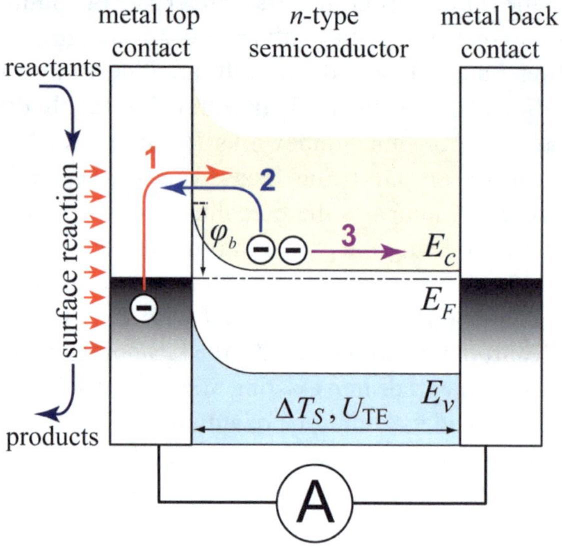

Figure 3.15. *Charge carrier transport in a nanofilm metal-semiconductor structure.*

Source: https://www.researchgate.net/profile/Ievgen-Nedrygailov/publication/260355514/figure/fig5/AS:667122577465353@1536065825596/Color-online-Charge-carrier-transport-in-a-nanofilm-metal-semiconductor-structure-1.ppm.

The variation impacts optoelectronic applications quite dramatically, affecting how devices respond to light as well as electrical input. As such, an sensitive understanding of these recombination dynamics with diffusion profiles placed between them becomes essential in an effort to create strategies that can mitigate these losses and capitalize on the particular strengths afforded by the nanoscale carriers. The different interfaces—heterojunctions and multilayer structures—are other complicating factors for carrier transport dynamics within nanoscale materials. New studies reveal the degree to which these interface interactions influence electronic properties and transport behavior to a significant extent, and new functionalities are now made accessible, previously out of reach. Utilization of materials such as graphene and transition metal dichalcogenides (TMDs) has demonstrated such effects, with their incorporation into devices creating lateral and vertical heterostructures, enhancing transport properties. It is essential to understand the thermal and electronic transport across these interfaces, particularly with the increasing demand for efficient energy conversion and nano-electronic devices. The recent advancements in simulation methods, particularly with the application of non-equilibrium Greens functions and Wigner forms, emphasize the importance of modeling these systems efficiently to facilitate predictions that could guide the engineering of next-generation nanoscale devices to maximize carrier transport mechanisms, enhancing performance and efficiency in practical applications (Figure 3.15).

3.6.3. Applications in Quantum Dots and Nanowires

The coupling between carrier transport mechanisms and quantum confinement is of central importance in the application of nanowires and quantum dots. The dimensional reduction in nanoscale systems plays a deep impact on the electronic properties of the materials to enable one to have greater control over the generation, recombination, and transport of the charge carriers. For instance, quantum dots possess discrete energy levels that are based on spatial charge carrier confinement and thus display unique optical and electronic properties, making them model systems for photovoltaic devices. Additionally, their support of ballistic transport enhances the efficiency of charge carriers, minimizing the loss of energy through scattering and consequently device improvement. This engineered confinement allows enhanced optimization of carrier dynamics, which ultimately contributes to more efficient photodetectors and light-emitting devices (Qin et al., 2020). Nanowires, on the other hand, offer favorable channels for charge transport due to their one-dimensional character, which plays a significant role in carrier mobility. Their reduced cross-sectional area allows them to have high surface-to-volume ratios, which can help increase charge carrier interactions and offer enhanced mobility through engineered synthesis methods. The unique features of nanowires render them capable of application in various optoelectronic devices, such as high-efficiency solar cells and LEDs, where carrier lifetime should be maximized. Therefore, charge carrier density variations in these nanostructures can regulate recombination processes in play, influencing the overall efficiency of the device. This is accompanied by the understanding of high-field effects, which leads to emerging developments in next-generation electronic and optoelectronic technology (Shen et al., 2012). Furthermore, chemical tunability of nanowires and quantum dots enables functionality that can cause increased carrier transport through modification of their surface states. Tailoring could lead to low defect density, which further improves the recombination dynamics of charge carriers. For instance, with rising excitation intensities, as levels of generation increase, knowledge about transitions among competing recombination channels becomes important. Such data has a direct impact on the performance parameters of devices that employ these nanoscale materials, e.g., low-light detectors. Together, the utilization of the novel transport characteristics and optical properties of nanowires and quantum dots offers the window of opportunity for advanced applications in technology fields ranging from renewable energy to bioimaging, a demonstration of their excellence compared to classical semiconductor models (Qin et al., 2020).

SUMMARY

When one analyzes the multifaceted phenomenon behind the carrier transport mechanism, it is evident that various processes are linked with the efficiency and functionality of semiconductor devices. The interplay between drift, diffusion, recombination, and generation highlights not only the inherent physics responsible for charge carrier transport but also their significant effects on device optimization. For instance, as illustrated in recent research, the implementation of bilayer structures, e.g., In_2O_3/IGTO, has demonstrated notable improvements in carrier mobility and concentration. These improvements are based on strategic management of carrier dynamics that effectively leverage individual layer contributions as well as their coupling. Understanding these mechanisms allows researchers to design material compositions and structures to realize upgraded electronic functionalities, thus creating more effective devices suitable for different applications, particularly thin-film electronics and photodetectors, a focus of ongoing research interest (Han et al., 2024). Second, developments in carrier transport mechanism extend beyond conventional structures into the regimes of high-field effects and ballistic transport, which are especially important at larger electric fields. Features exhibited by carriers in such situations warrant close consideration since they have significant influences on device performance at actual operating conditions. Recent research on AlGaN-based photodetectors highlights the operation of multiple transport regimes, such as thermionic-field emission and Poole–Frenkel emission, under different bias conditions. Not only does this analysis of high-temperature carrier transport mechanisms imply the physical limitation of current material but also indicates the need for constant innovation in semiconductor design. Investigation of these effects is the precursor to device performance improvement, including increased responsivity and breakdown voltages optimized for specific application demands, maximizing the utility of these high-end materials in solar-blind UV detection and other applications (Gu et al., 2020). Quantum confinement effects in nanoscale devices introduce another critical component to understanding carrier transport mechanisms. With decreasing device size, quantum effects predominate, altering carrier behavior in drastic ways. This shift demands a rethink of traditional transport models, possibly ones that are incapable of handling the idiosyncrasies introduced by confinement. This paradigm shift highlights the need for cross-disciplinary research involving physics and material science. Through the correlation of quantum mechanics with actual engineering issues, researchers will be able to devise innovative problem-solving strategies that fit the needs of today's technology. Finally, an understanding of the full scope of carrier transport mechanisms and their quantum characteristics will give birth to the next generation high-performance semiconductor devices, thus forming a cornerstone point in material sciences and their applications in various sectors.

The involved charge carrier dynamics in semiconductor materials can be severely influenced by the many involved transport phenomena. Drift represents one of the elementary processes, with charge carriers moving under an electric field. The latter can significantly be enhanced in high-field regimes, with such high-field effects occurring and causing increased carrier velocities. In addition, diffusion is also driven by concentration gradient, and this illustrates how particles seek equilibrium within the material. The combination of these mechanisms is most critical in carriers' explanation of mobility because drift and diffusion between them govern the entire conductivity of the semiconductor. Knowledge of the processes is important, particularly in such applications as meticulous control of the material's electrical behavior is required, which is a basic field of physics in pure and applied physics. Generation and recombination are also as important to understanding carrier transport in semiconductor physics. Recombination is the process by which free charge carriers, i.e., holes and electrons, meet up and decrease conductivity. Generation is the electron-hole pair creation process, often caused by external triggers such as thermal energy or photon absorption. Both of these processes are directly related to carrier lifetime, a critical parameter in photovoltaic device efficiency characterization. The balance of recombination and generation rate determines solar cell performance, with minimization of recombination losses leading to a major improvement in the efficiency of a device. Therefore, a thorough understanding of such processes is the platform for enhancing carrier dynamics in energy conversion devices. Emerging technologies in nanostructured devices have brought forth the phenomenon of quantum confinement effects, which alter carrier transport behaviors greatly. In nanoscale devices, traditional semiconductor physics laws are modified according to quantization of energy levels, thereby influencing drift and diffusion phenomena. For example, in carbon nanotubes and quantum dots, the reduced size permits new electronic properties that enhance device performance. Furthermore, ballistic transport issues come to the forefront since carriers move short distances without being annihilated by scattering, thereby encouraging improved conductivity. This transition towards comprehension of carrier transport at the nanoscale not only reflects advances in device design but also poses new challenges at the material design and fabrication levels, as scientists seek to exploit fully the advantages provided by such new structures.

REFERENCES

1. Abalmassov, A., Abrahams, E., Adachi, S., Adelmann, C., Adelmann, M., Aeschlimann, M., Agnihotri, P., et al. (2010). Spin dynamics in semiconductors. http://arxiv.org/abs/1001.0606 (accessed on 27th September 2025).
2. Aeberhard, U. (2013). Simulation of nanostructure-based high-efficiency solar cells: Challenges, existing approaches and future directions. http://arxiv.org/abs/1304.1957 (accessed on 27th September 2025).
3. Ai, P., Tang, S., Bai, S., Wan, D., Li, X., Guo, W., Zheng, T., et al. (2025). Synergistically tuning thermoelectric properties of *BaCdPF* via strain engineering. *Langmuir: The ACS Journal of Surfaces and Colloids.* https://www.semanticscholar.org/paper/9953e7be3c6738ed0ac48e2d0d41e6b6c2377d1b (accessed on 27th September 2025).
4. Aksamija, Z., Kommini, A., & Majee, A. K. (2019). Electronic transport and thermopower in 2D and 3D heterostructures–A theory perspective. http://arxiv.org/abs/1901.10976 (accessed on 27th September 2025).
5. Bechinger, C., Di Leonardo, R., Löwen, H., Reichhardt, C., Volpe, G., & Volpe, G. (2016). Active particles in complex and crowded environments. *Reviews of Modern Physics, 88.* https://doi.org/10.1103/revmodphys.88.045006 (accessed on 27th September 2025).
6. Beenakker, C. W. J., Papaconstantopoulos, D. A., Ferry, D. K., Nicollian, E. H., Rhoderick, E. H., Haug, H., Henisch, H. K., et al. (2004). Scaling analysis of electron transport through metal-semiconducting carbon nanotube interfaces: Evolution from the molecular limit to the bulk limit. http://arxiv.org/abs/cond-mat/0405465 (accessed on 27th September 2025).
7. Bisri, S. Z. (2025). Charge carrier injection and extraction of colloidal quantum dot assemblies: A perspective for optoelectronic devices. *Journal of Information Display, 26*, 109–130. https://www.semanticscholar.org/paper/30814e4a816f6e300372b398b79ff7be5d79e588 (accessed on 27th September 2025).
8. Choy, W. C. H., Ren, X., Sha, W. E. I., Wang, Z., et al. (2017). Exploring the way to approach the efficiency limit of perovskite solar cells by drift-diffusion model. http://arxiv.org/abs/1703.07576 (accessed on 27th September 2025).
9. Connolly, J. P., & Mencaraglia, D. (2013). III–V solar cells. http://arxiv.org/abs/1301.1278 (accessed on 27th September 2025).
10. Deng, Z., Chen, S., Yang, K., Song, Y., Xue, S., Yao, X., Yang, L., et al. (2024). Tailoring interfacial structures to regulate carrier transport in solid-

state batteries. *Advanced Materials, 36.* https://www.semanticscholar.org/paper/69efb14aaba2ce1a0bb1b95eb4a4548ac283d199 (accessed on 27th September 2025).

11. Furukawa, H., Cordova, K. E., O'Keeffe, M., & Yaghi, O. M. (2013). The chemistry and applications of metal-organic frameworks. *Science, 341.* https://doi.org/10.1126/science.1230444 (accessed on 27th September 2025).
12. Gasparini, N., Salvador, M., Heumueller, T., Richter, M., Classen, A., Shrestha, S., Matt, G. J., et al. (2017). Polymer:Nonfullerene bulk heterojunction solar cells with exceptionally low recombination rates. *Advanced Energy Materials, 7.* https://doi.org/10.1002/aenm.201701561 (accessed on 27th September 2025).
13. Giri, A., & Hopkins, P. E. (2019). A review of experimental and computational advances in thermal boundary conductance and nanoscale thermal transport across solid interfaces. *Advanced Functional Materials, 30.* https://doi.org/10.1002/adfm.201903857 (accessed on 27th September 2025).
14. Gu, Y., Ni, X., Fan, Q., Gu, X., Chen, G. Q., Yang, G., Danner, A., et al. (2020). Analysis of high-temperature carrier transport mechanisms for high Al-content $Al_{0.6}Ga_{0.4}N$ MSM photodetectors. *IEEE Transactions on Electron Devices, 67*, 160–165. https://www.semanticscholar.org/paper/bbdc50cc07bc87691bac47d45fa4dadf06151783 (accessed on 27th September 2025).
15. Han, Z., & Abliz, A. (2024). Investigation of the electrical performance and carrier transport mechanism for heterostructured bilayer $In_2O_3/InGaSnO$ thin-film transistors. *Applied Physics Letters.* https://www.semanticscholar.org/paper/dfb61b429223c971127b3eaf8099c18af6d16a48 (accessed on 27th September 2025).
16. He, F., & Li, Y. (2022). Advances on theory and experiments of the energy applications in graphdiyne. *CCS Chemistry, 5*, 72–94. https://doi.org/10.31635/ccschem.022.202202328 (accessed on 27th September 2025).
17. Herrera-González, M., Herrera-May, A. L., Nolasco-Montaño, J. C., Hernández-Quiroz, T., Mabil-Espinosa, P., & Martínez-Castillo, J. (2024). Investigation of graphene field-effect transistor behavior using a semiclassical transport model and Monte Carlo simulation. *2024 IEEE International Conference on Engineering Veracruz (ICEV).* https://www.semanticscholar.org/paper/31c52234be3313bc7907b33e835655b23ffa8568 (accessed on 27th September 2025).

18. Hirohata, A., Yamada, K., Nakatani, Y., Prejbeanu, L., Diény, B., Pirro, P., & Hillebrands, B. (2020). Review on spintronics: Principles and device applications. *Journal of Magnetism and Magnetic Materials, 509*, 166711–166711. https://doi.org/10.1016/j.jmmm.2020.166711 (accessed on 27th September 2025).

19. Huang, C. Y., Li, H., Wu, Y., Lin, C. H., Guan, X., Hu, L., Kim, J., et al. (2022). Inorganic halide perovskite quantum dots: A versatile nanomaterial platform for electronic applications. *Nano-Micro Letters, 15.* https://doi.org/10.1007/s40820-022-00983-6 (accessed on 27th September 2025).

20. Kerner, R. A., & Rand, B. P. (2017). Ionic–electronic ambipolar transport in metal halide perovskites: Can electronic conductivity limit ionic diffusion? *The Journal of Physical Chemistry Letters, 9*, 132–137. https://doi.org/10.1021/acs.jpclett.7b02401 (accessed on 27th September 2025).

21. Khan, I., Saeed, K., & Khan, I. (2017). Nanoparticles: Properties, applications and toxicities. *Arabian Journal of Chemistry, 12*, 908–931. https://doi.org/10.1016/j.arabjc.2017.05.011 (accessed on 27th September 2025).

22. Langpoklakpam, C., Liu, A. C., Chu, K. H., Hsu, L. H., Lee, W. C., Chen, S. C., Sun, C. W., et al. (2022). Review of silicon carbide processing for power MOSFET. *Crystals, 12*, 245–245. https://doi.org/10.3390/cryst12020245 (accessed on 27th September 2025).

23. Lee, H. Y., Yatsu, K., Kim, T. I., Kwon, H., & Park, I. (2022). Synthesis of vacancy-controlled copper iodide semiconductor for high-performance p-type thin-film transistors. *ACS Applied, 14*(50), 56416–56426.

24. Li, Z., Senanayak, S. P., Dai, L., Kusch, G., Shivanna, R., Zhang, Y., Pradhan, D. D., et al. (2021). Understanding the role of grain boundaries on charge-carrier and ion transport in $Cs_2AgBiBr_6$ thin films. *Advanced Functional Materials, 31.* https://doi.org/10.1002/adfm.202104981 (accessed on 27th September 2025).

25. Liu, W., & Bai, S. (2019). Thermoelectric interface materials: A perspective to the challenge of thermoelectric power generation module. *Journal of Materiomics, 5*, 321–336. https://doi.org/10.1016/j.jmat.2019.04.004 (accessed on 27th September 2025).

26. López-Varo, P., Frégnaux, M., Regaldo, D., Puel, J. B., Béchu, S., Hajhemati, J., Kleider, J., et al. (2024). Characterization of the performance of perovskite solar cells by operando X-ray photoemission spectroscopy. *2024 IEEE 52nd Photovoltaic Specialist Conference (PVSC).* https://www.semanticscholar.org/paper/35cd37385c180e923674ac565c7634459941c201 (accessed on 27th September 2025).

27. Low, J., Yu, J., Jaroniec, M., Wageh, S., & Al-Ghamdi, A. A. (2017). Heterojunction photocatalysts. *Advanced Materials, 29.* https://doi.org/10.1002/adma.201601694 (accessed on 27th September 2025).

28. Ma, X. X., Shen, X., Chen, X., Fu, Z., Yao, N., Zhang, R., & Zhang, Q. (2022). The origin of fast lithium-ion transport in the inorganic solid electrolyte interphase on lithium metal anodes. *Small Structures, 3.* https://doi.org/10.1002/sstr.202200071 (accessed on 27th September 2025).

29. Manchon, A., Železný, J., Miron, I. M., Jungwirth, T., Sinova, J., Thiaville, A., Garello, K., et al. (2019). Current-induced spin-orbit torques in ferromagnetic and antiferromagnetic systems. *Reviews of Modern Physics, 91.* https://doi.org/10.1103/revmodphys.91.035004 (accessed on 27th September 2025).

30. Mateo, D., Cerrillo, J. L., Durini, S., & Gascón, J. (2020). Fundamentals and applications of photo-thermal catalysis. *Chemical Society Reviews, 50*, 2173–2210. https://doi.org/10.1039/d0cs00357c (accessed on 27th September 2025).

31. Novoselov, K. S., Mishchenko, A., Carvalho, A., & Castro Neto, A. H. (2016). 2D materials and van der Waals heterostructures. *Science, 353.* https://doi.org/10.1126/science.aac9439 (accessed on 27th September 2025).

32. Ozawa, T., Price, H. M., Amo, A., Goldman, N., Hafezi, M., Lü, L., Rechtsman, M. C., et al. (2019). Topological photonics. *Reviews of Modern Physics, 91.* https://doi.org/10.1103/revmodphys.91.015006 (accessed on 27th September 2025).

33. Pang, T., Jia, R., Wang, Y., Sun, K., Hu, Z., Zhu, Y., Luan, S., et al. (2018). Giant zero-drift electronic behaviors in methylammonium lead halide perovskite diodes by doping iodine ions. *Materials, 11*, 1606–1606. https://doi.org/10.3390/ma11091606 (accessed on 27th September 2025).

34. Qin, J., Liu, X., Yin, C., & Gao, F. (2020). Carrier dynamics and evaluation of lasing actions in halide perovskites. *Trends in Chemistry, 3*, 34–46. https://doi.org/10.1016/j.trechm.2020.10.010 (accessed on 27th September 2025).

35. Scaccabarozzi, A. D., Basu, A., Aniés, F., Liu, J., Zapata-Arteaga, O., Warren, R., Firdaus, Y., et al. (2021). Doping approaches for organic semiconductors. *Chemical Reviews, 122*, 4420–4492. https://doi.org/10.1021/acs.chemrev.1c00581 (accessed on 27th September 2025).

36. Shen, J., Zhu, Y., Yang, X., & Li, C. (2012). Graphene quantum dots: Emergent nanolights for bioimaging, sensors, catalysis and photovoltaic devices. *Chemical Communications, 48*, 3686–3686. https://doi.org/10.1039/c2cc00110a (accessed on 27th September 2025).

37. Song, Y., Qiu, C., & Wei, S. H. (2024). Mechanism of E'_{γ} defect generation in ionizing-irradiated *a-SiO₂*: The nonradiative carrier capture-structural relaxation model. https://www.semanticscholar.org/paper/2703f0d25c1f19ec602e990db7c95d73abf88c95 (accessed on 27th September 2025).

38. Sorrell, S. (2015). Reducing energy demand: A review of issues, challenges and approaches. *Renewable and Sustainable Energy Reviews, 47*, 74–82. https://doi.org/10.1016/j.rser.2015.03.002 (accessed on 27th September 2025).

39. Stassen, I., Burtch, N. C., Talin, A., Falcaro, P., Allendorf, M. D., & Ameloot, R. (2017). An updated roadmap for the integration of metal–organic frameworks with electronic devices and chemical sensors. *Chemical Society Reviews, 46*, 3185–3241. https://doi.org/10.1039/c7cs00122c (accessed on 27th September 2025).

40. Tian, L. Q., Pan, C., Shi, W., Pan, Y. K., Ran, E. Z., & Li, C. X. (2023). Mechanism of avalanche charge domain transport for nonlinear mode of *GaAs* photoconductive semiconductor switches. *Acta Physica Sinica.* https://www.semanticscholar.org/paper/2a7d4d517bb1ab378e352d1d9380191efef6fa88 (accessed on 27th September 2025).

41. Tu, D., & Fabiano, S. (2020). Mixed ion-electron transport in organic electrochemical transistors. *Applied Physics Letters, 117.* https://doi.org/10.1063/5.0012599 (accessed on 27th September 2025).

42. Veshkini, A., Kühn, C., Dengler, F., Bachmann, L., Liermann, W., Helm, C., Ulrich, R., et al. (2024). Cryptosporidium parvum infection alters the intestinal mucosa transcriptome in neonatal calves: Impacts on epithelial barriers and transcellular transport systems. *Frontiers in Cellular and Infection Microbiology, 14.* https://www.semanticscholar.org/paper/314893407a73358ee6e05efab5496694c5106448 (accessed on 27th September 2025).

43. Waqas, M., Khadka, D. B., Khan, A. H. H., & Wang, Y. C. (2025). Fullerene-driven photocarrier processes in perovskite solar cells: Recent advances. *Nanoscale.* https://www.semanticscholar.org/paper/d321d8d12b3d112cc478c318adb598cb031a67c0 (accessed on 27th September 2025).

44. Wijesinghe, A., Ou, Y., Rathore, A., Edirisinghe, C., Adhikari, P., Chen, A. H., Gilbert, D., et al. (2025). Quantum confinement effect in *Sb* thin films. https://www.semanticscholar.org/paper/0ac53c4c04796596a45c8fade30cb69fca4e9dbc (accessed on 27th September 2025).

45. Xiao, K., & Schmidt, O. G. (2021). Light-driven ion transport in nanofluidic devices: Photochemical, photoelectric, and photothermal effects. *CCS*

Chemistry, 4, 54–65. https://doi.org/10.31635/ccschem.021.202101297 (accessed on 27th September 2025).

46. Xu, Y., Chen, X., Zhou, D., Ren, F., Zhou, J., Bai, S., Lu, H., et al. (2019). Carrier transport and gain mechanisms in *β–Ga_2O_3*-based metal–semiconductor–metal solar-blind Schottky photodetectors. *IEEE Transactions on Electron Devices, 66*, 2276–2281. https://www.semanticscholar.org/paper/edbfbd7934e87907ada03dc582946677a4c74d56 (accessed on 27th September 2025).

47. Yanagida, T. (2018). Inorganic scintillating materials and scintillation detectors. *Proceedings of the Japan Academy Series B, 94*, 75–97. https://doi.org/10.2183/pjab.94.007 (accessed on 27th September 2025).

48. Yang, X., Guo, A., Yang, J., Chen, J., Meng, K., Hu, S., Duan, R., et al. (2024). Halogenated-edge polymeric semiconductor for efficient spin transport. *Nature Communications, 15.* https://www.semanticscholar.org/paper/ed8c1e65090b6072611fa10a1604a427a97657ca (accessed on 27th September 2025).

49. Zeng, H. (2025). A study on the photovoltaic conversion mechanism of high-efficiency lead halide perovskite solar cells. *Applied and Computational Engineering.* https://www.semanticscholar.org/paper/63e315d254022af62c537ca235cec7d0486bcdf3 (accessed on 27th September 2025).

CHAPTER 4

MOSFET FUNDAMENTALS AND SCALING CHALLENGES

Contents

Introduction 133

4.1. Basic Principles of Metal-Oxide-Semiconductor Field-Effect Transistor (MOSFET) Operation 138

4.2. Short-Channel Effects 144

4.3. Scaling of Metal-Oxide-Semiconductor Field-Effect Transistors (MOSFETs) 149

4.4. Advanced Metal-Oxide-Semiconductor Field-Effect Transistor (MOSFET) Architectures 154

4.5. Subthreshold Slope and Leakage Control 159

4.6. Future Trends in Metal-Oxide-Semiconductor Field-Effect Transistor (MOSFET) Technology 164

Summary 169

References 172

INTRODUCTION

In the extensive field of semiconductor technology, the evolution of MOSFET devices has been a turning point towards the current age of electronics. With the world striving to meet progressively escalating requirements on performance, scalability, and power consumption, scaling MOSFETs is an insurmountable challenge. The continued miniaturization of these devices, a requirement to maintain Moore's Law, has created the short-channel effects, which degrade performance with decreasing channel lengths. Minimization of these degrading effects becomes a necessity to sustaining device reliability, necessitating thinking outside the box in transistor design and architecture. Several researchers are already looking into advanced structures such as FinFETs, GAAFETs, and nanosheet transistors, which promise improved electrostatics and reduced leakage. Apart from these structural advancements, acquaintance with the quintessential parameters such as subthreshold slope is quintessential in ensuring prolonged efficiency as well as in controlling leakage current effectively. Creating effective methods to combat the complexities brought by scaling necessitates special attention to both the physical dimensions as well as the electrical properties of transistors. Most recent advancements in device architecture, such as FinFET technology, enhance gate control of the channel, thereby minimizing short-channel effects that limit conventional planar MOSFETs. Such 3D devices hold the potential to minimize leakage currents significantly without sacrificing greater drive currents, which are vital for high-performance applications. Also, the development of Gate-All-Around Field-Effect Transistors (GAAFETs) leverages the advantages of nanosheet technology by providing greater electrostatic control. With the addition of these new geometries, it is possible to decrease the subthreshold slope along with lessening the impact of leakage currents, which is crucial in low-power applications. This innovation presents a significant response to the scaling restrictions created by continuing to scale, offering a blueprint for future transistor technology. As we continue to navigate the constraints of MOSFET scaling, increasingly it appears that innovative solutions must be found if advances in integrated circuits are to continue. The implications of improved transistor performance extend beyond mere speed or power savings; they also result in expanding technological uses across an array of applications from consumer electronics to high-performance computing systems. In addition to architectural breakthroughs, materials science is central to bridging the limitations of traditional silicon-based technologies. The discovery of new materials, for example, two-dimensional materials and new dielectrics, can herald a new era of performance optimization. As such, studies over time need to be focused on building robust frameworks that navigate the subtle dance of electrical performance and material potential. As the evolution of MOSFET continues, it promises to redefine the face of electronics for the coming generations.

Definition of Metal-Oxide-Semiconductor Field-Effect Transistor (MOSFET)

The MOSFET is a block of modern electronics, and it forms the core of integrated circuit design and operation. Its source, gate, and drain terminal configuration allows for efficient modulation of electrical current from source to drain by applying a voltage across the gate terminal. This working mechanism makes MOSFETs ideal to be used as both switches and amplifiers and hence are of prime importance in applications from microprocessors to power regulation systems. Scaling of MOSFETs has triggered large performance and energy advantages but has also triggered complex challenges, particularly in managing short-channel effects. These are done in reduced devices, and scientists have hunted forward transistor structures such as FinFETs and GAAFETs to guarantee performance but with fewer problems from smaller device size. With continuing shrinking of semiconductor dimensions, the limitations of classical MOSFETs become more prominent. These limitations are further more important in the context of subthreshold slope, a parameter which is immensely important in the context of leakage current management in ultra-scaled transistors. Tight control of the subthreshold behavior is critical in the context of upholding energy efficiency, a critical parameter in contemporary electronic systems. Transistor innovations, such as nanosheet transistors, are being developed to address these issues. Nanosheet MOSFETs utilize the advantage of a three-dimensional structure for enhancement in electrostatic control of the channel and thus improvement of subthreshold slope, minimizing leakage (Figure 4.1).

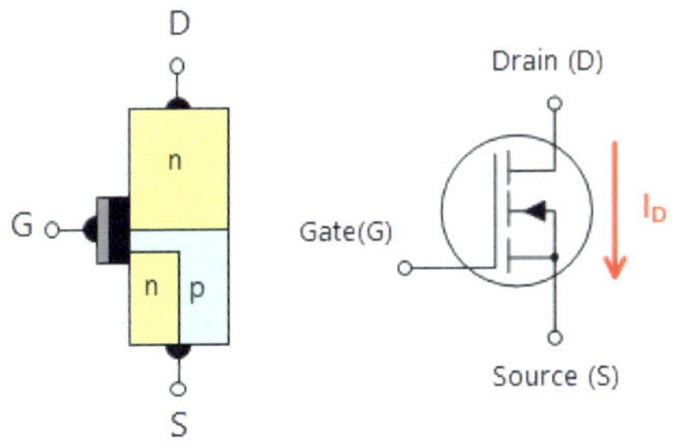

Figure 4.1. *Function of metal-oxide-semiconductor field-effect transistor (MOSFET).*

Source: https://toshiba.semicon-storage.com/content/dam/toshiba-ss-v3/master/en/semiconductor/knowledge/e-learning/discrete/chap3-5-1_en.png.

This improvement underscores the ongoing transformation of MOSFET technology in accordance with the requirements of modern application, demanding understanding not only of the fundamental operation of MOSFETs but also of scaling's impact on performance metrics. Continuing research into material and architecture brings profound implications for the future of MOSFETs, with innovation fueled by advanced computation techniques in fields such as artificial intelligence (AI). These approaches promise to optimize the design and functionality of new materials, further pushing the boundary of what is possible in semiconductor technology. The use of machine learning integration as an example can enable the discovery of new material discovery that increases transistor efficiency and miniaturization potential. The complexity of material development challenges that can be effectively embedded within MOSFET architectures highlights the intersection between theoretical prediction and pragmatic engineering outcomes. With electronics reaching so wide and deep, the quest for new solutions that maximize MOSFET performance is poised to shape the semiconductor technology of the future, significantly touching various industries from consumer electronics to renewable energy systems (Bauer et al., 2023; Cefe López, 2022).

Importance in Modern Electronics

The evolution of electronic devices has given rise to an imperative need for materials and structures that can effectively control short-channel effects, and these effects have increasingly been given more stress as the size of the transistor shrinks. With the discovery of the most recent device architectures, such as FinFETs, GAAFETs, and nanosheet transistors, the landscape of modern electronics has been revolutionized by minimizing these effects. These new designs enhance electrostatic control over the channel to help sustain device performance in spite of reduced dimensions. Such capabilities are needed not only to improve computational efficiency but also to address power utilization in increasingly portable and high-performance applications. Because the industry pushes transistors to be smaller and more power-efficient, the development and implementation of such higher-level structures mirror the path electronics are going today, resulting in solutions which allow one to enter a high-density device era. Moreover, subthreshold slope and leakage current control are of the highest priority in maximizing the performance of modern transistors. Traditional MOSFETs suffer from massive leakage at smaller scales, leading to wasted power and lower efficiency. To counter this, emerging methods utilize materials like gallium nitride (GaN), which carry inbuilt traits that significantly reduce these leakage currents while allowing for fast switching, as evidenced from the latest advancements in GaN technology (Meneghini et al., 2021). Steep subthreshold slopes realized in new transistor structures are crucial to maintaining performance and increasing overall energy efficiency. This is critical in applications as wide-ranging as consumer devices to big data-processing systems, where energy conservation is critical. As demands for improved performance and lower power consumption continue to rise, focus on the tailoring of subthreshold and leakage behavior will persist in driving research and development in high-end semiconductor technology. The widespread significance of such advances to today's electronics cannot be overstated. Strategies that address the problem of scaling and device performance improvement are crucial in meeting the demands of current technology, where efficiency and reliability take precedence. Ongoing advancements in materials and device structures by focused research, as exemplified by ongoing investment by institutions like the U.S. Department of Energy (Bauer et al., 2023), ensure that the electronics industry can innovate against changing customer needs and emerging technological applications. As MOSFETs improve together with new materials and structure designs, the field is experiencing miniaturization, performance, and efficiency come together. This transformation not only propels next-generation device technology but also provides the foundation for an increasingly integrated world, with the function of modern electronics only increasing in complexity and effect on society.

Overview of Scaling Challenges

The relentless miniaturization of MOSFETs has created a plethora of challenges, primarily stemming from short-channel effects that appear when device dimensions shrink. As transistors reduce in size, the electrostatic control of the gate over the channel decreases, and leakage currents rise, together with a deterioration in device performance. These short-channel effects impact various critical parameters such as the subthreshold slope, which ought theoretically to be approaching 60 mV/decade, but actually varies significantly due to these effects. Such a degradation in performance requires out-of-the-box solutions, and thus researchers are investigating more advanced device structures like FinFETs and more recently Gate-All-Around (GAA) Nanosheet (NS) transistors. GAA NS transistors have been identified as a promising direction for devices at the 3 nm node and beyond because of their improved electrostatic control and overall efficiency (Zhou, 2025). Proper control of short-channel effects continues to be important to realize optimum performance nonetheless. Transition to next-generation transistor technologies like FinFETs and GAA NS transistors is a critical countermeasure to the challenges posed by the scaling of traditional MOSFETs. The new fabrications target the enhancement of gate control over the channel and a much reduced leakage current. Three-dimensional structure allows FinFETs to possess an enhanced effective gate area and improved electrostatic confinement. This innovation reflects a significant move towards alleviating the adverse short-channel effects plaguing traditional planar MOSFETs. GAA NS transistors go even further in this by encapsulating the channel on all sides, further enhancing gate control (Zhou, 2025). With such changing technology, continued research into their geometrical and material components becomes increasingly important, as variation can have a direct bearing on their reliability and performance enhancement. Such developments point to the necessity for global scaling rather than mere dimensional scaling. Concurrent with the new transistor geometries, dealing with leakage current is also a critical challenge in the MOSFET scaling problem. Subthreshold leakage can severely impact total power consumption, leading to inefficiencies in today's power-sensitive world of electronics. There is normally a compromise between performance and leakage control with traditional scaling, so new ways are being sought to increase device efficiency without compromising its functionality. Novel materials like high-k dielectrics and low-bandgap semiconductors are under investigation that may aid in the improvement of the electrostatic properties of novel architectures. Moreover, the application of artificial intelligence offers encouraging prospects for enhancing semiconductor materials and structures by analyzing huge collections of data and identifying trends that might yield leaps in performance. Lastly, resolving these intricate problems will entail collaboration in research and development activities to bring in the next generation of high-performance, reliable MOSFET technology.

4.1. BASIC PRINCIPLES OF METAL-OXIDE-SEMICONDUCTOR FIELD-EFFECT TRANSISTOR (MOSFET) OPERATION

Increased demand for more compact and energy-efficient electronic devices has created the need to advance semiconductor technology. Most significant advancements in the semiconductor industry are the advancements in the Metal-Oxide-Semiconductor Field-Effect Transistor (MOSFET). The basic operating mode of a MOSFET relies on controlling the flow of electrons over a semiconductor channel through voltage applied to the gate terminal, effectively manipulating the drain-source current. However, as the scaling progresses, short-channel effects become increasingly dominant, adversely affecting device performance and reliability. Such effects include threshold voltage roll-off and increased leakage current, both of which can severely limit the efficacy of standard MOSFET architecture in ultra-scaled devices. To mitigate against these problems, emerging architectures such as FinFETs, Gate-All-Around FETs (GAAFETs), and nanosheet transistors have emerged, significantly reducing the adverse effects associated with short-channel operation. The development of non-planar MOSFET structures is a key strategy to enhance device performance without the compromise of silicon-based technology benefits. FinFETs utilize a three-dimensional structure to further electrostatic channel control through the deployment of a gate that wraps around multiple sides of the fin structure. Geometry suppresses short-channel effects, making scalability better. GAAFETs also have more control with full enveloping of the channel, exhibiting greater efficiency and lower power consumption. Nanosheet transistors offer a differentiated approach in that they support adjustable width of the channel and improved performance characteristics (Figure 4.2).

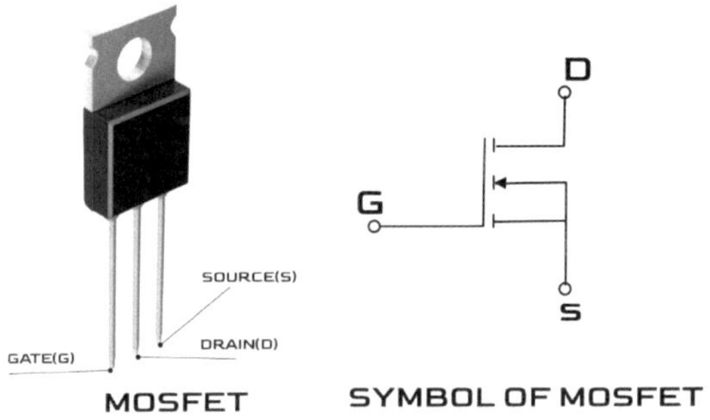

Figure 4.2. *Principle of metal-oxide-semiconductor field-effect transistor (MOS-FET).*

Source: https://media.licdn.com/dms/image/v2/D5612AQE6P0ROYEHyDg/article-cover_image-shrink_720_1280/article-cover_image-shrink_720_1280/0/1700217460591?e=2147483647&v=beta&t=VL14qhUvIlWub8UU_vym8zca5V23YpG4H-vV8_YRIWbQ.

These emerging device structures are vital to meet the ciad and operating demands of next-generation technology in the context of increasing necessity for more frequencies and better power efficiency in applications in most areas, including mobile communication and high-speed computation. Concurrent with structural innovations, understanding the subthreshold slope is vital to continue improving the performance of MOSFETs. The subthreshold regime is when the transistor is turned off but still lets some leakage current pass through, causing subthreshold leakage problems. Leakage control is essential, and it will become even more important as device scaling continues. Being higher in subthreshold slope would be desirable because that translates to lower subthreshold currents, which mean improved overall efficiency and performance. Recent progress focuses on material and device engineering that can achieve a subthreshold slope less than the typical 60 mV/decade minimum, which would be especially useful for ultra-low-power applications. These techniques serve as the continued effort to overcome MOSFET operational obstacles, which are still at the forefront of future technology achievement, as we continue ahead towards beyond-next-generation systems and solve the issue of scaling that lies ahead of us(Bauer et al., 2023; Tataria et al., 2021).

4.1.1. Structure and Components of a Metal-Oxide-Semiconductor Field-Effect Transistor (MOSFET)

In the context of microelectronics, structural complexity in MOSFETs plays an important part in their operational efficiency. They usually possess a source, drain, gate, body, and oxide layer with the gate being a significant one controlling the flow of current in an adjacent semiconductor region. With decreasing device size, issues like short-channel effects are encountered, and conductivity control and leakage currents become more challenging. For instance, the subthreshold slope defining the capability of the current to switch off becomes more critical at smaller dimensions. To combat these problems, new devices such as FinFETs have emerged, which entail a three-dimensional structure to control electrostatics more effectively, minimize short-channel effects, and enhance performance parameters quite a lot. Physical modifications directly map to the MOSFET's capability to perform well in more compact electronic devices. Enhanced MOSFET design has given rise to new architectures such as Gate-All-Around FETs (GAAFETs) and nanosheet transistors. These devices offer a superior gate control mechanism in that they essentially surround the channel on all sides and hence become easily controlled electrostatically as compared to planar MOSFETs. This is due to the fact that the technology continues to move towards smaller technology nodes where it becomes all the more challenging to exercise gate control over progressively shorter channel lengths (Figure 4.3).

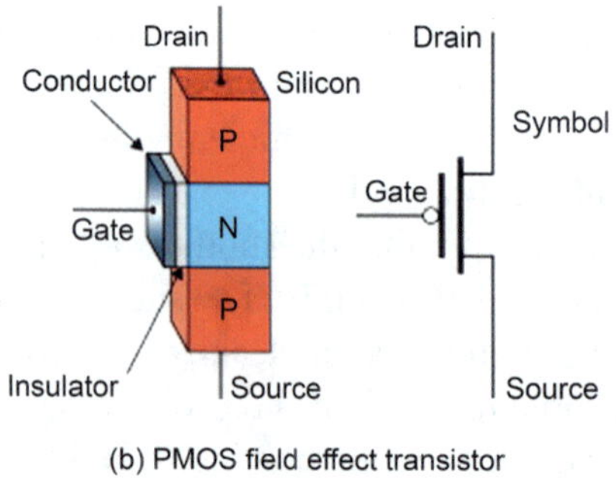

Figure 4.3. *Structure of metal-oxide-semiconductor field-effect transistor (MOSFET).*

Source:https://ars.els-cdn.com/content/image/3-s2.0-B9781856175074000048-gr9b.jpg.

The formation of these multilayered structures aims to offset the adverse impacts of short-channel behavior that are likely to yield increased leakage current and reduced overall efficiency. GAAFETs and nanosheet technologies therefore not only address near-term performance issues but also extend the lifespan of MOSFET technologies, setting them in good stead for the future. The second key aspect of the MOSFET structure is mechanisms for leakage control and subthreshold slope effects. Good leakage control is encouraged by the increasing need for low power devices, with subthreshold leakage accounting for a significant percentage of the overall integrated circuit power dissipation. As device size further decreases, changes in design rules and material selection become necessary to sustain performance operation. High materials like high-k dielectrics can help gate capacitance improvement and leakage paths reduction. According to recent discussions, developments in FinFET-based structures have been shown to offer promise in addressing leakage problems while offering better control, which is essential in enabling operational efficiencies required in modern electronic devices (Karimi et al., 2024). The shift towards the MOSFET design paradigm presents a response to not only current performance demands but also to future scalability challenges embedded in semiconductor technology.

4.1.2. Operating Modes: Enhancement and Depletion

The study of MOSFET operation modes encompasses two prominent configurations: enhancement and depletion. These modes significantly affect device performance, particularly with decreasing transistor sizes during the age of advanced scaling. For an n-channel MOSFET in enhancement mode, it operates by creating a conductive channel only under a positive gate voltage, and it facilitates an easy flow of electrons from the source to drain. This operating characteristic is particularly prone to short-channel effects such as Drain-Induced Barrier Lowering (DIBL) and reduced threshold voltage. The more the transistor is scaled down, the broader these effects will make leakage currents and influence the overall device reliability. Depletion mode transistors, however, already have the capability to conduct current at zero gate voltage due to a pre-formed channel that is present beforehand, decreasing with negative gate voltage, offering designers some leeway in analog applications where linearity and power efficiency are the issues. Modern device architecture has introduced FinFETs and GAAFETs as a solution to the scaling problem of traditional planar MOSFETs. FinFETs utilize a three-dimensional structure that provides improved electrostatic control over the channel, boosting performance significantly with lower short-channel effects. It is essential in cases where gate lengths approach a few nanometers, where issues such as threshold voltage fluctuations and increasing leakage currents become problems.

Similarly, GAAFETs make use of innovative geometries to shrink device dimensions even further and drive the performance parameters to the limit, thereby gaining traction in new circuit design regimes. As geometries go innovative, behavior under operation in enhancement as well as depletion modes also needs to keep up to ensure devices remain functional as required while operating in aggressive environments. The transition from traditional to innovative MOSFET architectures brings about challenges related to subthreshold slope management and leakage control. The subthreshold slope, which dictates the rate with which drain current increases in response to gate voltage, can be optimized by applying new material and geometries. Enhanced channel control mechanisms in devices like FinFETs allow for improved subthreshold slope, thus minimizing the leakage currents that typically sabotage power efficiency. Apart from that, the presence of multi-gate architectures, as is seen in both FinFETs and recent SGrFETs, provides independent gate control that can be used strategically to allow for peak performance when working in a variety of circuit architectures. Through thorough simulations and experimental implementations, scientists are studying these advancements to more accurately evaluate the impact on enhancement and depletion modes in upcoming devices, taking into account the chronic scaling concerns in the semiconductor industry (Asenov et al., 2001; Shadrokh et al., 2010).

4.1.3. Key Electrical Characteristics: Threshold Voltage and Transconductance

MOSFET technology evolution has necessitated a better understanding of key electrical parameters, particularly transconductance and threshold voltage, in light of diminishing device size. As the scaling of transistors toward nanoscale regimes significantly increases the prevalence of short-channel effects, a significant gate-drain terminal interaction is the result. This interaction is a detrimental factor affecting the threshold voltage, and therefore, controlling this parameter with high precision is crucial to ensure proper device operation. High threshold voltage is crucial as it has a direct effect on power consumption and switching speed. Innovative device architectures like FinFETs and GAAFETs are being used in an effort to mitigate these issues. These new-generation devices enhance electrostatics that allow a steeper subthreshold slope and hence reduce leakage currents while maintaining a healthy threshold voltage, thus optimizing overall device performance. Transconductance, i.e., the change of drain current with respect to the change of gate voltage, is a key parameter for assessing MOSFET performance. Scalability problems arise since traditional planar MOSFETs do not deliver adequate transconductance values at severe scaling constraints. As the transition into three-dimensional geometries like nanosheet transistors, gate control of the channel is much more significant, and thus the values of transconductance are even greater. Higher transconductance is preferable as it ensures quicker switching and greater drive current, which is desirable for high-performance applications. The new geometries also exhibit better short-channel effects, further enhancing the transconductance response. Thus, there is a need to comprehend the relationship between transconductance and threshold voltage so that maximum advantage can be taken of such advanced transistors in advanced electronic circuits. In rendering MOSFETs more efficient, leakage current management and subthreshold slope optimization became emerging challenges while increasing efficiency. A perfect steeper subthreshold slope can suppress leakage, a source of trouble that becomes aggravated in devices that are shrunk.

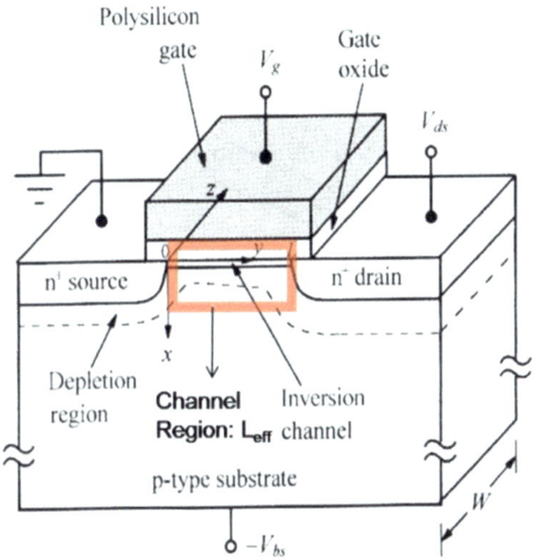

Figure 4.4. *(a) Metal-oxide semiconductor field effect transistor (MOSFET), and (b) Schematic describing scaling laws applied to MOSFET where scaling down by a factor λ implies increasing the density of devices on IC by a factor of λ 2 and decreasing the gate-length or CD by 1/λ. The cost reduction is a function of scaling factor, decreasing with the increase in λ.*

Source:https://www.researchgate.net/profile/Piyush-Pathak-2/publication/241232559/figure/fig3/AS:648970510544897@1531738035150/a-Metal-Oxide-Semiconductor-Field-Effect-Transistor-MOSFET-and-b-Schematic.png.

Devices must sacrifice performance against reduced leakage to acquire practical applications in battery-operated and high-density circuits. Current investigations have shown that novel materials, as discussed on innovations around GaN technology, have a significant impact on such electrical parameters. Being capable of maintaining a stable threshold voltage and reducing leakage without sacrificing speed again highlights the importance of these electrical parameters. As the virtual world advances, adapting to these parameters—threshold voltage and transconductance—will be vital in defining the future of MOSFET technology such that scalability will not compromise performance and energy efficiency (Figure 4.4).

4.2. SHORT-CHANNEL EFFECTS

Modern semiconductor technology is faced with unprecedented challenges because scaling down devices creates a variety of phenomena that fall under the umbrella of short-channel effects. As MOSFETs become smaller in size, the electrostatic control of the channel weakens and initiates several key issues such as increased leakage currents and lower subthreshold slope. In traditional planar MOSFETs, the edge between off-state and on-state becomes less steep, which means dependence on a variety of unintended factors like channel length modulation.

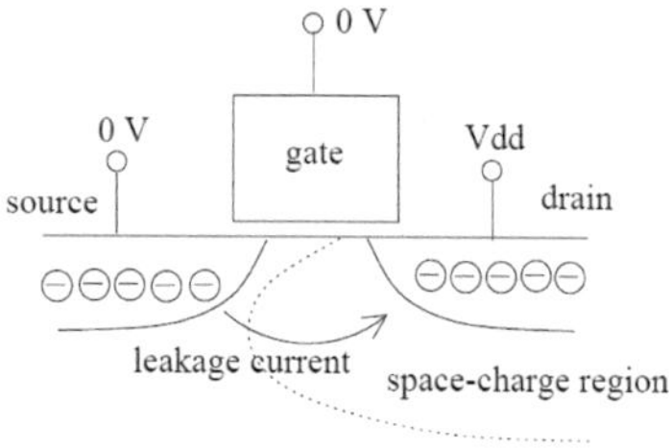

Figure 4.5. *Short channel effects in a metal-oxide-semiconductor field-effect transistor (MOSFET).*

Source: https://www.researchgate.net/profile/Ashish-Kumar-75/publication/321149694/figure/fig1/AS:562082748866560@1511022378201/Short-Channel-Effects-in-a-MOSFET.png.

Therefore, effective device operation in scaled-down designs calls for novel strategies to compensate for these adverse effects. Surmounting such issues is imperative as the technology evolves in electrical characterization of short-channel devices with the knowledge that the conventional practice may no longer be sufficient to maintain device performance against increasing demands of emerging applications. As a response to challenges raised by short-channel effects, new designs of transistors have been introduced, notably FinFETs and Gate-All-Around FETs (GAAFETs). These structures are more complex and offer better electrostatic control due to their three-dimensional geometries, which help to reduce the instability issues of the standard flat-channel devices.

Greater surface area of the channel allows better control of current and reduced leakage, thus improving the devices on-off behavior. As a result, the devices show a significant advantage in that they can achieve sub-VTH control, thus reducing some of the problems brought about by short-channel effects. Also, the progress in nanosheet transistors reflects ongoing research efforts to explore new directions towards optimal performance improvement (Figure 4.5).

These advancements are critical in guaranteeing device dependability as technology is advancing towards smaller nodes and increased integration densities, naturally plagued with short-channel problems. Besides, subthreshold slope optimization and leakage control are essential factors affecting the performance of emerging MOSFET technologies with short-channel challenges. Equally important is attaining a steeper subthreshold slope because it affects power consumption and switching directly. Advanced material systems and device structures play pivotal roles in maximizing the current flow, and as a result, in limiting leakage currents generated by classical scaling techniques. Consequently, the employment of high-k dielectrics and new semiconductors such as GaN assists in enhancing performance overall while allowing the designer to benefit from complexity in short-channel dynamics. As research looks deeper into the electrical and physical characteristics of such materials and devices, it will become increasingly critical to develop solutions that balance invention with pragmatism, ensuring that MOSFET technologies continue to develop despite the challenges presented by reducing device sizes.

4.2.1. Definition and Significance of Short-Channel Effects

As silicon technology continues to advance, miniaturization of transistors has become a requirement and an issue. It has led to the inherent formation of short-channel effects (SCEs) due to this relentless scaling, which gravely disables the functionality of Metal-Oxide-Semiconductor Field-Effect Transistors (MOSFETs). SCEs manifest as undesirable effects in the form of increased leakage currents and compromised electrostatic control over the channel. These effects arise when the channel length of the transistor is of a similar order as the depletion region, resulting in distortions to the intended behavior of the device. Not only does this effect complicate the stability of the performance but it also has huge implications for power consumption, an issue of utmost importance for modern applications in very large scale integrated (VLSI) technologies requiring high speed and efficiency (Wang et al., 2025). Groundbreaking answers have emerged to address the SCE challenges, particularly by using enhanced transistor architectures like FinFETs, Gate-All-Around FETs (GAAFETs), and nanosheet transistors. These architectures enhance gate control and electrical performance by increasing the surface area with the vertical geometry. FinFETs, for instance, provide improved scalability and effectively clamp SCEs through a three-dimensional structure that facilitates better electrostatic confinement of the channel charge (Figure 4.6).

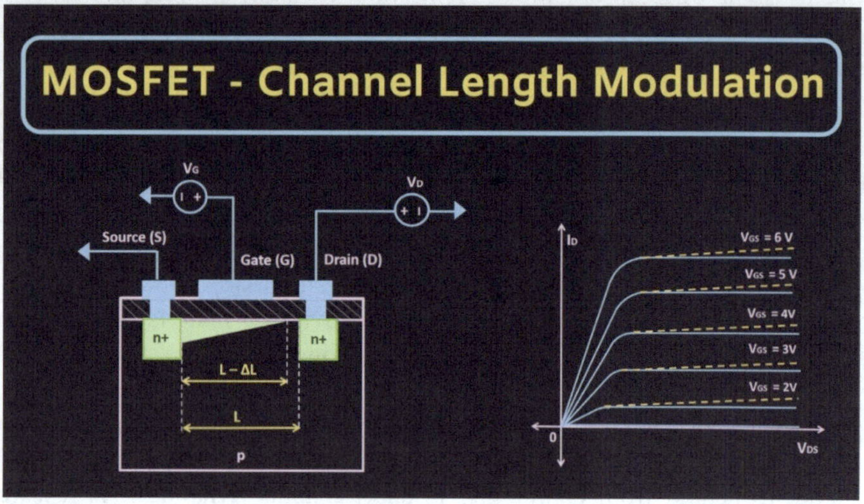

Figure 4.6. *The miniaturization of short-channel effects.*

Source:https://media.licdn.com/dms/image/v2/D4D12AQFnvkjSoE9CsA/article-cover_image-shrink_720_1280/article-cover_image-shrink_720_1280/0/1737121661924?e=2147483647&v=beta&t=46oshRBBsabCzHZYnkr5G0lPkw1Nw_A-2j-J6r6kXb4.

Similarly, GAAFETs benefit from a fully wrap-around gate structure, which results in improved leakage current control and subthreshold slope control. The intrinsic benefits provided by these new forms of field-effect transistors make them to be the focal solutions in addressing the limitations created by conventional MOSFETs in response to the demands of next-generation electronics (Wang et al., 2025). Besides, the importance of subthreshold slope and leakage control cannot be overstated in addressing short-channel effects. Traditional MOSFETs possess a theoretical subthreshold swing of 60 mV/decade, which brings inherent limitations to scaling the device and to operational efficiency. Emergence of novel transistor geometries has the potential to break these barriers to provide subthreshold swings much better than traditional expectations in scaling. For instance, advances in material sciences and geometry of transistors have resulted in devices that can approach an ultra-low leakage regime, thus enhancing the viability of power-efficient design. In order to achieve effective control of leakage is of extreme importance, as it has a direct bearing on the energy efficiency and thermal integrity of devices with increasing circuit densities. Thus, adding augmented designs in combination with effectively managing short-channel effects is a prime challenge in improving the efficacy and sustainability of next-generation MOSFET technologies (Mohammad et al., 2022).

4.2.2. Impact on Device Performance and Reliability

The unchecked scaling of MOSFETs has resulted in the evolution of a number of high-end architectures like FinFETs, GAAFETs, and nanosheet transistors. All of these technologies address the concerns regarding short-channel effects that become more prominent as device size decreases. In FinFETs, the three-dimensional structure ensures better electrostatic control over the channel, thus reducing leakage currents and improving subthreshold slope behavior. This advancement not only enhances performance but is also essential to enabling reliability by minimizing susceptibility to random variations that are prone to accompany diminished geometries. At the other extreme, emerging technologies like GAAFETs are in the pipeline to continue enhancing by utilizing a gate-all-around configuration, having yet tighter channel control, and optimizing electrical performance. These developments are essential in maintaining device efficiency against increased demands of modern electronics, where reliability plays a key role in maintaining operational integrity. With regards to these developments, the impact on device performance and reliability cannot be overstated. High-frequency operation enabled by these new transistor geometries introduces tremendous power efficiency improvement but is accompanied by complex issues regarding thermal management and electromagnetic interference. Although the mobility gains seen in materials like gallium nitride (GaN) permit faster switching and reduced resistive loss, the subtleties of trapping and potential degradation remain concerns which require thoughtful investigation and understanding. Device structures must therefore be tailored not only for enhanced performance characteristics but also to address reliability concerns. Continuing innovation in the quest for materials and structures that do not deteriorate under operating loads is alluding to a multidisciplinary effort between materials science and electrical engineering. New technologies such as the use of solid-state transformers (SST) also demonstrate the complementarity of new devices with existing electrical infrastructures, yet again suggesting the requirement for system behavior reliability. SSTs integrate power electronic converters with traditional transformer operations, providing more control and flexibility in AC and DC applications. The reliability of SSTs depends significantly on the resilience of MOSFET performance, especially under varying operating environments. The simultaneous evolution of control strategies and research on various SST topologies highlight the necessity for SSTs to support the evolving needs of electronic systems. As the technology advances, consideration must be made of the interaction between device performance and reliability and system-level integration because the future of electronics will largely rely on being able to efficiently deal with these interdependencies for top performance.

4.2.3. Techniques to Mitigate Short-Channel Effects

With the advancement in semiconductor technology, short-channel effects become more critical, and innovative design methods are required to improve device performance. Among the hopeful technology to overcome these problems are FinFETs, Gate-All-Around Field Effect Transistors (GAAFETs), and nanosheet transistors. FinFETs, owing to their three-dimensional structure, effectively control short-channel effects by promoting better electrostatic confinement of the channel. This structure causes improved subthreshold slope and reduced leakage current, parameters critical to maintaining device performance as scaling persists. On the other hand, GAAFETs are taking control to a higher level by surrounding the gate around the channel material to gain more control over electrostatics compared to traditional structures.

The three-dimensional geometries of these new transistors do not only suppress short-channel effects but also result in less power consumption, ultimately paving the way for more efficient integrated circuits suitable for different applications. Another major way to manage short-channel effects is by leveraging new materials that resist such phenomena as Drain-Induced Barrier Lowering (DIBL) and enhanced leakage currents. Employing materials such as In0.53Ga0.47As as the channel material in GAAFETs has proven to improve modulation of electric fields along the channel, with a drastic elimination of short-channel effects and improvement in on-current features. This is more than silicon-based technologies because such III-V compound semiconductors guarantee enhanced electron mobility and lower effective mass, leading to improved performance characteristics. Nevertheless, new material integration is faced with thermal challenges, particularly self-heating effects. Effective thermal management strategies, together with device architectural optimization, are imperative to realizing the complete potential of such innovative transistor structures for future electronic devices (Khrisshna et al., 2024).

Apart from that, attention on leakage current management and optimization of subthreshold slopes as a technique to neutralize short-channel effects cannot be overemphasized. Techniques such as the use of negative-capacitance concepts or asymmetrical doping of channels can significantly alter the gate-channel junction electrostatics, thus improving performance for short-channel devices. Optimum doping profiles and high-k gate dielectrics can be used to minimize the routes of leakage effectively. These developments enable power-performance trade-offs to be more easily controlled, thus facilitating MOSFET scaling. The scenario is evolving rapidly as there is a race among researchers trying out combinations of material science and architectural ingenuity to design avenues that not only suppress short-channel effects but also enhance the overall resilience of scaled MOSFET technologies for next-generation computing models (Supriya, 2025).

4.3.SCALINGOFMETAL-OXIDE-SEMICONDUCTOR FIELD-EFFECT TRANSISTORS (MOSFETS)

Advanced transistor design techniques are essential as conventional scaling techniques are severely constrained by short-channel effects. With reducing gate length of MOSFETs, these effects cause an increase in leakage current and reduction in transistor performance. The cause behind this is the fact that the channel is not long enough to be controlled by the gate electrostatically, resulting in the transistors inability to switch off completely. To address these problems, new transistor geometries, including FinFETs and Gate-All-Around FETs (GAAFETs), are being studied.

These are intended to enhance electrostatic control by providing a multi-gate architecture that surrounds the channel, effectively preventing leakage currents and improving subthreshold slope values. Improved understanding of these short-channel effects and reasons for concern to device performance is necessary in addressing the present scaling challenges of MOSFET technology, which has prompted the exploration of newer structures and materials to further enable beneficial scaling enhancements. Aside from the traditional planar structure, FinFETs are a significant advancement towards MOSFET scaling. By incorporating a three-dimensional shape, FinFETs offer better control over the gate on the channel, with reduced short-channel effects and enhanced drive current capability.

The devices utilize a vertically rising thin silicon fin where the gate can envelop multiple sides, effectively enhancing the effective channel width without resorting to bulky dimensions. Moreover, the transition to nanosheet transistors and GAAFET technology also highlights the value of innovative designs to meet aggressive scaling demands. The new geometries offer improved control of carrier transport and improved electrostatics, which are essential since the industry is focusing on driving the device dimensions below 5 nm. Deploying such advanced architectures can usher in a paradigm shift in addressing the needs of scaling, with performance benchmarks being constantly pushed higher (Figure 4.7).

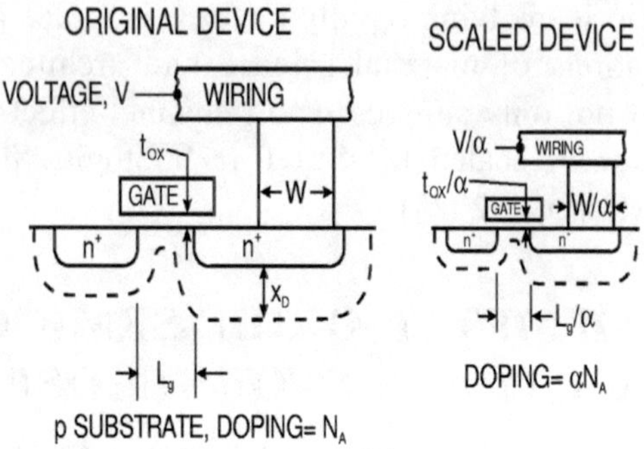

Figure 4.7. *Schematic illustration of the scaling of a Si metal-oxide-semiconductor field-effect transistor (MOSFET).*

Source: https://www.researchgate.net/profile/Subhra-Dhar-3/publication/256595549/figure/fig1/AS:297684926779396@1447985025728/Schematic-illustration-of-the-scaling-of-a-Si-MOSFET-by-a-factor-alpha-7.png.

Subthreshold slope and leakage current control is a significant area of MOSFET scaling since these parameters directly influence energy efficiency and the overall performance of the device. As transistors continue to reduce in size, retaining a steep subthreshold slope becomes increasingly challenging, though it is significant in power electronics for leakage protection during off-state operation. Recent development has seen focus on wider bandgap semiconductor material such as Gallium Nitride (GaN) and Silicon Carbide (SiC), which hold great potential in power electronics as evidenced by their performance ratings compared to silicon (Paul et al., 2017). These materials offer better performance at high voltages, which is useful in high-power devices. Thus, transcending the intrinsic limitation of current materials as well as exploring novel device geometries like FinFETs and GAAFETs can significantly improve MOSFET performance and play a vital role in resolving the scaling challenges of today's electronics (Amoroso et al., 2014).

4.3.1. Historical Perspective on Metal-Oxide-Semiconductor Field-Effect Transistor (MOSFET) Scaling

MOSFET technology development has been an important factor in driving the digital revolution as the need for continued miniaturization has spurred a tremendous amount of innovation within the semiconductor industry. The first studies recognized the need to eliminate short-channel effects that were emerging as devices were miniaturized, and this had a direct impact on performance and efficiency. When transistors grew smaller, phenomena such as increasing leakage current and diminishing control over the channel grew prevalent, which demanded inventive solutions. This was the start of new transistor architectures that would redress these short-channel effects. One such early move away from traditional alternatives was the turn towards alternatives such as the FinFET, enabling better electrostatic gate control and significant performance metric reduction. MOSFET history illustrates a continuous process within the engineering world to improve and adapt technology in response to challenges posed by physically shrinking dimensions (Bauer et al., 2023). Because there are advancing demands for performance and efficiency, history in new transistor technology has been dotted with further developments such as Gate-All-Around FETs (GAAFETs) and nanosheet transistors. These new architectures are a leap in scaling capabilities, enhancing gate control by wrapping the channel on all four sides. This extreme design has better electrostatic control, not only diminishing leakage but also enabling a steeper subthreshold slope, thus making switching more efficient. As the demand for more high-end computers and improved-performing phones intensifies, the architectures lay the foundation for breathtaking reductions in power consumption and device density. It is important to realize the context of these advancements historically because they were part of a widespread consensus within the industry that scaling not only was desirable but was necessary for near-term technology requirements (Tataria et al., 2021). Also, the MOSFET scaling history emphasizes the fine balancing act between device size and effectiveness. When engineers got over problems coming from shrinking dimensions, every move in transistor design advanced the boundaries of dominating manufacturing processes and materials science. The conversation between scaling and performance continues over into research and development work directed at understanding the inherent limits of transistor operation, including dielectric scaling and surface scattering problems. This delicate dance of invention and refinement reflects a dynamic situation within semiconductor technology, in which each solution is crafted with foresight toward future needs. Finally, continued study of scaling technologies not only enhances current device performance but also lays the groundwork for disruptive applications that will shape the electronics future (Bauer et al., 2023).

4.3.2. Physical Limitations of Traditional Scaling

The relentless pursuit of MOSFET technology miniaturization has uncovered a plethora of physical challenges that accompany traditional scaling approaches. Short-channel effects like increased leakage current and reduced electrostatic control become evident when devices are scaled below 5 nm. These effects in typical planar MOSFET architectures result in suboptimal performance since gate control of the channel degrades, which poses serious issues to maintaining performance characteristics such as threshold voltage stability. This necessitates reassessment of design paradigms to reverse such constraints, as simply reducing dimensions cannot guarantee enhanced performance or efficiency any longer. New design paradigms must thus be considered in order to address these challenges to pave the way for proper integration of scaling practices that can preserve performance as device dimensions keep going further into the nanoscale regime. To overcome decreasing returns of traditional scaling, innovative transistor architectures like FinFETs, GAAFETs, and nanosheet transistors have been introduced as viable alternatives. They have multiple gates that enhance electrostatic control over the channel even further, which is essential to controlling short-channel effects and leakage currents well (Figure 4.8).

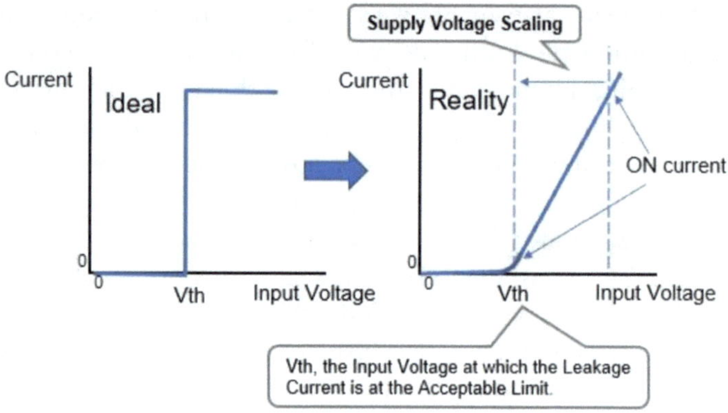

Figure 4.8. *Semiconductor miniaturization: limitations of metal-oxide-semiconductor field-effect transistor (MOSFET) scaling.*

Source:https://blog.nisshinbo-microdevices.co.jp/hs-fs/hubfs/blog/process/vol_8_fig02_supply_voltage_limitation.jpg?width=549&height=304&name=vol_8_fig02_supply_voltage_limitation.jpg.

With the application of a three-dimensional transistor structure, FinFETs increase the effective gate area, thereby improving device performance without sacrificing size. GAAFETs and nanosheet transistors continue by allowing even more electrostatic gate control through their new shapes. These advanced architectures are increasingly being used as the industry continues grappling with the limitations of two-dimensional planar devices. All of these innovations are a step in fulfilling the necessary performance specifications with adherence to principles of continuous scaling and miniaturization. Further, control of subthreshold slope and leakage current become an essential feature of transistor performance due to scaling-related problems. Planar MOSFETs have a thermal voltage-limited subthreshold slope that inherently impacts the switchability of the device. As the technology of semiconductors is being driven to ultra-scaled devices, it is necessary to maximize the subthreshold slope to achieve lower power modes of operation while retaining robust switching behavior. Innovations in gate material, channel engineering, and architectural innovations like that in FinFETs offer new ways to enhance control on these critical parameters. Research indicates a move towards using nanomaterials that could further aid in developing more favorable electrostatic conditions and improved on-off ratios, which are critical for the optimal operation of future scaled devices amidst an increasingly competitive market scenario(Higashiwaki et al., 2018; Naresh et al., 2021).

4.3.3. Performance Trade-Offs in Scaled Devices

As semiconductor devices continue to scale down, several problems arise that affect performance and efficiency directly. The most critical issue is the short-channel effects, which are faced when gate length is diminished. These effects result in diminished gate control of the channel, lower electrostatic integrity, and larger leakage currents. The leakage is not only responsible for energy wastage but also impacts the overall reliability of the devices. As a consequence, novel transistor structures like FinFETs and GAAFETs have come into picture to tackle these challenges through higher electrostatic control and better performance figures. The short-channel effects are decreased better by these novel structures, but they also demand careful handling of complexity in manufacturing and cost factors involved, thus facilitating another set of trade-offs that have to be considered in terms of performance scaling. The transition towards emerging transistor technologies has been accompanied by spectacularly enhanced device performance at the expense of some performance trade-offs that must be catered to in the context of scaling. An example is the nanosheet transistor, where they allow improved electrostatic control because of their three-dimensionality but at the expense of additional fabrication complexity and more complex material. Moreover, as the size of chips diminishes, subthreshold slope characteristics become important in dictating energy efficiency and power performance. Unfavorable subthreshold devices can cause drastic increases in power consumption at low voltages, ultimately limiting the payoff on scaling activities. Therefore, a very thin line has to be walked between architecture complexity and energy efficiency to facilitate the sustainability of device scaling while optimizing for performance gain. Another important consideration comes in testing leakage control in scaled devices. Currents of leakage can actually negate gains made through downsizing, particularly in battery devices where the power consumption has to be kept low. Minimizing the subthreshold slope of scaled MOSFETs, FinFETs, and GAAFETs is a must to curb unwanted leakage but usually results in trade-offs in other performance parameters such as drive current and speed of switching. For instance, the application of advanced materials or geometries to gate control for optimizing it can have the unforeseen effect of causing variability in performance from one manufacturing lot to another. Furthermore, competitiveness in performance is obtained by having continuous innovation in device structure, e.g., the integration of approaches proposed for maximum blood pressure monitoring (Devadasan et al., 2025) or user authentication systems (Wang et al., 2025), highlighting the necessity for flexible, scalable technologies. Hence, controlling these performance compromises is crucial in the evolution of semiconductor technologies while remaining compliant to the demands of modern electronic applications.

4.4.ADVANCEDMETAL-OXIDE-SEMICONDUCTOR FIELD-EFFECT TRANSISTOR (MOSFET) ARCHITECTURES

MOSFET architecture evolution is essential in fighting the battle against short-channel effects, particularly with shrinking device sizes in the world of semiconductors. As devices shrink in size, traditional planar MOSFETs degrade with diminishing electrostatic control, thereby leading to increased leakage currents and degradation of the subthreshold slope. In response to these constraints, alternative architectures such as FinFETs and Gate-All-Around Field-Effect Transistors (GAAFETs) have emerged. These architectures offer enhanced gate control and improved short-channel performance due to their three-dimensional layouts which offer superior electrostatic integrity. Thus, FinFET and GAAFET technology adoption is a significant milestone in maintaining device performance while furthering scaling limits effectively mitigating some of the inherent difficulties of mainstream planar technologies with cutting-edge semiconductor nodes. Outside of FinFETs and GAAFETs, nanosheet transistors have become yet another new architecture that also holds the promise of next-generation applications. This new architecture maximizes the advantages of stable gate control and reduced short-channel effects by using multiple stacked sheets, enabling further scaling without decreasing performance. The number of sheets can be varied to enable tailoring of the electrical characteristics of the devices with versatile configurations for various operation requirements. Such versatility is critical as industries shift towards more power-constrained and high-performance applications (Figure 4.9).

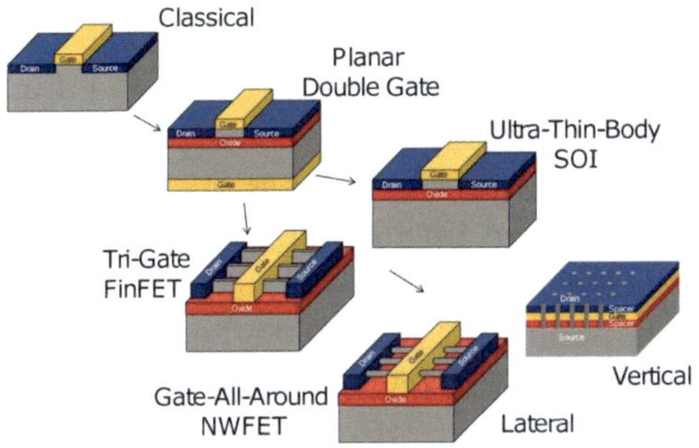

Figure 4.9. *Evolution of metal-oxide-semiconductor field-effect transistor (MOSFET) from classical to nanowire FETs.*

Source: https://www.researchgate.net/profile/Christoforos-Theodorou/publication/312022427/figure/fig4/AS:669039655395351@1536522892895/Evolution-of-MOSFETs-from-classical-to-nanowire-FETs-CopyrightSciRep12.jpg.

Furthermore, the nanosheet technology calls for the establishment of fabrication methods, particularly with respect to micron-sized dimensions and precision patterning, which are all highly important for the effective integration into broader semiconductor technologies (Lanzillo et al., 2025). With these new architectures of MOSFETs ongoing, subthreshold slope and leakage current control remain an overarching concern in the design process. Improved gate control in FinFETs, GAAFETs, and nanosheet transistors significantly enhances the electrostatic performance and subsequently decreases off-state leakage—a key determinant in prolonging battery life in handheld devices. However, with the utilization of these newer structures comes the challenge; balancing improvements in performance with increases in complexity in manufacturing processes requires innovative solutions. Power, performance, area, and cost (PPAC) intersections establish the feasibility of these architectures in practical applications. Continuous R&D will be required in the search for new materials and techniques that can be employed to enable these new designs, enabling the semiconductor industry to continue to meet the growing demand for high-density, efficient integrated circuits (Sarangi et al., 2024).

4.4.1. Introduction to FinFET Technology

With the relentless drive to shrink semiconductors, it has become clear that the traditional planar MOSFET architectures become progressively less efficient at the nanoscale. Such inefficiency of these smaller devices is, for the most part, due to short-channel effects, which happen with decreasing device size, resulting in issues like increased leakage currents and loss of gate control.

These lead to calls for innovations in transistor design, particularly those aimed at improved electrostatic control of the channel region. FinFET, or Fin Field-Effect Transistor, technology is one such solution, providing a three-dimensional control scheme that successfully obliterates such short-channel effects. By employing a vertically stacked fin structure, FinFETs facilitate a greater gate capacitance, hence increasing electrostatic integrity and further improving device performance at reduced sizes. Upgrading FinFET architecture, other architectures such as GAAFETs (Gate-All-Around FinFETs) and nanosheet transistors are the future of transistor scaling. GAAFETs surround the channel region on all four sides with more control over subthreshold slope and leakage currents, which are key parameters in characterizing the overall behavior of a transistor (Figure 4.10).

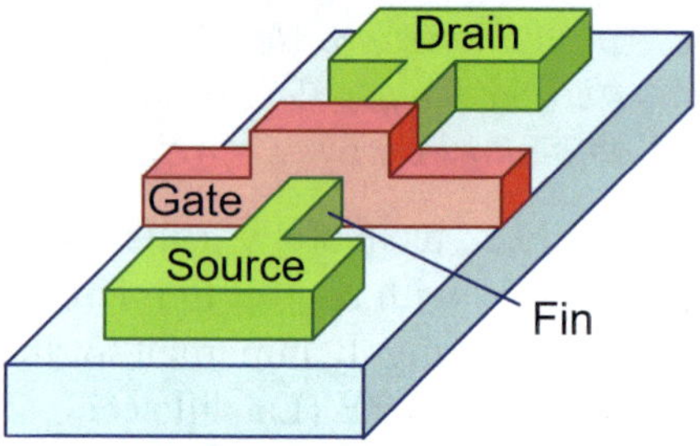

Figure 4.10. *Fin field-effect transistor.*

Source: https://upload.wikimedia.org/wikipedia/commons/7/79/Doublegate_Fin-FET-en.svg.

With gate leakage reduced and performance improved, GAAFETs represent a paradigm shift that heralds even better improvements compared to traditional FinFET designs. Nanosheet transistors, which use a similar three-dimensional architecture but thicker channels, demand equally huge future scaling potential. Such higher-end architectures underscore the need for continued innovation in transistor design to narrow the gap between the requirements for performance and physical issues encountered during scaling CMOS technology to newer nodes. The unification of these innovations results in a total strategy for fighting problems brought about by MOSFET technology scaling. Key to the implementation of such new transistor structures is the ability to achieve strong control over leakage currents with optimum on-state current maximization of the devices. The ability to manage the scaling effects in such advanced devices comes at the cost of significant power reduction and enhanced performance. Power device efficiency, particularly with respect to high voltage and temperature stability, remains in the center of attention for industry and academic experts alike. Materials such as diamond and wide-bandgap semiconductors are also at stake, as they guarantee better reliability and overall efficiency, particularly at high-performance levels. Further research in these fields is important, with serious implications for semiconductor technology's future in a broad spectrum of industries.

4.4.2. Overview of GAAFETs (Gate-All-Around FETs)

The uncompromising pursuit of miniaturization in semiconductor technology has led to a turning away from traditional MOSFET structures to more advanced alternatives in minimizing short-channel effects. Among these developments, GAAFETs have been a game-changer. By encasing the channel in gate material on all four sides, GAAFETs provide superior electrostatic control compared to their FinFET equivalents. This greater control helps in suppression of leakage current, ultimately leading to improved performance at lower dimensions. The three-dimensional aspect inherent in GAAFET design allows devices to scale more efficiently as it makes the maintenance of optimal subthreshold slope properties feasible despite shrinking transistor sizes. Thus, the GAAFETs are likely to overcome the limitations of intrinsic scaling issues in conventional MOSFETs' architecture to pave the way for improvements in the electronics world. Field-effect transistors have seen massive technology overhauls, particularly with the invention of nanosheet transistors that serve as a fulcrum between FinFETs and GAAFETs. Nanosheet transistors are an intermediate technology that utilizes thicker channel structures to boost drive current with small short-channel effects (Figure 4.11).

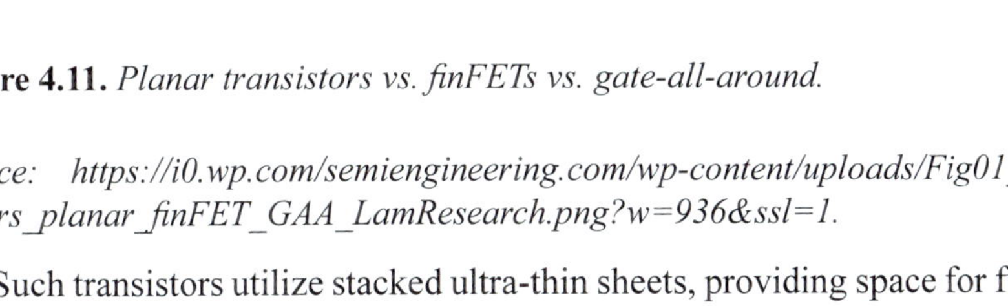

Figure 4.11. *Planar transistors vs. finFETs vs. gate-all-around.*

Source: https://i0.wp.com/semiengineering.com/wp-content/uploads/Fig01_transistors_planar_finFET_GAA_LamResearch.png?w=936&ssl=1.

Such transistors utilize stacked ultra-thin sheets, providing space for further reductions in footprint with effective control over the channel. This transition to nanosheet architectures represents a turning point toward GAAFET implementation because enhanced electrostatic integrity corresponds linearly with enhanced device performance. As the semiconductor industry is grappling with rising design complexities and physical limitations, GAAFET structures' utilization appears unavoidable on the condition that they can blend the virtues of FinFET and nanosheet architectures while resolving problems associated with performance inherent in miniaturization methods. The demand for efficient leakage management and subthreshold slope control on an emergent basis has fueled motivation towards GAAFET technology as an encouraging answer for next-generation integrated circuits. GAAFETs' all-encompassing gate architecture fosters not just greater electrostatic confinement but also tends to stabilize the electrical characteristic of the transistor under various operating regimes. Such stabilization is crucial in keeping the adverse effects of short-channel behavior low, thus allowing for improved power efficiency and performance. In addition to more sophisticated material alternatives, such as Ge1-xSnx alloys, GAAFETs can leverage their unique characteristics further to fine-tune the device structure. The convergence of GAAFET technology with innovative materials is hand in hand with the industry's aim to break through the intricate challenges that scaling presents, thus propelling the technology of MOSFET application to ever more advanced electronic systems (Cao et al., 2021; Galluccio et al., 2020).

4.4.3. Nanosheet Transistors and Their Advantages

Given that the semiconductor industry has confronted the limitations of traditional scaling methods, new transistor structures like nanosheet transistors have been overwhelmingly given prime importance. These advanced devices effectively address the issues created by short-channel effects, which significantly impact the performance of conventional planar metal-oxide-semiconductor field-effect transistors (MOSFETs). By employing a vertical structure consisting of multiple nanometer-thick sheets, nanosheet transistors enhance electrostatic control of the channel and, therefore, suppress leakage currents and improve subthreshold slopes. This superior level of electrostatic integrity allows enhanced performance at smaller sizes, allowing further scaling without loss of properties. In addition, design freedom provided by nanosheet configurations allows tailored device performance parameters, which becomes ever more important as the drive towards smaller, more efficient transistors continues within the industry. After FinFETs and gate-all-around field-effect transistors (GAAFETs), nanosheet transistors represent the latest frontier in device development. Their new design permits a larger gate area to channel length ratio, which provides enhanced electrostatic control. This kind of architecture promotes reduced batch-to-batch variability in device performance during manufacturing, alleviating some of the statistical problems that existed with earlier technology. Moreover, the ability to highly control the thickness of nanosheets provides engineers a powerful tool for device property tuning to meet precise application requirements. These advances not only facilitate improved on-current performance but also minimize off-state leakage, thereby improving power efficiency. As such, transistors fabricated from nanosheets are set to drive future electronics growth, particularly in high-performance and power-efficient application environments. The advantages of nanosheet transistors extend far beyond simply breaking current shorts in channel effects; they also hold promise to reduce broader challenges associated with scaling. As continuous miniaturization continues, conventional scaling techniques have begun to collapse, and techniques such as vertical transistor stacking are being explored for higher performance. This technology, which enables multiple nanosheets within a single device footprint, permits greater transistor density without compromising or improving electrostatic control. The latest developments discussed in the literature illustrate the potential of these devices to realize future nodes of technology beyond the resolution limit of current fabrication technology. In particular, the structure's ability to be easily augmented with new material makes it easy to overcome evolving semiconductor requirements (Liu et al., 2024; Zhang et al., 2024). These advantages all come together to account for how nanosheet transistors can be key to finding ways to satisfy Moores Law in the face of new scaling challenges.

4.5. SUBTHRESHOLD SLOPE AND LEAKAGE CONTROL

The relentless push for miniaturization in MOSFET technology has led to the invention of a number of sophisticated device architectures, including FinFETs, GAAFETs, and nanosheet transistors. Such architectures are particularly designed to mitigate short-channel effects that worsen with decreasing transistor size. As conventional planar MOSFETs scale down, they are affected by increasing leakage currents, primarily resulting from insufficient electrostatic control of the channel. This leakage not only reduces the performance of devices but also contributes significantly to power dissipation, a particular concern in modern electronics that need high energy efficiency. Device design advances have utilized multi-gate structures, enhancing gate control and reducing the subthreshold slope—thus enhancing leakage control. This architectural shift is an essential transition toward tackling the shortcomings of conventional MOSFETs so that generations of integrated circuits in the future will be able to satisfy performance and energy usage requirements. The study of the effects of subthreshold slope on MOSFET scaling unveils its central function in device efficiency. Theoretical subthreshold slope limit is set at 60 mV/decade; however, it becomes harder to approach as devices shrink (Figure 4.12).

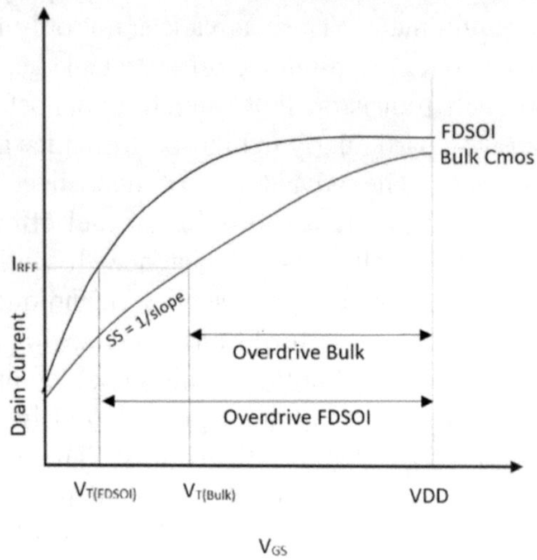

Figure 4.12. *Subthreshold Slope in FDSOI FD-SOI has a more vertical subthreshold slope (SS) than bulk CMOS (Vth) and requires less subthreshold voltage to forward bias the transistor in Thus, FD-SOI can match bulk CMOS performance with less voltage.*

Source: https://www.researchgate.net/profile/Chamak-Ganguly-2/publication/378534761/figure/fig2/AS:11431281228524775@1710061251800/Subthreshold-Slope-in-FDSOI-FD-SOI-has-a-more-vertical-subthreshold-slope-SS-than-bulk.jpg.

The physical attributes that control current transport mechanisms like drift and diffusion start to fail, and there is a need for new solutions in the form of tunnel field effect transistors (TFETs). Unlike traditional MOSFETs, TFETs have the capability of producing steep subthreshold slopes, improving the control over leakage current. Such a feature can provide better device performance at reduced subthreshold voltages, providing critical benefits in applications where power efficiency takes precedence. Therefore, the study of subthreshold steep devices, particularly together with new materials, is a promising means to challenge the challenges caused by traditional MOSFET technologies (Fahad, 2017). During this revolution, new materials and device architectures offer alternative means to leak efficiently and optimize the subthreshold slope. For instance, the utilization of two-dimensional (2D) materials such as graphene or transition metal dichalcogenides has been regarded as an extremely efficient approach to amplifying electrostatic control for short-channel devices. These materials possess some unique electronic properties that can significantly boost gate efficiency and potentially render subthreshold slopes smaller than the conventional limit. In this, the study of junctionless tunnel effect transistors (JTETs) introduces new interlayer tunneling mechanisms with it, leading to an intriguing new field in transistor structure and performance. With increased studies of these emerging technologies, the potential for new transistor structures to provide improved leakage control while sustaining rigorous performance requirements becomes more feasible, with the prospect of a new era in the formation of low-energy integrated circuits (Chen et al., 2017).

4.5.1. Importance of Subthreshold Slope in MOSFET Performance

Scaling of MOSFETs has transformed the scenario of semiconductor technology by a tremendous amount but has introduced complex issues involving short-channel effects. As transistor sizes reduce, the gating over control of the effect on the channel decreases, causing leakage currents and non-ideal behavior in the subthreshold region to increase. The subthreshold slope increasingly matters during this stage because it quantifies how well a transistor can switch from the off to the on state. A steeper subthreshold slope indicates tighter control of leakage currents, as required in power efficiency for current applications. Therefore, understanding the inherent limitation of the subthreshold slope is essential in enhancing the performance parameters of MOSFETs and addressing the scaling problems (González Márquez et al., 2019). New transistor devices such as FinFETs, GAAFETs, and nanosheet transistors are designed to address the limitations of traditional planar MOSFETs, particularly subthreshold performance. These new architectures enhance electrostatic control of short-channel effects by maximum gate surface area and minimum short-channel effect. For instance, FinFETs achieve improved drive current due to their three-dimensional structure that provides the maximum gate control over the channel. Despite such enhancement, the subthreshold slope is still a critical parameter that must be well controlled in order to minimize leakage. An ideal subthreshold slope below the theoretical limit of 60 mV/decade is an objective, and ongoing research is crucial in developing these technologies further to make them suitable for low-power applications (Haensch et al., 2010). As industries require more energy-efficient devices, controlling the subthreshold slope has become a research and development area for MOSFET technology. The need for a low subthreshold slope is apparent from its direct impact on leakage performance, crucial for battery-powered systems. An optimal subthreshold slope minimizes the off-state leakage currents that cause power dissipation, therefore extending battery life and enhancing overall system energy efficiency. Additionally, maintaining the balance between short-channel effects and the need for transistors to be fabricated more aggressively in scaling is vital. Introduction of devices that violate the 60 mV/decade constraint, such as tunneling field-effect transistors, is evidence of ongoing attempts to push performance limits while addressing issues of power dissipation (Haensch et al., 2010).

4.5.2. Mechanisms of Leakage Current in Scaled Devices

As device dimensions are scaled down in an effort to improve performance, issues caused by leakage current are becoming more apparent. Short-channel effects, which are taking place mainly as the transistors are scaling down, induce degradation of electrostatic control of the channel. This results in higher off-state leakage currents, which spoil device performance and also lead to power consumption. The impact of these effects is particularly conspicuous in typical planar MOSFETs, where the reduction of the gate length significantly degrades the transistors ability to well shut off between on-state and off-state operation. Consequently, leakage currents can increase, and the effective management of these effects is an important subject of study for further enhancing the energy efficiency of scaled devices. As a countermeasure to the issues introduced by short-channel effects, new transistor architectures such as FinFETs, GAAFETs, and nanosheet transistors have emerged. These structures employ multiple gates to achieve enhanced electrostatic control of the channel and thus lower leakage current. Through enhanced gate-channel coupling and promoting improved subthreshold behavior, such novel structures address intrinsic shortcomings of conventional MOSFETs. In particular, such sophisticated configurations promote superior subthreshold slope, enabling steeper off-on transitions. This transition not only assists in leakage control but also significantly decreases the operational inefficiency of modern-day semiconductor devices. The use of such technologies is a promising avenue for going beyond the intrinsic limits of scaling traditional planar transistors. Leaking current control lies very much integrated with the intricate balance of many operating parameters in such scaled-down structures. The subthreshold slope, which is a critical parameter that rules the switching between conduction and leakage, is of utmost significance in sustaining efficient performance at lower voltages. In particular, the best subthreshold slope must be achieved in order to reduce leakage as little as possible while maintaining drive current characteristics. The challenge therefore is not only to achieve a steep subthreshold slope but also stability across varying conditions. Emerging materials and engineering methods can be employed to reduce this slope and manage leakage. Besides, the incorporation of diversified sensing mechanisms, as evident in fields such as biosensor technology, creates greater alertness with regard to leakage control. Intelligent structures such as proposed in Kaisti (2017) and Naresh et al. (2021) indicate a viable way toward attaining such measures, eventually rectifying the performance issues typical of scaled devices.

4.5.3. Strategies for Leakage Control in Advanced Transistors

Transistor technology has called for creative techniques to manage leakage currents as a result of short-channel effects, whose severity increases with decreasing device size. In the traditional MOSFET, sub-threshold leakage raises very critical problems that degrade performance by creating static power. By the implementation of advanced device structures such as FinFETs and Gate-All-Around FETs (GAAFETs), researchers have come a long way in addressing these issues. These structures manage the electrostatic potential efficiently, thereby minimizing the short-channel effects while maintaining the functionality of transistor operation at smaller dimensions. The combination of these multi-gate configurations has improved controllability over channel areas to offer low leakage currents even at lower voltages, which is extremely crucial to enhance energy efficiency in future digital systems. In addition, nanosheet transistor design introduces one additional layer of potential and complexity of handling leakage. Nanosheet structure provides larger effective gate width with low profile, a means to lower the subthreshold slope. Subthreshold region is critical to the efficiency of the transistors, particularly for low-power applications where leakage directly affects battery life as well as device efficiency. With the capability to provide steeper subthreshold slope, nanosheet transistors show better switching characteristics, thereby presenting a possible solution for leakage control.

This optimization not only improves performance but also allows for even more extreme scaling, thus paving the way for generations of transistors ahead to be burdened with meeting increasingly burgeoning demands. With research in materials and device structure continuing, leakage control remains a number one priority issue, particularly as it relates to device reliability and life. The research of GaN and other alternative semiconductors offers potential advantages from their material properties, offering lower leakage rates than current silicon-based technologies. Applying GaN within improved transistor structures would allow designers to exploit its higher bandgap and higher critical electric field strength, leading to more efficient running devices. However, with GaN now only a recent addition to the list of well-established semiconductor materials, further work will be needed in order to fully understand its long-term reliability and degradation characteristics in real application conditions. Overcoming these problems will be crucial to enabling the successful incorporation of GaN into next-generation high-performance MOSFET designs and facilitating leap-ahead innovation in power electronics and beyond.

4.6. FUTURE TRENDS IN METAL-OXIDE-SEMICONDUCTOR FIELD-EFFECT TRANSISTOR (MOSFET) TECHNOLOGY

As semiconductor technology has improved, the ongoing search for more performance has placed pressures on novel device structures to assist in alleviating scaling-based intrinsic issues. Among the leading candidates in the bid to become the next-generation transistors are FinFETs, Gate-All-Around (GAAFETs), and nanosheet transistors, some of which have dominated developments in the area of MOSFET technology.

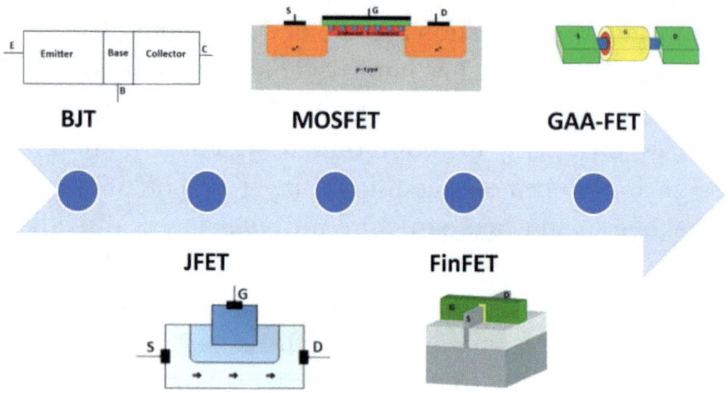

Figure 4.13. *The enhancement property of metal-oxide-semiconductor field-effect transistor (MOSFET).*

Source: https://cdnintech.com/media/chapter/1176382/1751010453-1539820664/media/F1.png.

The structures provide improved electrostatics with the ability to manage short-channel effects and excellent gate control of the channel. As an example, FinFETs' three-dimensional architecture improves the effective channel length, resulting in better performance of the device without any compromise on area. Here, the shift to GAAFETs is noteworthy, as they yield even larger decreases in subthreshold slope and leakage currents, reflecting the move towards more advanced architectures in semiconductor devices promising to extend the shelf life of Moores Law. One of the most significant aspects of the future of MOSFET technology is dealing with leakage currents as well as subthreshold slope improvement. These are extremely vital parameters in improving the overall power efficiency of the transistors when they shrink down even smaller. Material and device structure advancements assist in achieving significant improvements in leakage management, which is extremely vital in facilitating energy-efficient electronics. New materials such as high-k dielectrics and two-dimensional material are also being investigated for the potential to enhance the electrical properties of transistors while minimizing power loss (Figure 4.13).

Furthermore, ongoing work on hybrid structures that combine traditional silicon with other semiconductors sets the industry towards a new generation of power-efficient devices at no performance cost, thus responding to the challenge the growing demand for reduced-power electronics poses in an energy-conscious world (Harame, 2022). As the technology of semiconductor manufacturing evolves, the use of next-generation materials and novel design will play a pivotal role in setting the direction for the future of MOSFET technology. Researchers are exploring the possibility of using silicon together with III-V compounds for utilizing their high electron mobility, which can bring about significant enhancements in MOSFET performance. Additionally, the application of modular architectures and heterogeneous integration has the potential to ease manufacturing processes and achieve efficient coupling of disparate technologies. These advancements not only aim to push the boundary of device performance but also aim to offer systems capable of supporting complex applications like artificial intelligence and Internet of Things (IoT) devices. In short, with the progressive development of technology, the constant advancement of device structures and materials will have a profound influence on MOSFET design, and initiate groundbreaking changes in numerous applications in the semiconductor industry (Luryi et al., 2007).

4.6.1. Emerging Materials and Their Potential Impact

Transistor technology has largely underscored the limitations of traditional silicon-based materials, particularly the issue that is associated with short-channel effects for MOSFETs. When transistors scale down to the nanoscale, these effects become dominant, degrading performance parameters like subthreshold slope and leakage current. Novel materials, e.g., gallium nitride (GaN) and graphene, provide a potential path to eliminate the same. GaN, e.g., boasts good electron mobility and large bandgap, making it well suited for high-power and high-frequency applications. This advantage could bring more efficient electronic devices that produce less heat and elevate overall performance. By reducing short-channel effects, GaN-on-Si power electronics can not only increase functionality but also reduce operating costs compared to traditional silicon devices, hence offering a competitive edge in the semiconductor industry (Gao et al., 2019). In addition to GaN, flexible electronics innovations are revolutionizing wearable technology and medical monitoring devices. The ability to integrate wearable flexible sensors that can be curved around the body to monitor health in real time is essential for personalized medicine. Materials science innovations, particularly the development of soft flexible electronics, enable the user-device interface to be as natural as possible. These on-body sensors can record physiological data, such as temperature and heart rate, with the ability to minimize motion artifacts that typically confound measurements. Additionally, the incorporation of liquid metals in flexible electronics enhances conductivity and stretchability, enabling real-time monitoring of multiple health parameters, like hydration and blood glucose. With these technologies evolving, the prospect of noninvasive medicine becomes increasingly feasible, and the role of future materials in transforming healthcare delivery (Gao et al., 2019). Shrinking down transistors has been the problem, which has given rise to various architectures, such as FinFETs, Gate-All-Around FETs (GAAFETs), and nanosheet transistors. These architectures not only assist in offering better electrostatic control of the channel but also significantly reduce leakage current, leading to the overall improvement in efficiency of semiconductor devices. The novel materials play a major role in making these next-generation architectures possible by offering better electrical properties and mechanical flexibility. For instance, the use of two-dimensional materials, like transition metal dichalcogenides, in their manufacture could lead to further performance improvements as well as sheet scaling that is beyond the limit of traditional silicon. This advancement towards new materials and architectural designs is pivotal in finding resolution to the inherent constraints presented by current scaling methods to cater for the continued development of semiconductor technology despite growing demands for high-performance, low-loss electronic devices (Chen et al., 2017).

4.6.2. Integration of Artificial Intelligence (AI) and Machine Learning in Design

The evolution of semiconductor design techniques has been significantly influenced by the application of AI and machine learning. Such advanced computational techniques are increasingly used to address scaling constraints of MOSFETs, particularly as devices reach the nanoscale regime. Transistor scaling is a cause of short-channel effects, and such effects adversely influence device operation. AI can assist in predictive modeling and optimization by simulating various transistor architectures, such as FinFETs and GAAFETs, intended to minimize such unfavorable effects. The addition of AI-enabled design approaches allows effective space exploration of design spaces and its ability to find configurations that deliver the optimal subthreshold slope and leakage behavior and thereby enhance overall device performance and reliability in emerging applications. Hence, the use of machine learning at the design stage not only resolves existing limitations but also inspires innovation in transistor development. Additionally, the invention of nanosheet transistors represents a revolution in device design, where AI technologies are employed to assist their design and optimization. By employing algorithms modeling and predicting the performance of these novel configurations, engineers can better regulate the interaction between short-channel effects and scalability of the device. The nanosheet structural complexity needs sophisticated design methods, where AI facilitates ease of process through simulations to examine various physical effects on electrical properties. This convergence of smart systems facilitates the better tuning of leakage and subthreshold-related parameters, ultimately yielding devices that meet the stringent high-performance specifications requested by a wide range of applications, for example, artificial intelligence chips in the theme of future semiconductor technologies. Lastly, the integration of AI and machine learning into MOSFET design represents a synergistic partnership with the potential to enhance performance as well as to introduce new solutions to longstanding scaling challenges. As the world of semiconductor design forges its future, the application of predictive maintenance methods, developed nearly entirely on AI, stands forth as a foundation stone to quality and effectiveness. The ability to predict component failure and realign maintenance schedules enhances the overall integrity of semiconductor manufacturing processes. This outcome identifies the potential for AI to not only revolutionize design methodologies but also operational strategies in semiconductor manufacturing. Consequently, as advances in AI such as data analysis and advanced modeling develop the future of transistors, they bring forth the potential of overcoming the inherent limitations faced by traditional silicon-based technologies, maintaining a sustainable path for continued progress in the electronics sector.

4.6.3. Predictions for the Next Generation of Metal-Oxide-Semiconductor Field-Effect Transistors (MOSFETs)

With the silicon industry hitting the wall of traditional scaling, new approaches are imperative for sustaining the performance improvements of MOSFET technology. The future MOSFET architectures' predictions point to the transition away from planar geometry towards advanced three-dimensional designs such as FinFETs and Gate-All-Around Field-Effect Transistors (GAAFETs). The new geometries address short-channel effects, which are becoming progressively more destructive as the devices are miniaturized. FinFETs overcome these problems effectively by their vertical channel profile, giving better electrostatic control of the channel and more drive current capability.

Additionally, GAAFETs, with their completely surrounding gate, also hold the promise of reduced leakage currents, further boosting energy efficiency without compromising speed. With traditional planar MOSFETs in diminishing returns, these new architectures represent a breakthrough towards robust and efficient electronic devices that meet the relentless quest for more performance despite scaling challenges (Asenov et al., 2006). In conjunction with these new transistor architectures are advances in subthreshold slope engineering that are central to future MOSFETs development (Figure 4.14).

Figure 4.14. *Performance improvements in metal-oxide-semiconductor field-effect transistor (MOSFET) technology.*

Source: https://www.magnachip.com/wp-content/uploads/2022/01/600V-SJ-MOSFET-4-e1642112730938.jpg.

Control of the subthreshold slope is crucial for reducing leakage currents, which are significant power-dissipation components, especially in low-power designs. Traditional MOSFET structures are bounded by Boltzmann statistics, which determine the subthreshold slope to at least on the order of about 60 mV/decade. However, by the adoption of new material and device concepts, researchers are searching for substitutes that might render subthreshold slopes less than this threshold. For instance, the incorporation of quantum-dot structures or the use of two-dimensional materials can potentially bring with them new physical phenomena to transcend the traditional limits. These advances do more than optimize energy efficiency but create the potential for devices that can operate with reduced supply voltages, thus catering to the demands of future applications (Estribeau et al., 2009). The implications of these advancements extend far beyond simple performance enhancements; they portend a paradigm shift in the semiconductor ecosystem. As scaling issues increase and traditional MOSFET architectures reach physical boundaries, the industry needs to revolutionize its electronics design strategy entirely. The predicted advances in materials science, such as the insertion of new-generation dielectrics and high mobility materials, together with innovative transistor geometries, promise a future where integrated circuits can continue performing as well even as sizes shrink even further. Furthermore, radiation hardening techniques for future devices highlight the industry's emphasis on environmental resistance, an argument which is extremely relevant to aerospace and high-energy physics applications (Asenov et al., 2006; Estribeau et al., 2009). With the new technologies spreading their influence, they not only promise to provide answers to current problems but also create the foundation for ongoing innovation in the semiconductor industry.

SUMMARY

The relentless drive towards innovation in semiconductor technology has necessitated deep understanding of the scaling issues of MOSFETs, especially short-channel effects. As devices shrink, phenomena such as increased off-state leakage and subthreshold slope degradation set in, constituting a major bottleneck in performance enhancement. All these necessitate innovative solutions that breakthrough device architecture. Emerging transistor architectures, such as FinFETs and GAAFETs, show much promise in minimizing these detrimental effects through better electrostatic control of the channel. The three-dimensional nature of the architectures allows for improved electron manipulation, eventually leading to improved leakage control and sustaining performance despite further scaling. The integration of these emerging technologies into high-volume manufacturing highlights the importance of design flexibility in semiconductors as the industry aims to decrease feature size and boost performance. The strategic transition towards the adoption of advanced transistor architectures, like nanosheet transistors, is a critical watershed in addressing the unavoidable scalabilities facing conventional MOSFETs. As dimensions get smaller, maintaining compelling electrical performance and minimizing power dissipation become increasingly difficult. Nanosheet transistors deliver improved scaling performance with excellent short-channel control using their new structures that inherently suppress parasitic capacitance and improve electrostatic integrity. With these new structures, the semiconductor industry can better address the dual demand of delivering maximum performance while saving energy. Thus, scaling of MOSFET is not merely shrinking device size, but it is about redesigning paradigms to ensure that future devices are capable of meeting the rigorous requirements of applications without compromising reliability or efficiency. Overall, overcoming the scaling challenge of MOSFET is a multi-pronged approach that spans innovation in material, architecture, and fabrication techniques. As has been indicated, short-channel effects cause performance degradation, making it imperative to pursue alternative transistor technologies such as FinFETs and nanosheet geometries. Tackling such issues not only demands technical innovation but a firm methodology for failure analysis as well. The results of the present researches underscore necessity in using integrated failure assessment approaches that can diagnose and correct issues early enough in the scaling process. With evolving terrain in the semiconductors, building collaboration across the industry can spur innovation in actual real-world solutions to make sure the potential of high-end MOSFET technology can be leveraged for the next generation of electronic devices (Ahmad et al., 2024). Besides, lexicographic extension in language models as observed in Tao et al. (2024) can model scaling techniques that can be adapted to semiconductor design, promoting the overall insight across disciplines into optimization.

Transistor miniaturization has led to the emergence of short-channel effects, with vital impacts on device action as the dimensions dwindle. Such effects, comprising a reduction in gate control over the channel, may lead to increased leakage currents and threshold voltage variability. Hence, there is a need to manage such short-channel effects to maintain the functional integrity of MOSFETs. Transistor architecture innovation such as FinFETs, GAAFETs, and nanosheet transistors introduces efficient solutions against such problems. By the use of well-placed multiple gates or three-dimensional structures, these newer setups enhance electrostatic control, thereby enhancing performance metrics such as drive current and subthreshold slope. Such designs have essential functions in achieving desired energy efficiency and maintaining device scaling despite physical limitations that traditional bulk MOSFETs encounter. Besides reducing short-channel effects, advances in transistor technology also focus on reducing subthreshold slope and controlling leakage currents, both of which are extremely critical for power-efficient operation. Subthreshold slope, which measures the quality of a transistor to switch from an off-state to an on-state, is an extremely critical contributor to the overall power consumption of MOSFETs. An ideality of close to theoretical 60 mV/decade subthreshold slope is optimal in that it allows for steeper current transitions, and thus leakage minimization. Innovations such as GAAFETs contribute in this regard by allowing tighter electrostatic control over the channel, minimizing leakage currents without a compromise in switching speed. Thus, the achievement of an ideal subthreshold slope not only enhances the performance of individual devices but also to the overall efficiency of integrated circuits, especially in processing systems that are increasingly reliant on low-power operations. In addition to the dramatic enhancement of transistor structures and behavior, the continued improvement in semiconductor materials is necessary to address the challenges that MOSFET scaling brings. The development of materials such as GaN illustrates the possibility of a dramatic performance enhancement based on such attributes as wide energy gaps and high saturation velocity. GaN technology is particularly applicable for high-voltage applications, where the materials' properties enable more compact and efficient devices, thereby overcoming silicon-based process constraints. With increased power needs, particularly in wireless and mobile technologies, innovations in circuit design and optimization in the future will rely on the incorporation of new materials as well as creative transistor structures to drive these advances in momentum.

REFERENCES

1. Ahmad, I. H., Arifen, N. I. B. M., Mohamad, M. F., Ahmad, A. R. S., & Mohamad, M. F. B. (2024). Identification of recovering Idss failure mode via novel comprehensive failure analysis technique for MOSFET device. *2024 IEEE International Symposium on the Physical and Failure Analysis of Integrated Circuits (IPFA).* https://www.semanticscholar.org/paper/92371f47ec8584f8734e19c99715bba6f024f255 (accessed on 27th September 2025).
2. Amoroso, S., Asenov, A., & Georgiev, V. (2014). 3D multi-subband ensemble Monte Carlo simulator of FinFETs and nanowire transistors. https://core.ac.uk/download/42356098.pdf (accessed on 27th September 2025).
3. Asenov, A., Brown, A. R., Davies, J. H., Saini, S., et al. (2001). Increase in the random dopant induced threshold fluctuations and lowering in sub-100 nm MOSFETs due to quantum effects: A 3-D density-gradient simulation study. https://core.ac.uk/download/1393558.pdf (accessed on 27th September 2025).
4. Bauer, D., Khazdozian, H., Mehta, J., Nguyen, R. T., Severson, M., Vaagensmith, B., Toba, A. L., et al. (2023). *2023 Critical Materials Strategy.* https://doi.org/10.2172/1998242 (accessed on 27th September 2025).
5. Cao, L. (2021). Multi-objective digital VLSI design optimization. https://core.ac.uk/download/482023452.pdf (accessed on 27th September 2025).
6. Chen, K. J., Häberlen, O. D., Lidow, A., Tsai, C. L., Ueda, T., Uemoto, Y., & Wu, Y. (2017). GaN-on-Si power technology: Devices and applications. *IEEE Transactions on Electron Devices, 64*, 779–795. https://doi.org/10.1109/ted.2017.2657579 (accessed on 27th September 2025).
7. Chow, T. P., Higashiwaki, M., Kawarada, H., Omura, I., & Pala, V. (2017). Smart power devices and ICs using GaAs and wide and extreme bandgap semiconductors. https://core.ac.uk/download/147427115.pdf (accessed on 27th September 2025).
8. Devadasan, A. V., Sengupta, S., & Masum, M. (2025). Deep learning for non-invasive blood pressure monitoring: Model performance and quantization trade-offs. *Electronics.* https://www.semanticscholar.org/paper/125a2902b25baefbf50cd1afb8e427558cb6840d (accessed on 27th September 2025).
9. Donato, N., Rouger, N., Pernot, J., Longobardi, G., & Udrea, F. (2019). Diamond power devices: State of the art, modelling, figures of merit and

future perspective. *Journal of Physics D: Applied Physics, 53*, 093001–093001. https://doi.org/10.1088/1361-6463/ab4eab (accessed on 27th September 2025).

10. Estribeau, M., Goiffon, V., & Magnan, P. (2009). Overview of ionizing radiation effects in image sensors fabricated in a deep-sub micrometer CMOS imaging technology. https://core.ac.uk/download/12041448.pdf (accessed on 27th September 2025).
11. Fahad, M. S. (2017). Modeling of two-dimensional graphene and non-graphene material-based tunnel field effect transistors for integrated circuit design. https://doi.org/10.31390/gradschool_dissertations.4489 (accessed on 27th September 2025).
12. Galluccio, E. (2020). GeSn semiconductor for micro-nanoelectronics applications. https://core.ac.uk/download/334778909.pdf (accessed on 27th September 2025).
13. Gao, W., Ota, H., Kiriya, D., Takei, K., & Javey, A. (2019). Flexible electronics toward wearable sensing. *Accounts of Chemical Research, 52*, 523–533. https://doi.org/10.1021/acs.accounts.8b00500 (accessed on 27th September 2025).
14. González Márquez, R. (2019). A study on fundamental physical limitations of CMOS technologies. https://core.ac.uk/download/483360299.pdf (accessed on 27th September 2025).
15. Haensch, W., Koester, S. J., & Koswatta, S. O. (2010). On the possibility of obtaining MOSFET-like performance and sub-60 mV/decade swing in 1D broken-gap tunnel transistors. http://arxiv.org/abs/1011.5241 (accessed on 27th September 2025).
16. Hannan, M. A., Ker, P. J., Lipu, M. S. H., Choi, Z. H., Rahman, M. S. A., Muttaqi, K. M., & Blaabjerg, F. (2020). State of the art of solid-state transformers: Advanced topologies, implementation issues, recent progress and improvements. *IEEE Access, 8*, 19113–19132. https://doi.org/10.1109/access.2020.2967345 (accessed on 27th September 2025).
17. Harame, D. (2022). Perspectives on how the "SiGe, Ge, & related compounds: Materials, processing, and devices" field has changed over the last 20 years. *ECS Meeting Abstracts.* https://www.semanticscholar.org/paper/15861cee6879b4382402e0fbd0db168e6414b85e (accessed on 27th September 2025).
18. Higashiwaki, M., & Jessen, G. H. (2018). Guest editorial: The dawn of gallium oxide microelectronics. *Applied Physics Letters, 112.* https://doi.org/10.1063/1.5017845 (accessed on 27th September 2025).

19. Ji, H., Wang, J., Ma, J., Cheng, H. M., & Zhou, G. (2023). Fundamentals, status and challenges of direct recycling technologies for lithium-ion batteries. *Chemical Society Reviews.* https://www.semanticscholar.org/paper/1c77ce9faab67326520241240c86cbb52dbb3128 (accessed on 27th September 2025).
20. Kaisti, M. (2017). Detection principles of biological and chemical FET sensors. *Biosensors and Bioelectronics, 98*, 437–448. https://doi.org/10.1016/j.bios.2017.07.010 (accessed on 27th September 2025).
21. Karimi, K., Fardoost, A., & Javanmard, M. (2024). Comprehensive review of FinFET technology: History, structure, challenges, innovations, and emerging sensing applications. *Micromachines, 15.* https://www.semanticscholar.org/paper/b493dfd404e13349468f9161608426b4e5550974 (accessed on 27th September 2025).
22. Khrisshna, N. R., Kumar, B. P., Bhol, K., Tayal, S., & Jena, B. (2024). $In_xGa_{1-x}As$ stacked multichannel MOSFET design to improve the electrostatic performance by self-heating effect reduction. *IEEE Access, 12*, 196567–196575. https://www.semanticscholar.org/paper/f8c1b5545925f54d3b44651edf6fd633b9a5b727 (accessed on 27th September 2025).
23. Lanzillo, N. A., Khan, S., Mazza, J., Bajpai, U., & Motoyama, K. (2025). A perspective on interconnect scaling challenges in the NanoStack transistor era: AP/DFM: Advanced patterning / design for manufacturing (design-technology co-optimization). *2025 36th Annual SEMI Advanced Semiconductor Manufacturing Conference (ASMC).* https://www.semanticscholar.org/paper/1a6ff494705146508485ba5b08726b2748be4162 (accessed on 27th September 2025).
24. Liu, A., Zhang, X., Liu, Z., Li, Y., Peng, X., Li, X., Yue, Q., et al. (2024). The roadmap of 2D materials and devices toward chips. *Nano-Micro Letters, 16.* https://doi.org/10.1007/s40820-023-01273-5 (accessed on 27th September 2025).
25. López, C. (2022). Artificial intelligence and advanced materials. *Advanced Materials, 35.* https://doi.org/10.1002/adma.202208683 (accessed on 27th September 2025).
26. Luryi, S., Xu, J., & Zaslavsky, A. (2007). Future trends in microelectronics: Up the nano creek. https://www.semanticscholar.org/paper/ffc6ac857e3796b7a95d26b3b1fac5ac6832f868 (accessed on 27th September 2025).
27. Mahesh, A., Bharatiraja, C., & Mihet-Popa, L. (2021). Inductive wireless power transfer charging for electric vehicles–A review. *IEEE Access, 9,*

137667–137713. https://doi.org/10.1109/access.2021.3116678 (accessed on 27th September 2025).

28. Meneghini, M., De Santi, C., Abid, I., Buffolo, M., Cioni, M., Khadar, R. A., Nela, L., et al. (2021). GaN-based power devices: Physics, reliability, and perspectives. *Journal of Applied Physics, 130.* https://doi.org/10.1063/5.0061354 (accessed on 27th September 2025).
29. Mohammad, H. P., Hadimani, H. C., Yedukondalu, U., & Udara, S. R. (2022). Distinct ρ-based model of silicon N-channel double gate MOSFET. *International Journal of Reconfigurable and Embedded Systems (IJRES).* https://www.semanticscholar.org/paper/4da8fc988f46a986f3fdcd38fa5fe2306a9d4a89 (accessed on 27th September 2025).
30. Naresh, V., & Lee, N. (2021). A review on biosensors and recent development of nanostructured materials-enabled biosensors. *Sensors, 21*, 1109–1109. https://doi.org/10.3390/s21041109 (accessed on 27th September 2025).
31. Sarangi, S., & Kanungo, S. (2024). Quantum computing integration with multi-cloud architectures: Enhancing computational efficiency and security in advanced cloud environments. *World Journal of Advanced Engineering Technology and Sciences.* https://www.semanticscholar.org/paper/fd903b36b5529b1940945f74e28606fe07329031 (accessed on 27th September 2025).
32. Shadrokh, Y., & Shadrokh, Y. (2010). Benchmarking the screen-grid field effect transistor (SGrFET) for digital applications.
33. Supriya, D. V. (2025). Exploration of CNTFET-oriented approximate full adders' techniques, challenges, and future prospects. *Journal on Electronic and Automation Engineering.* https://www.semanticscholar.org/paper/db977eec4987911a5b5f3b76eb295043b3afe7c9 (accessed on 27th September 2025).
34. Tao, C., Liu, Q., Dou, L., Muennighoff, N., Wan, Z., Luo, P., Lin, M., et al. (2024). Scaling laws with vocabulary: Larger models deserve larger vocabularies. https://www.semanticscholar.org/paper/96686b8a1cd281584b1f4812a124a434dbda1c17 (accessed on 27th September 2025).
35. Tataria, H., Shafi, M., Molisch, A. F., Döhler, M., Sjöland, H., & Tufvesson, F. (2021). 6G wireless systems: Vision, requirements, challenges, insights, and opportunities. *Proceedings of the IEEE, 109*, 1166–1199. https://doi.org/10.1109/jproc.2021.3061701 (accessed on 27th September 2025).
36. Uçar, A., Karaköse, M., & Kırımça, N. (2024). Artificial intelligence for predictive maintenance applications: Key components, trustworthiness,

and future trends. *Applied Sciences, 14*, 898–898. https://doi.org/10.3390/app14020898 (accessed on 27th September 2025).

37. Wang, K., Huang, M., Dai, Z., Yan, S., Guo, Z., Xiang, H., Gong, X. G., et al. (2025). Silicon nanowire gate-all-around cold source MOSFET with ultralow power dissipation: A machine-learning-Hamiltonian accelerated design. *Advanced Functional Materials.* https://www.semanticscholar.org/paper/316d81fea1bb18d18fc812f55a1c8b99e13d45ea (accessed on 27th September 2025).
38. Wang, Y., Wu, C., Liao, Y., & You, M. (2025). Optimizing mouse dynamics for user authentication by machine learning: Addressing data sufficiency, accuracy-practicality trade-off, and model performance challenges. https://www.semanticscholar.org/paper/66e8aca7a612fcc580eceaa2a5616b809f3831ed (accessed on 27th September 2025).
39. Zhang, Q., Zhang, Y., Luo, Y. N., & Yin, H. (2024). New structure transistors for advanced technology node CMOS ICs. *National Science Review, 11.* https://doi.org/10.1093/nsr/nwae008 (accessed on 27th September 2025).
40. Zheng, Y., Xu, H., Li, Z., Li, L., Yu, Y., Jiang, P., Shi, Y., et al. (2025). Artificial intelligence-driven approaches in semiconductor research. *Advanced Materials.* https://www.semanticscholar.org/paper/b54f0af12f8140abd8b846466fb3788a4deb3cf5 (accessed on 27th September 2025).
41. Zhou, H. (2025). An overview of hot carrier degradation on gate-all-around nanosheet transistors. *Micromachines, 16.* https://www.semanticscholar.org/paper/0d1255ce3ec4caeea7e3bf97df4b6947598b113d (accessed on 27th September 2025).

CHAPTER 5

ADVANCED TRANSISTOR ARCHITECTURES

Contents

Introduction 177
5.1. Historical Context 182
5.2. Tunnel Field-Effect Transistors (TFETs)..188
5.3. Negative Capacitance Field-Effect Transistors (NC-FETs) 192
5.4. Vertical Transport Devices 197
5.5. Emerging Trends in Transistor Architectures 203
5.6. Challenges and Limitations...207
Summary...212
References...214

INTRODUCTION

The electronics face has undergone dramatic transformations, and most of these result from transformations in transistor design. While the limits of traditional silicon-based technologies are yet to be exploited, new technologies are emerging that aim at enhancing performance at lower energy levels. This is also fueled by the emergence of devices such as Tunnel FETs (TFETs), which rely on quantum tunneling phenomena to achieve lower voltage operation compared to their traditional counterparts. The transition from traditional silicon designs to yet more sophisticated designs is a breakthrough not merely in device building but in the fundamental principles of electronic functioning. As scientists and engineers exploit the limits of the potential within these new architectures, a knowledge of their theoretical foundations and real-world implementation becomes increasingly vital. At the same time, NCFETs represent a new frontier of transistor development. With the addition of ferroelectric materials to the gate stack, NCFETs can achieve subthreshold slope values above the traditional 60 mV/decade limit at room temperature. This can facilitate enhancement in switching characteristics and power reduction required for next-generation devices with emphasis on energy efficiency (Figure 5.1).

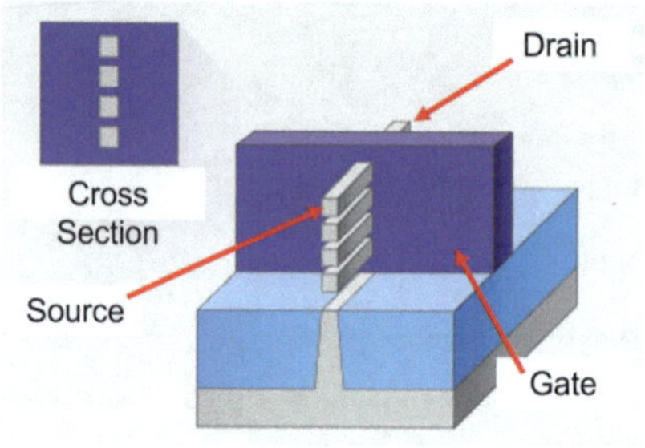

Figure 5.1. *The technical advancements in transistor architectures.*

Source: https://www.nist.gov/sites/default/files/images/2024/03/23/Causal%20Greens%20function%20simulations%20of%20phonons%20for%20multiscale%20multiphysics%20modeling%20of%20thermal%20transport%20in%20gate%20all%20around%20transistors.png.

The marriage of semiconductor physics and materials science provides a rich soil for innovation, as researchers explore how the new structures can better be made compatible with existing manufacturing techniques. Every step towards NCFET integration represents the relentless pursuit of efficiency, performance, and sustainability of modern electronics, and constitutes a compelling case for their more extensive adoption in future electronic devices. Furthermore, research into vertical transport devices presents the importance of architecture in making transistors more capable. Such devices, which utilize a vertical structure to maximize electron transport, have the potential to provide significant advancements in speed and power efficiency. With increasing demand for high-speed computing and mobile usage, demands for new device structures are urgent. By reshaping the spatial geometry and working dynamics of transistors, researchers are capable of taking advantage of advantages that traditional planar geometries may not be able to offer. The integration of vertical transport concepts with technologies like TFETs and NCFETs is a comprehensive approach for creating transistor structures so that future electronic technology will have the capability to meet the surging computational demands of society. Thus, the transition towards advanced transistor architectures is not merely a technological revolution but an imperative necessity in the unfolding saga of technological advancement.

Definition of Transistor Architectures

The unveiling of advanced transistor architectures presents a multifaceted arena driven by the primal principles of electronic engineering and materials science. TFETs represent some of the seminal developments in this field, which function on the quantum tunneling effect in place of the traditional channel conduction. This effect has the potential to lead to lower power consumption with high speeds of operation, addressing one of the main problems with conventional transistors. With the application of an electrostatic control mechanism, TFETs have the potential to produce low-voltage and ultra-scaled electronics, essential for future devices that demand greater energy efficiency and better performance. Hence, TFET development represents a vital adaptive innovation to shifting needs in modern technology, particularly in those applications that demand severe power control and thermal stability. Subsequent advancements have led to NC-FETs, with ferroelectric materials that minimize gate voltage requirements. The design allows for the scaling of transistors above the scale limit of regular designs, significantly increasing the speed and efficiency of operations within circuits. In addition to increasing subthreshold swing in transistors, the integration of negative capacitance mechanisms provides a method for power saving at an unprecedented level. As device sizes shrink, NC-FETs offer a major solution outside the inherent thermal limit of semiconductor technology. The use of ferroelectric material with standard semiconductor processing is one of the major paradigm changes, thereby ensuring that technology keeps up with the rapidly evolving computational requirements in the age of digitization. Vertical transport devices are also a frontier in transistor structure that has the benefits of several advantages over traditional planar structures. With the use of the three-dimensional device structure, vertical transistors can offer improved current densities and thermal efficiency, both being significant for high-performance devices of today. Vertical orientation allows for reducing channel length and consequently improved switching speed and reduced power consumption. In response to the growing demand for smaller yet high-performance electronic systems, devices such as vertical-gate FinFETs demonstrate the trend of multiscale architectures that optimize space and performance. In addition, these designs open new possibilities for the integration of complex functionalities into a reduced volume, significantly impacting the performance requirements of devices in most fields of electronics (Rasheed et al., 2020; Tataria et al., 2021).

Importance in Modern Electronics

Innovations in transistor technology are paramount in shaping the landscape of modern electronics, particularly with the advent of advanced architectures such as TFETs. TFETs present a possible alternative to traditional field-effect transistors, primarily due to their unique tunneling mechanism that enhances performance at the cost of greatly reduced power consumption. This capability has a direct correlation with the increasing demand for power-efficient devices in mobile and IoT applications, thereby making these transistors an imperative building block for future electronics. Besides, as researchers work to integrate TFETs into the existing semiconductor production processes, the way to their commercialization becomes more and more evident, and this is an important achievement in the solution of energy issues of contemporary electronic devices (Luo et al., 2023).

Parallel with this, the creation of negative capacitance FETs is yet another breakthrough in the history of transistor design. By the use of negative capacitance, these transistors enable voltage to be cut below the present requirement for operation, utilizing less power but without a corresponding rise in performance. This phenomenon not only enhances device efficiency but also admits of immense scalability in the case of CMOS technology, which is important as the semiconductor industry works toward increasingly minute nodes. The integration of negative capacitance devices may thus lead to a new generation of low-power electronics, appropriate for consumer demand and regulatory concerns regarding energy consumption. The promising aspects of the FETs reinforce their position in the development of modern electronic devices (Zhang et al., 2022). Vertical transport devices are increasing in significance too, as they handle the call for high-performance applications that require greater efficiency and smaller size. These transistors utilize a three-dimensional structure that allows for better current drive and reduced short-channel effects, significant as transistor sizes shrink. Vertical transport arrangements not only enhance device performance but also provide the potential for new functionality, e.g., multi-gate architecture and in-built sensors. With increasing demand for smaller and more efficient electronics across various areas, e.g., automotive and telecom, the significance of vertical transport devices will surely grow. Together with TFETs and negative capacitance FETs, vertical transport architectures are a concerted move towards revolution and advancement in electronic systems in the face of exponential technology growth (Luo et al., 2023).

Overview of Advanced Architectures

Advanced development of transistor architecture marks the milestone for semiconductor technology because of the need for increased energy efficiency and performance. Among these new devices, TFETs are a significant innovation since they employ novel operating principles based on quantum mechanical tunneling. As opposed to the conventional MOSFETs relying on thermionic emission, TFETs can offer a steep subthreshold swing, which helps in achieving lower operating voltages and lower power consumption. This aspect is particularly crucial in ultra-low power electronics and circuits of the future, where performance needs to be maintained while reducing energy expenditure. Hence, studies towards optimizing TFET materials as well as the processing techniques continue to build momentum, pinning emphasis on the transformational character of the technology in the production of electronic devices. Another key innovation in next-generation architectures is the NCFET that includes ferroelectric material for gaining increased electrostatic control over the channel. This design utilizes negative capacitance to effectively amplify the gate voltage at low voltages, reducing static power losses that are typical in traditional devices. NCFET's ability to maintain a preferred subthreshold swing while operating at lower voltages can lead to spectacular gains in performance metrics such as read/write speed and energy efficiency (Figure 5.2).

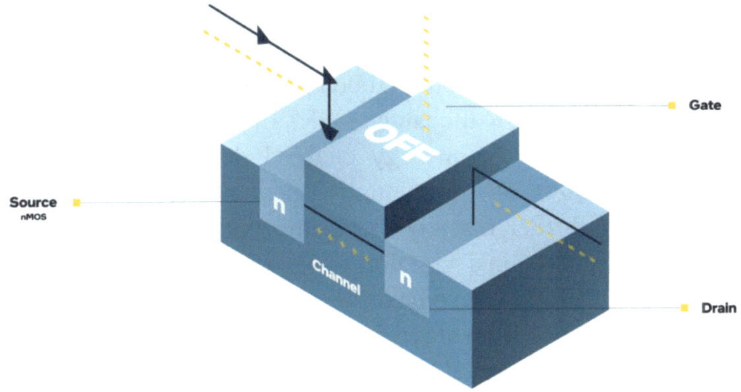

Figure 5.2. *The evolution of advanced transistor architectures.*

Source: https://newsroom.intel.com/wp-content/uploads/2025/03/tech101_transistors_vis5_v01-1024x510.gif.

As businesses strive to meet the demands of increasingly complex applications, the potential impact of NCFETs may drive the future of low-power electronics and high-performance computing systems. Thus, the evolution of NCFET technology represents a classic example of how newer architectures can rise above the issues of semiconductor performance. Vertical transport devices are yet another disruptive technology in the field of novel transistor architectures, challenging conventional lateral designs. By incorporating a vertical aspect, these devices can reduce current density in an efficient manner, improve scalability, and pack more transistors into a constrained footprint, making them extremely sought after in dense applications. Vertical transistors utilize epitaxial growth techniques and innovative material selection to enable performance while consuming very little power. Besides, with increased emphasis on green technologies, vertical architectures offer sustainable solutions for overcoming environmental problems associated with semiconductor processing. With further research in this area, semiconductor manufacturing companies will be able to use vertical transport devices to not only improve current technologies but also promote green practices in the industry. The ongoing research and development in the frontier architectures highlight the critical role of interdisciplinary work in the development of innovation capable of dictating the future of electronic systems.

5.1. HISTORICAL CONTEXT

Transistor technology has been progressing with sweeping revolutions in underlying design and structure, both in accordance with the requirements of the marketplace and technological breakthroughs in materials science. The initial transistors were largely built on bulk silicon substrates, and those were largely limited by the inherent physical characteristics of silicon. The complementary metal-oxide-semiconductor technology enabled device miniaturization, and this led to a dramatic increase in computing power and energy efficiency. As technology advanced further, researchers began working with new materials and structures, paving the way for technologies such as Tunnel FETs (TFETs) and negative capacitance FETs. These developments demonstrate a trend towards more sophisticated designs from classical approaches towards utilizing quantum mechanical effects for enhanced performance at lower power consumption, demonstrating how past research into the development of transistors leads to current innovation in semiconductor design (Figure 5.3).

Scientifically and technologically, the technical and scientific challenges of emerging transistor architectures have required rigorous research and development. For more than a decade, the U.S. Department of Energy (DOE) has committed substantial funds to learning about the important materials that make these developments possible, overcoming the underling S&T hurdles that arise with emerging architectures, including vertical transport devices. This emphasis on ground-level research is a testament to the realization of not just pushing current technologies further but also investigating new mechanisms that would break the conventional semiconductor operation paradigm. Emerging techniques in doping technologies, particularly for sophisticated devices like FinFET and gate-all-around (GAA) transistors, underscore the adaptability required in the industry.

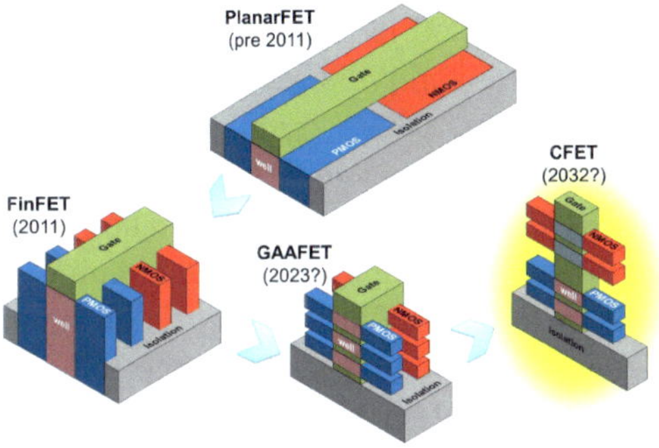

Figure 5.3. *The Evolution of Transistors: From PLANAR MOSFETs to FINFETs to NANOSHEET FETs.*

Source:https://techovedas.com/wp-content/uploads/2024/01/1_zEx-zyOsd9mTYH-HiK53Gg.webp.

New techniques to analyze and measure material characteristics are required to exploit fully the benefits of such emerging designs, portraying the dynamic interaction between the past and future of semiconductor technology. Besides, the use of metrology methods has become a critical aspect of maintaining the pace of transistor architecture advancement. With geometries shifting from 3D to 2D materials, conventional test methods struggle to keep pace with emerging demands associated with device performance parameters. The implications of such a change can be seen starkly in the methodology used for dopant profiling in association with next-generation structures. As defined in the literature for these new architectures, traditional figures-of-merit such as impurity concentration and activation profiles are less applicable, inducing the need for new measurement techniques. The growing reliance on less abusive doping methods and advanced characterization techniques is a feature of an even broader tendency in semiconductor research, since the nature of new architectures demands continued advancement of analysis tools, placing modern breakthroughs solidly in an important scientific context of investigation and technological progress (Bauer et al., 2023; Маргарита Георгиева et al., 2023).

5.1.1. Evolution of Transistor Technology

The development of transistor technology has ever been towards circumventing the constraints of conventional MOSFET structures. With increasing size constraints, new architectures such as Tunnel Field-Effect Transistors (TFETs) have been viewed as potential substitutes. TFETs allow for subthreshold conduction through tunneling and facilitate significant reduction in operating voltage and power. This capability is particularly important in the ultra-nanoscale device environment, where traditional MOSFETs are confronted with leakage current and short-channel effects. Additionally, the unique electrostatics of TFETs enable improved performance parameters at reduced dimensions, rendering their inclusion in advanced circuit designs an imperative endeavor. Thus, research into TFET technology is an important milestone in the development of transistor technology, hoping to follow the pattern of ever greater performance established in the semiconductor industry. TFETs are studied in combination with other research into Negative Capacitance Field-Effect Transistors (NC-FETs). This new architecture tries to take advantage of the process of negative capacitance, which would be capable of further enhancing gate control over the channel charge carriers, thus enabling subsequent scaling of the transistor size without sacrificing substantial leakage. This structure promises a path to reduced power operation with high performance and gives solutions to the biggest issues of energy efficiency in modern electronics.

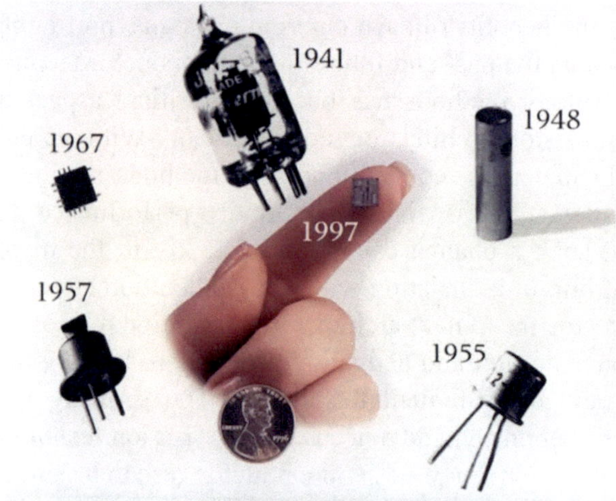

Figure 5.4. *The historical development in transistor.*

Source:https://www.allaboutcircuits.com/uploads/articles/The_transistor_has_evolved_in_sizes_and_types.jpg.

Furthermore, NC-FETs can be designed to enhance device figures such as speed and scalability without losing the inherent advantages of conventional MOSFETs. Using the incorporation of ideas from ferroelectric materials into the design of the transistor, NC-FETs are a cutting-edge next step, with focus on sustainability as well as performance in the next-generation applications of electronics. Another breakthrough in transistor technology is the development on vertical transport devices, which shift the conventional planar design paradigm. These devices provide more current paths using vertical structures, with highest space utilization and potential for higher integration densities. Taking advantage of the three-dimensional space, vertical transport devices not only alleviate the performance limitations seen in interconnects, but also minimize the device fabrication area. This alternate paradigm enables higher scalability and efficiency, which is essential to meet the increasing demands of small form factor yet high-performance electronic systems. Along with advanced materials and design in architecture, vertical transportation systems can potentially lead the way for progress in circuit design, system overall performance such as high-frequency RF applications. All these advancements are bringing into prominence important milestones to the evolution of transistor technology, holding promise for a robust future for semiconductor devices in a more competitive market.

5.1.2. Key Milestones in Transistor Development

Transistor technology has reached a watershed moment with the advent of Tunnel Field Effect Transistors (TFETs). TFETs are amazing in that their mechanism of operation capitalizes on quantum tunneling to enable sub-threshold conduction. TFET's ability to work at lower voltages significantly reduces power consumption, thus addressing the pressing need for energy-efficient electronics. This advance not only enhances the performance of digital circuits but also facilitates miniaturization of chips in line with Moore's Law. Based on cutting-edge materials and new doping methods, TFETs can revolutionize applications in high-speed and low-power electronics. Their compatibility with existing semiconductor technologies is an exhilarating frontier to be discovered by research and development that can lead to the implementation of TFETs in upcoming integrated circuits and be the key to sustainable technology in a power-constrained era. Another breakthrough in transistor technology is the evolution of Negative Capacitance FETs (NC-FETs) that make use of the ferroelectricity of materials to achieve reduced demands of gate voltage. The use of ferroelectric materials in the gate stack of traditional FETs allows for such devices to tap into a negative capacitance effect, resulting in better electrostatic control over the channel. This breakthrough defies the constraints brought about by short-channel effects typical with traditional silicon devices, thereby allowing for the design of smaller, faster, and more efficient transistors. NC-FETs are the focus of research to create scalable schemes, paving the way for the enhancement of low-power digital circuits, as advocated by varied research that highlights their potential to go beyond Silicon CMOS technology scaling limits. Their use flexibility to integrate upcoming semiconductor procedures also validates the use of NC-FETs in future electronics with an emphasis on power efficiency and performance. Vertical transport devices are another critical breakthrough in the wake of transistor technology, inviting interest for their three-dimensional architecture that offers improved electron transport. By positioning the channel vertically, these devices can reduce the distance carriers must travel by a significant amount, leading to improved performance parameters like lower delay times and reduced power consumption. Vertical transportation geometries, such as vertical-TFETs and vertical MOSFETs, come to prominence more and more due to their compatibility with integration into state-of-the-art 3D integrated circuits. Research indicates that their distinct structure allows for higher current densities, thereby providing compact designs while maintaining efficiency. While vertical transport devices are being enhanced with novel materials, such new architectures are already well on their way towards becoming quintessential facilitators for future electronic systems, which will transform a variety of applications in computing and communications technology.

5.1.3. Transition from Bipolar to Field-Effect Transistors

The advancement in transistor technology has made a deep influence in the electronic world, particularly with the transition from bipolar junction transistors (BJTs) to field-effect transistors (FETs). This transformation is marked not only by the advancements in device performance but also by the pressing requirements of modern applications. FETs, due primarily to their improved input impedance and lower power consumption, provide a welcome alternative to BJTs, especially for battery devices. Among the new architectures emerging are tunnel FETs (TFETs), which leverage quantum tunneling to achieve sub-threshold swing behavior that is beyond the capabilities of traditional thermal devices. Through the employment of TFETs, device performance can be enhanced, and the operational voltage can be reduced concurrently, aligning with the recent push for energy-efficient technology-(Tataria et al., 2021). Such innovation points to the wide-ranging impact of FET architectures in consumer electronics and industrial use. In addition to the promising performance of tunnel FETs, the application of negative capacitance FETs (NC-FETs) has garnered interest due to their potential to overcome the intrinsic constraints of conventional FETs.

NC-FETs employ ferroelectric materials to induce a negative capacitance effect, which can significantly reduce the subthreshold swing and optimize operational efficiency. This technology provides a credible way of minimizing power consumption without decreasing speed and is very compatible with integrated circuit design advancements. Moreover, as technology advances toward more robust field-effect devices, vertical transport architectures are being researched for optimizing electrical performance through the reduction of channel length while effectively managing current density (Figure 5.5).

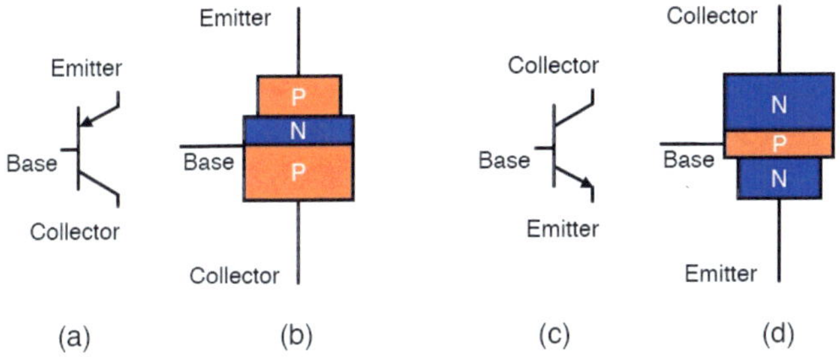

Figure 5.5. *Interplay between bipolar junction and field effect transistors (BJTs and FETs).*

Source: https://www.allaboutcircuits.com/uploads/articles/4.4_Transistors1_.jpg.

These transistor architectures are a pioneering step towards the achievement of better performance and lower power transistors in future semiconductor applications-(César Thadeo de Lima et al., 2021). The study of these next-generation transistor architectures is inevitable to address the evolving requirements of telecommunications and computing. Particularly, with the communications landscape shifting towards more bandwidth-demanding applications, FETs, including TFETs and NC-FETs, emerge as pillar technologies that are capable of taking the challenge head-on. The anticipated growth of sixth-generation (6G) networks highlights the demand for next-generation devices with heretofore unseen performance and power efficiency in a broad spectrum of frequency bands. The integration of machine learning protocols and artificial intelligence in devices will revolutionize their capacity to react to real-time scenarios and dynamically adjust their performance. With the evolution of these transistor technologies, not only do they promise to enable electronic performance but also new uses that redefine data connectivity and processing in an increasingly digitized ecosystem-(Tataria et al., 2021).

5.2. TUNNEL FIELD-EFFECT TRANSISTORS (TFETS)

The demand for low-energy electronic devices has long been there, driving innovation beyond the utilization of conventional transistors. Leading such innovations are Tunnel Field-Effect Transistors (TFETs), and they are a preferable solution due to their unique working mechanism. TFETs leverage band-to-band tunneling, and the potential to achieve subthreshold slopes lower than the conventional 60 mV/decade threshold is a significant improvement for low-power applications. This characteristic is in stark contrast with conventional MOSFETs, which suffer from short-channel effects and enhanced leakage currents when device size diminishes. While manufacturers venture out in search of nanoscale technology, the evident advantages provided by TFETs become progressively more noteworthy, making them an option worthy of exploration for upcoming integrated circuit designs that seek to be power-efficient and performance-driven (Agopian et al., 2025). As far as application is concerned, TFETs are particularly useful in low-power applications, considering that they can also operate efficiently at lower voltages. Their recent applications point to analog devices, particularly operational transconductance amplifiers and voltage regulators with low-dropout (Kumar et al., 2023). These arrangements leverage the high voltage gain characteristic of TFETs to make them the perfect choice in applications where power efficiency matters most. As semiconductor technologies continue to evolve, integrating TFETs into analog and digital circuits not only promises to maintain with existing performance levels but also mitigate the increasing power density challenges that are intrinsic in traditional MOSFET technologies (Figure 5.6).

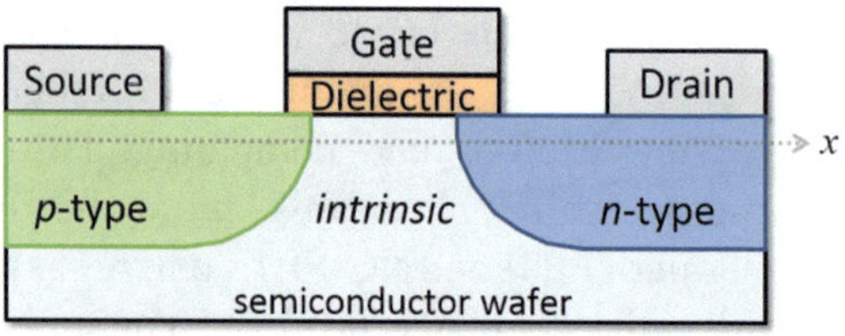

Figure 5.6. *Tunnel field-effect transistor (TFET): At a sufficient gate bias a tunnel current flows across the device.*

Source: https://www.researchgate.net/publication/320905812/figure/fig8/AS:558059325407232@1510063119021/Tunnel-field-effect-transistor-TFET-At-a-sufficient-gate-bias-a-tunnel-current-flows.png.

In essence, TFETs provide a route for the development of devices that are not only efficient but also sync with the increasing need for sustainability and energy efficiency (Agopian et al., 2025). Further, although TFETs exhibit several benefits, certain limitations also exist, especially regarding their ambipolar nature and consistency of performance under different conditions. To address these likely drawbacks, research focuses on device architecture optimization, for instance, the deployment of asymmetric structures that can enhance conduction behavior and prevent leakage issues. For instance, the implementation of double-gate TFETs has shown promise in the avoidance of performance limitation by enabling better control over tunneling and channel behavior (Kumar et al., 2023). As the field of advanced transistor designs expands, new technologies aim to further advance the use of TFETs so that it will compete adequately with not only traditional MOSFETs but also other emerging designs like negative capacitance FETs and vertical transport devices. As research evolves and becomes more refined, TFETs will be capable of ruling the future landscape of semiconductor technologies.

5.2.1. Operating Principles of Tunnel Field-Effect Transistors (TFETs)

The complexity of modern electronic devices has necessitated the exploration of new architectures for higher performance and power savings. Of these newer types, TFETs are unique due to their different operation mechanisms based on quantum tunneling for the transportation of charge. Unlike traditional MOSFETs, which are based on large energy barriers to modulate current, TFETs use a band-to-band tunneling mechanism. This enables them to achieve a subthreshold swing that can be less than the theoretical 60 mV/decade limit, ultimately resulting in lower power consumption when in operation. Through their capability to drive technology needs for increased performance, TFETs offer a hopeful foundation for addressing rising power dissipation and scaling issues, as highlighted by advances in transistor devices demonstrating the importance of overcoming short-channel effects as well as rising leakage currents (Agopian et al., 2025). Aside from the power concerns being addressed by the architecture of TFETs, they also introduce defining improvements in functionality and device integration (Figure 5.7).

With the application of materials such as transition metal dichalcogenides (TMDs), researchers have developed new TFET structures with improved electrical characteristics. The incorporation of locational controls and heterostructures in TFET structures allows for tunable performance parameters, thus expanding the potential application area, including low-power logic and analog circuits. The onset of options like local-control gates (LCGs) has been promising to filter defect-induced effects that can hinder the performance of transistors, along with enhancing peak-to-valley current ratios. These developments mark a milestone in design philosophy for future electronics, where efficient usage of TFETs may lead to revolutionary leaps in energy efficiency and functionality (Yu et al., 2024). Besides, TFETs possess excellent attributes for specific applications, particularly for low-dropout voltage regulators (LDO). The low quiescent current of the device combined with high voltage gains enables improved power management in integrated circuits, and as such, they are especially well-suited to power-sensitive designs. This is especially important in the case of miniaturized devices where power efficiency assumes a very central position. Moreover, the multivaluedness of TFETs is carried over to multivalued logic arrays, hence serving as a significant contribution to the development of wiser, capable circuits. As research and development are focused on demystifying the fundamental transport mechanisms and increasing scalability, TFETs are going to be significant drivers in the future trajectory of advanced transistor architecture, indicative of the marriage of performance, energy efficiency, and technological innovation (Agopian et al., 2025).

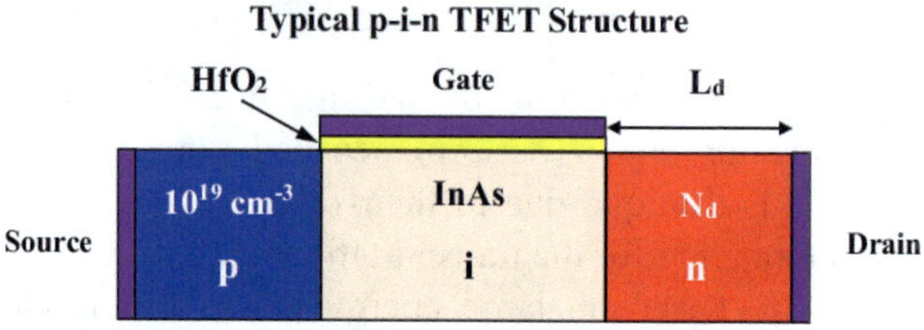

Figure 5.7. *Schematic structure of a typical tunnel field-effect transistor (TFET).*

Source: https://www.researchgate.net/profile/Nguyen-Dang-Chien/publication/292981595/figure/fig1/AS:583930126303233@1516231199679/Schematic-structure-of-a-typical-tunnel-field-effect-transistor-using-lowbandgap.png.

5.2.2. Advantages Over Traditional Field-Effect Transistors (FETs)

Transistor technologies have advanced over the years with a series of high-end architectures offering better performance than traditional FETs. An example par excellence is the Tunnel FET (TFET) based on a new tunneling mechanism to deliver low operating voltage while offering high drive currents. While traditional FETs employ thermal energy for generating charge carriers, TFETs exploit quantum tunneling for switching more effectively. This property not only enables reduced power consumption but also achieves greater switching speed, which is of significant concern with today's electronics demanding greater performance per energy efficiency. TFETs' reduced subthreshold swing, a measurement of significant concern regarding transistor performance, enables them to provide better control over leakage current, hence making them well-suited to low-power applications, providing a means to introduce further enhancements in small and energy-efficient devices. Negative capacitance FETs or NC-FETs are yet another innovation to enhance the performance of transistors. The new design consists of a ferroelectric material within the gate dielectric that provides a capacitance that can reduce the switching voltage. By taking advantage of the negative capacitance effect, NC-FETs decrease the gate voltage thresholds, thereby effectively equivalent to much less power consumption when operational. Besides, the ability of these devices to exhibit strong electrostatic control reduces short-channel effects, which have proven a persistent issue for traditional FET designs as device scales continue to reduce. With enhanced control of the electrostatic environment, NC-FETs allow more significant drive currents without sacrificing leakage, offering an appealing advantage to circuit designers seeking efficient but powerful electronic solutions. Such enhancements vindicate the potential of NC-FETs in applications in future semiconductor technology. Vertical transport devices also show the gains achieved with innovative transistor structures, greatly improving device scalability and performance. Using a three-dimensional approach, such devices optimize material usage and maximize electrostatic control. Vertical devices such as Vertical Gate-All-Around (V-GAA) transistors have improved scalability, enabling greater device density while reducing the area utilized for circuitry fabrication. This design evolution not only suppresses the short-channel effects of miniaturization but also supports increased drive currents and reduced access resistance, critical to high-frequency operations. V-GAA transistors are referenced in studies as efficient, with significant performance improvement to conventional planar configurations, having high logic functionality while minimizing size. Collectively, these new devices are a leap in performance and efficiency, an undisputable paradigm shift compared to their conventional counterparts, pushing the frontiers in semiconductor technologies (Pelloquin et al., 2025; Wu et al., 2025).

5.2.3. Applications in Low-Power Electronics

Low-power electronics have been evolving fast lately, particularly through novel device concepts like Tunnel Field Effect Transistors (TFETs), driven by advancements in the architecture of transistors. These devices operate on the quantum tunneling process to enable effective current conduction at lower voltages, which becomes necessary in minimizing power consumption in various applications. Their capability to handle high on-off ratios of current even at subthreshold voltages makes TFETs better than traditional MOSFETs in applications requiring energy efficiency. The application becomes significantly important in mobile applications, where battery life becomes very precious. Furthermore, ongoing research on enhancing TFET structures tries to improve them even further, making them a strong contender for upcoming low-power integrated circuits and clever electronics, practically addressing the increasing demand for green technology solutions. Thus, TFETs are rewriting the landscape of low-power electronics and play an important role in fostering advancement in the field. Another groundbreaking technique of semiconductor design is the use of negative capacitance FETs (NCFETs) in power-constrained electronics. Leaping on the concept of negative capacitance, these technologies have the ability to significantly reduce power consumption while boosting switching speed, thus breaking the limitations of traditional transistors. NCFETs provide steep subthreshold swing with improved electrostatic control over the channel, allowing lower operation voltages without degrading performance. This property is crucial in prospective applications of emerging markets such as the Internet of Things (IoT), where energy efficiency is vital in extending battery life for devices under connectivity. As NCFET technology research evolves, its potential of advancing the efficiency and performance of low-power electronics in a positive direction represents a positive path towards meeting the demands of high-functionality advanced digital systems based on extended battery life, advancing smart and sustainable electronic solutions. Vertical transport devices are also gaining prominence in low-power electronics primarily due to their distinct geometry and operational characteristics that favor effective current conduction. Unlike planar transistors of traditional design, vertical devices ensure a more effective channel length with less short-channel effect and superior overall device performance. This structural benefit enables the creation of miniaturized devices well suited to high-density integration applications like system-on-chip (SoC) technology. Furthermore, innovations in materials, such as the combination of gallium nitride (GaN) with Gate-All-Around technology, are the promise of these devices for next-generation low-power applications. The new properties of GaN, including high electron mobility and improved thermal management, put vertical transport devices at the forefront of activity toward energy-efficient solutions across a range of electronic platforms. As they further develop, they have the potential to bring substantial value to the field of low-power electronic applications in the future.

5.3. NEGATIVE CAPACITANCE FIELD-EFFECT TRANSISTORS (NC-FETS)

Within the context of state-of-the-art transistor technologies, the pursuit of higher performance figures of merit and, specifically, those in low-power applications is a principal theme for research and innovation. A key advancement in this area is the development of NC-FETs that utilize the special properties of ferroelectric materials to enable improved gate control. While traditional transistors are bulked by the scaling limit with leakage currents and bad electrostatic control, the NC-FETs mechanism relies on the intrinsic positive feedback cycle by the ferroelectric layer. This mechanism not only reduces the subthreshold swing below the traditional 60 mV/decade but also enhances the on-current, offering a potential solution to the challenges of additional device miniaturization in complementary metal-oxide-semiconductor (CMOS) technology (Hueting, 2018). Studies of NC-FETs reveal an intricate interplay between material parameters and device functionality. The ferroelectric material selection is vital; to achieve optimal performance, a hard ferroelectric material such as hafnium zirconate (HZO) is usually applied, which possesses a large polarization-field loop. Such a characteristic results in a high switchable polarization, which is vital for sustaining the negative capacitance effect and optimizing the transistors operation efficiency (Figure 5.8).

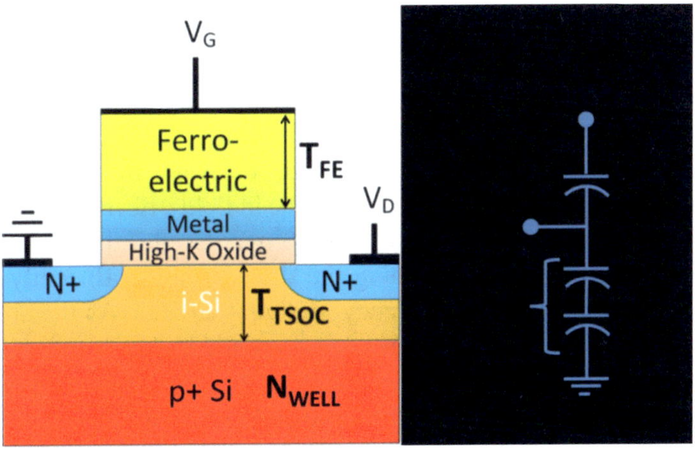

Figure 5.8. *Negative capacitance field effect transistor (NC-FETs).*

Source: https://media.licdn.com/dms/image/v2/D5612AQGyFm6tjSMjpw/article-cover_image-shrink_600_2000/article-cover_image-shrink_600_2000/0/1681923466164?e=2147483647&v=beta&t=igNvZwCGGCGx66IBd8IYBJaPiArles9tW7rFYnAit7k.

On the other hand, the piezoelectric field-effect transistor (π-FET) targets increasing the piezoelectric coefficient of the material to achieve optimal polarization-field interaction. While NC-FETs also have great advantages, particularly at low power, their speed may be penalized by the substantial polarization change required for good switching. Thus, understanding and potential hybridization of these architectures can give rise to the next generation of transistor technology through leveraging the strengths of both the technologies (Hueting, 2018). The area of two-dimensional materials, particularly transition metal dichalcogenides (TMDs), also holds promising prospects for integration with NC-FET technologies. The inherent electronic properties of TMDs, e.g., MoS_2 and WS2, provide the avenues for new functionality and device miniaturization. As progress continues in overcoming the fabrication and scaling limitations of defect-free 2D material fabrication, their incorporation into NC-FET design can potentially significantly enhance performance parameters. Thin layers reduced to atomic scales while retaining the desired electrical properties closely aligns with the power efficiency and performance goals of present electronics. This confluence of emerging materials with innovative transistor structures holds a thrilling course for near-term development in nanoelectronics, including ubiquitous application in energy storage, sensing, and optoelectronics, thus paving the way to revolutionary applications.

5.3.1. Concept of Negative Capacitance

The relentless pursuit of enhanced transistor performance has prompted creative explorations of a variety of architectural styles, some of which incorporate negative capacitance. Through redesign of the conventional definition of capacitance, negative capacitance in field-effect transistors (FETs) is a potential paradigm shift towards sub-60 mV/decade subthreshold slopes. It holds the significance of switching the devices more efficiently and thus improving the overall power efficiency in digital circuits. As the architecture of transistors evolves towards decreasing power and higher speed applications, negative capacitance is the requisite feature that enables this accomplishment by merging properties of materials science and quantum mechanics to further enhance the performance of devices under ever more stringent requirements for design. The integration of ferroelectric materials into the gate stack of transistors forms the basis for the realization of negative capacitance with the behavior of capacitance being optimized beyond conventional limits. Not only does this technology relieve the traditional constraints on metal-oxide-semiconductor FETs (MOSFETs), but it also makes possible new device structures, e.g., negative capacitance FETs (NC-FETs), which leverage the unique characteristics of ferroelectric materials to multiply gate voltage and thereby improve the steepness of the subthreshold switch. These assertions are confirmed by recent research, which indicates that the incorporation of ferroelectric layers increases sensitivity in devices such as DNA biosensing. The combination of greater sensitivity and reduced power consumption means that negative capacitance will transform the operation of future sensors and FETs across a range of technological applications (Tamersit et al., 2024). Moreover, negative capacitance devices are also not restricted to traditional FET structures but also encompass novel paradigms like tunnel FETs (TFETs) and vertical transport devices. For instance, introduction of negative quantum capacitance in MoS_2-FETs has unveiled extremely promising opportunities by exhibiting steep subthreshold slopes as low as 31 mV/decade. This improvement is due to the new electron band structure of two-dimensional materials such that the quantum effects can be explored that increase the device efficiency even further (Zhu et al., 2023). As more studies on negative capacitance mechanisms are carried out, the knowledge gained on these new transistor structures offers a roadmap to build highly efficient low-power electronics devices required for future applications in data processing and communication (Figure 5.9).

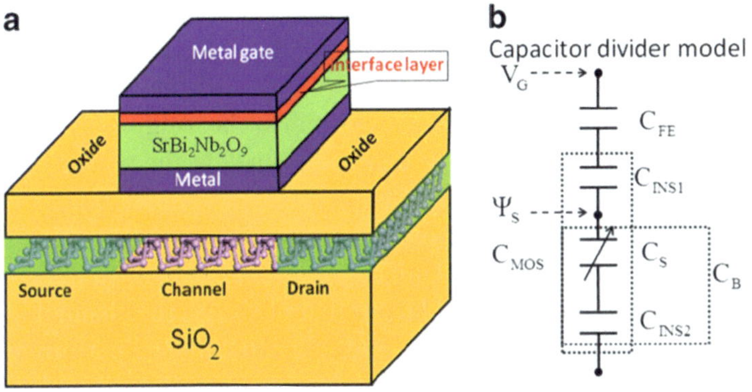

Figure 5.9. *The negative capacitance (NC) transistor (FET) based on monolayer black phosphorus (ML-BP).*

Source: https://www.researchgate.net/publication/305663758/figure/fig3/AS:391239133351936@1470290086133/The-negative-capacitance-NC-transistor-FET-based-on-monolayer-black-phosphorus.png.

5.3.2. Impact on Power Consumption

The history of transistor design has seen a dramatic shift of attention towards improving power efficiency in semiconductor devices. Tunnel Field-Effect Transistors (TFETs), in this regard, have shown themselves to be a viable competitor to traditional MOSFETs, primarily owing to their operation mode allowing subthreshold conduction with reduced power consumption. TFETs work by taking advantage of quantum tunneling to create a current, which by its nature has the capability of reducing operating voltage. This decrease in voltage directly translates to reduced static power consumption, an important consideration as chips move to smaller nodes where leakage current can be a problem. With device scaling continuing to advance, the integration of TFETs in upcoming technology could trigger a radical reduction in total energy use, making them a suitable option for applications requiring high energy efficiency and long battery life in portable devices (Desai, 2025). Negative capacitance field-effect transistors (NCFETs) are yet another groundbreaking architecture that aims to combat power consumption. By the use of ferroelectric materials, NCFETs can achieve a negative capacitance effect whereby the voltage is reduced in order to receive a specific quantity of current. This characteristic leads to a low-power consumption reduction of the threshold voltage, thus lowering dynamic power consumption when the circuit is activated. As the transition towards leading-edge technology nodes persists, the integration of NCFETs can be instrumental in addressing the dangers posed by power density and thermal dissipation in high-end markets. In contrast to conventional approaches, NCFETs not only enhance energy efficiency but also support robust performance under varying conditions, bearing testament to their capability to meet the increasingly demanding needs for power optimization in modern microelectronics (Chowdhury, 2025). Vertical transport devices are also being developed as a power consumption enhancement technique for advanced transistor designs. Unlike traditional planar architecture, these devices utilize a three-dimensional layout that positions the channel in the most suitable place for current conduction, and this can enhance performance while decreasing voltage concurrently. Vertical architecture allows for short channel lengths, which provide high switching rates and minimal energy loss during transitions. Besides, their geometry minimizes the area for interconnects, which can result in lower parasitic capacitance and thus lower power consumption. The ability of vertical transport devices to optimize high performance and low power characteristics well makes them an essential technology in the pursuit of designing energy-efficient, high-performance semiconductor solutions (Desai, 2025).

5.3.3. Potential Applications in Logic Circuits

New transistor technologies will revolutionize logic circuits with the enormous gains in efficiency and performance. Among the new technologies, Tunnel Field-Effect Transistors (TFETs) are attracting attention because they can switch at lower voltages without compromising high switching speeds. TFETs use quantum tunneling to achieve subthreshold conduction and have very low leakage currents. This attribute proves to be especially beneficial in situations requiring high energy efficiency, like portable electronics and IoT applications.

Additionally, incorporating TFETs into present logic circuit frameworks has the potential to render the designs even smaller, eventually paving the way for innovations in multi-core processors and advanced computing systems. The flexibility of TFETs is such that they can be made to work alongside standard CMOS technology, which suggests the path to hybrid circuits that enjoy the optimum of both worlds for these two classes of transistors. The more recent innovation, the negative capacitance FET (NCFET), brings yet another alternative to logic circuits through exploiting the properties of ferroelectric materials. By integrating negative capacitance in the gate stack, NCFETs can be capable of having lower voltage thresholds directly, i.e., reduced power consumption. This aspect carries huge potential for future logic circuits, particularly those intended for energy-limited applications such as artificial intelligence and machine learning. With AI workloads being increasingly more complex, the ability to perform operations with reduced energy cost, while also growing faster simultaneously, is highly critical. In addition, NCFETs could replace conventional CMOS in some applications, offering an avenue to greener computing methods. Properly designed and deployed, NCFETs could provide new paces of performance, allowing intensely power-efficient logic circuits that can provide for the demands of high-performance computational processes. Vertical transport devices like next-generation architectures for transistors are a grand frontier in logic circuits. These transistors are supported by the vertical nature of their active layers, which enhances electron mobility and the performance of the devices. With the reduction of physical space and enhancement of current flow, vertical transistors can facilitate the creation of more highly integrated logic circuits that operate at higher frequencies. Such enhancements find particular applications in telecommunications and high-performance computing. Moreover, as technology trends shift toward reducing energy consumption while shrinking technology, vertical transport devices are expected to meet these challenges. Coupled with the incorporation of emerging materials, such as 2D materials, vertical transistors have the potential to transform logic circuit design into a new horizon, driving adoption levels and guaranteeing next-generation computing technology development.

5.4. VERTICAL TRANSPORT DEVICES

Transistor technology depends upon effective means of increasing device performance and minimizing energy usage. This shift has involved examining vertical transport devices that offer a drastic break from traditional planar geometries. One of the notable ones among them are Tunnel Field-Effect Transistors (TFETs), which are able to achieve subthreshold slopes above the traditional 60 mV/decade hurdle. This characteristic enables TFETs to operate at reduced supply voltages but still maintain reasonable drive currents. The vertical geometry of TFETs promotes effective charge conduction via tunneling effects, which eliminates the bottlenecks typically faced by conventional transistors. With researchers going on to further refine the structure and materials employed in TFET structures, these devices are poised to take center stage in facilitating future low-voltage and ultra-low-power electronics, as with the increasing demand for low-power solutions in semiconductor technology. Exploring further into the frontiers of novel transistor structures, NC-FETs represent another exciting avenue of vertical transport devices. The introduction of ferroelectric material into the structure of the transistor makes it possible to realize the utilization of negative capacitance, resulting in the reduction of switching voltages and increased energy efficiency. In contrast to traditional structures, NC-FETs can enhance electrostatic gating of the channel and consequently suppress short-channel effects typical of nanoscale transistors (Figure 5.10).

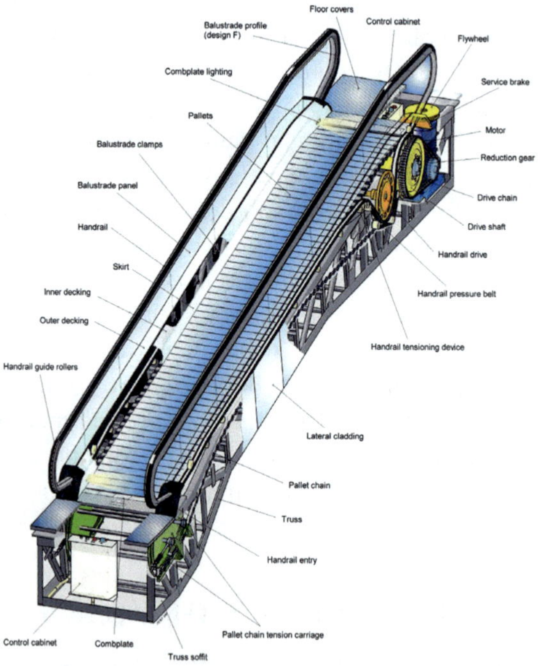

Figure 5.10. *The working principle of vertical transport devices.*

Source: https://archi-monarch.com/wp-content/uploads/2022/12/BASIC-OPERATING-GUIDELINES.webp.

This ability not only facilitates device size reduction without sacrifice of performance but also facilitates easier formation of complementary devices inherently more power-efficient. With ongoing research exploring several ferroelectric materials and integration with existing semiconductor processes, NC-FET architecture likely represents a substantial step toward future memory and logic uses, meeting the urgent need for devices capable of keeping pace with growing performance demands. Alongside NC-FETs and TFETs, vertical transport devices greatly improve the potential of semiconductor technology owing to their unique structures and operational advantages. The development of Vertical Gate-All-Around (V-GAA) transistors is reflective of this trend since these devices leverage the vertical integration to minimize device area while enhancing performance characteristics like drive current and electrostatic integrity. With scalable fabrication techniques, such as described in recent work, researchers have achieved outstanding success in constructing uniform and reliable logic cells, including inverters and NAND gates. The V-GAA transistor design can eventually pave the way for dense logic applications. While the vertical paradigms develop, they not only can optimize existing CMOS architectures but also facilitate integration of new concepts and materials into the overall scheme of post-CMOS technology, highlighting the pivotal role of vertical transport devices in charting the destiny of electronics.

5.4.1. Design and Structure of Vertical Devices

Advances in transistor technology have increasingly been directed towards the aim of seeking vertical device structures as a way of advancing performance at the expense of size. Tunnel Field-Effect Transistors (TFETs) have been a crucial solution in this context, taking advantage of band-to-band tunneling to achieve switching at lower voltage levels compared to conventional metal-oxide-semiconductor FETs (MOSFETs). This new operation mechanism makes TFETs more energy efficient, positioning them as front-runners for next-generation applications where power efficiency is the priority. The vertical architecture of TFETs also optimizes current flow by reducing the carriers' distance to travel, in effect amplifying speed and reducing device size overall. These advantages not only identify the potential of TFETs in future transistor technologies but also indicate the trend towards integration of vertical structures in semiconductor technologies in order to push the frontiers of miniaturization and efficiency. Following the ideology of ground-breaking transistor structures, Negative Capacitance FETs (NCFETs) provide yet another option that tries to exploit ferroelectric materials in order to achieve sub-threshold swing less than the conventional limit of 60 mV/decade. By incorporating a ferroelectric layer in the gate stack, these devices leverage negative capacitance to enhance electrostatic control of the channel, drive current, and reduce power requirements. The vertical structure of NCFETs allows for effective scaling without significant degradation of the device, which is crucial in high-performance computing applications. With the advent of vertical devices, the utilization of materials such as 2D transition metal dichalcogenides (Choi et al., 2017) also presents greater opportunities for optimizing these structures, creating design freedoms for NCFETs and bringing their viability into the spotlight for next-generation transistor designs. Blending advances from both TFETs and NCFETs, the world of vertical transport devices unveils the possibilities of enhanced functionality and efficiency in semiconductor devices. These geometries not only enable multiple modes of transport but also enable greater densities of integration, which are essential for today's electronics. As noted by the increasing demand for smaller, faster, and more power-efficient systems, the form and type of vertical devices become increasingly significant. Advanced fabrication techniques, along with thorough understanding of material behavior, favor the development of complex architectures that can be tailored to accommodate various electronic and optoelectronic functions, such as in 2D materials containing TMDs (Choi et al., 2017). Further studies in these areas hold the potential to disclose new horizons in device performance that will be suitably in accordance with the essential objectives that will shape the future of next-generation transistor architectures (Figure 5.11).

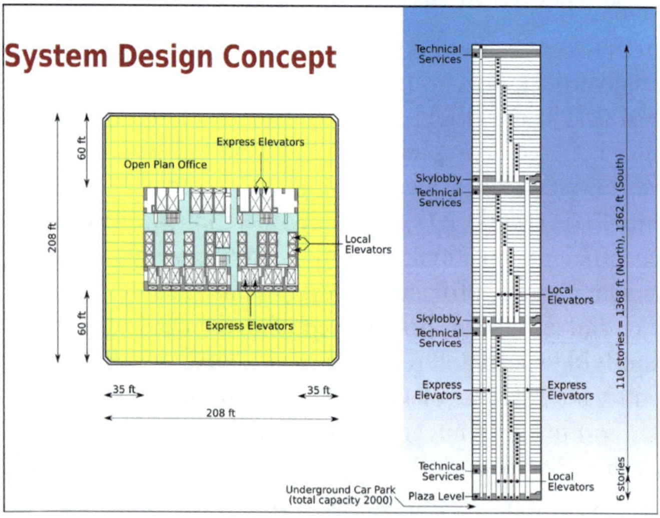

Figure 5.11. *The structure of vertical transportation system ★.*

Source: https://archi-monarch.com/wp-content/uploads/2022/12/SKY-LOBBY-ZONE-ONE.webp.

5.4.2. Benefits of Vertical Transport in Performance

The history of transistor technologies has been dominated by a quest for growing performance, i.e., increased speed and power efficiency. One of the important aspects of this development is the use of vertical transport architectures, which enable the carriers to travel vertically inside the device, so faster switching times can be achieved. Vertical transport technologies, such as Tunnel FETs (TFETs) and negative capacitance FETs, aim to cut down the distance a charge carrier must travel, and thus the intrinsic delay of horizontal transport. This approach benefits not only operation speed but also power consumption, and it has become an exceedingly promising solution in the fast-evolving domain of advanced transistor structures. These improvements are essential in meeting the stringent demands of tomorrow's applications with requirements for quick and efficient processing capability, now becoming critical as we transition into more data-drive systems. Moreover, the introduction of vertical transport concepts has the ability to efficiently resolve issues characteristically encountered by horizontal transistor architecture, such as short-channel effects and scaling power limitations. In contrast to traditional planar FETs, vertical transport designs allow for greater control of electrostatic effect, which can lead to improved device reliability and consistency in performance. Notably, these innovations leave opportunities for higher device density without compromising performance, an important factor as companies race to keep pace with increasing consumer demand. Also, the ability to function at lower voltages, a characteristic feature of vertical transport devices, enhances battery life of mobile and portable electronics. With technology trending toward miniaturization and increasing complexity, assuming vertical transport approaches presents an appealing way to push the frontiers of what is possible in cutting-edge transistor structures. The uses of vertical transport transistors are not just limited to the performance enhancement alone; they also find themselves at the forefront of driving communication technologies forward. The examination of future systems, i.e., 6G networks, centers on the requirement for devices to provide record-breaking data rates along with spectrum concerns. The unique features in the instance of vertical transportation equipment serve to fill these new requirements, as they have potential for lower latency and greater throughput which ultra-reliable communication systems demand. Moreover, with greater use of digital twins in system design and optimization, the flexibility of such vertical structures also offers immense value to simulation and modeling exercises. The interaction between innovative transistor structures and novel technologies not only advances the efficiency of electronic systems but also makes an important contribution to the ultimate objective of establishing strong, scalable, and high-performance networks capable of responding to future demands, as reported in Rasheed et al. (2020) and Tataria et al. (2021).

5.4.3. Comparison with Traditional Horizontal Devices

In the evaluation of the evolution of transistor technology, an emergent direction away from traditional horizontal devices goes toward new architectures like Tunnel Field-Effect Transistors (TFETs) and Negative Capacitance FETs (NC-FETs). Traditional horizontal devices tend to rely on long channel lengths and lateral transport mechanisms that limit efficiency and speed. TFETs use a tunneling mechanism that reduces energy required for switching and are therefore more suitable for low-voltage applications. This new technique has a major advantage of minimizing power consumption, hence supplementing the increasing demand for more energy-efficient electronic devices. Therefore, TFETs are a milestone in transistor technology that presents an improved alternative to conventional approaches, particularly with the future trend in electronic applications headed towards minimization and sustainability. As performance demand grows, vertical transport devices also illustrate a significant departure from horizontal architectures. These devices benefit from shorter channel lengths, which can contribute to bigger drive currents and improved electrostatic control. Vertical transport has the potential for improved thermal performance and reduced design compared to their horizontal counterparts.

These innovations not only permit high-level integration into modern semiconductor matrices but also address major issues created by increasingly diminishing circuit miniaturization. Furthermore, as high-frequency performance requirements are needed more, vertical devices offer a better method to scale while maintaining power dissipation low, a radical enhancement of transistor technology beyond limits imposed by traditional horizontal configurations. Parallel to that, the development of smart reflecting surfaces (IRS) represents a clear convergence of novel transistor structures and wireless communication systems. IRS can control signal propagation via a series of low-cost passive reflectors that maximize the effectiveness of signal paths to optimize communication performance.

This paradigm shift in network architecture is contrasted with traditional fixed transmission strategies, emphasizing the flexibility and cost benefits of integrating IRS with innovative transistor design. TFETs and IRS are two that are central to overcoming the future challenges of next-generation wireless systems, particularly with the emphasis of the sixth generation (6G) as the central area for technological advancement. Under such a situation, familiarity with the operational issues for IRS, such as reflection optimization and channel estimation, becomes pivotal in harmonious integration with next-generation networks (Tataria et al., 2021; Wu et al., 2021).

5.5. EMERGING TRENDS IN TRANSISTOR ARCHITECTURES

Transistor architecture evolution has pushed the discipline to devices beyond the limits of traditional semiconductor technologies. Another notable invention is the Tunnel Field-Effect Transistor (TFET) based on quantum tunneling to enable subthreshold operation and is well suited for ultra-low power applications. TFET exploits this mechanism to achieve high switching speeds with significantly lower leakage currents, and thus is a prime contender for future integrated circuits. Other researchers are exploring ways to enhance the on-state currents of TFETs through material selection and device scaling. The growing interest in TFETs is a result of the need for energy-conserving electronic devices in a rapidly accelerating technological landscape, propelling further research and development in this novel device structure. Alongside research in TFETs are Negative Capacitance Field-Effect Transistors (NC-FETs) as an innovative solution for improving subthreshold swing and subsequently minimizing power consumption.

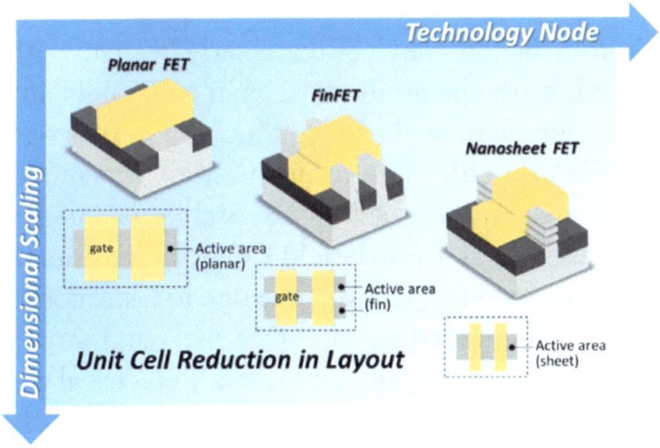

Figure 5.12. *The technical advancements in transistor.*

Source: https://www.researchgate.net/profile/Chih_Piao_Chuu/publication/348790742/figure/fig1/AS:1009257063395328@1617637043045/llustrative-diagram-of-transistor-architecture-evolution-For-the-lateral-dimension.jpg.

The fundamental concept in NC-FETs is the application of ferroelectric materials, which can potentially provide an extra capacitance that effectively reduces the operating voltage. This effect creates a highly promising route for achieving beyond the Boltzmann limit of conventional transistor operation. With the world heading towards the goal of reduced power consumption and better performance, NC-FETs are gaining more attention among researchers and industries alike. They hold the promise to drive energy-efficient computing and thereby pave the way to sustainable innovation in the semiconductor industry. Vertical transport device evolution is another basic trend in transistor architecture. Unlike standard planar devices, the vertical devices have the ability to achieve more efficient transport of the carriers to provide larger current densities and enable higher performance in small geometries (Figure 5.12).

The structures of vertical transistors, such as Vertical-TFETs and FinFETs, offer new possibilities for scaling device dimensions without compromising or even improving performance features. These advancements are particularly crucial in the environment of future sixth-generation (6G) wireless systems, which will need devices capable of operating between 100 GHz and 1 THz (Tataria et al., 2021). As the capabilities of devices are expanded by continued research, vertical transport structures can become the focus for providing the stringent requirements of next-generation high-speed communications and high-end computer systems.

5.5.1. Integration of Quantum Dots

The evolution of semiconductor devices has clearly been toward miniaturization and greater function, with quantum dots as focal points in this evolution. Quantum dots are nanometer-sized semiconductor particles that exhibit unique optical and electronic properties due to quantum confinement effects. Due to these characteristics, they are increasingly desirable for integration into future transistor architectures such as Tunnel Field Effect Transistors (TFETs) that will achieve lower power and higher performance due to a steeper turn-on behavior. Quantum dots can be employed to engineer the band structure in devices, thereby increasing tunneling currents and overall electrical performance. The integration of quantum dots into TFET geometries enables the realization of efficient compact devices at subthreshold voltage conditions, a critical energy requirement in modern electronics. Integration of quantum dots in such systems is a valuable route towards next-generation integrated circuit technology. Apart from their possible application to TFETs, quantum dots also hold promise for negative capacitance FETs (NC-FETs). NC-FETs focus on the use of negative capacitance to enhance subthreshold swing and enable sharper off-to-on state transitions. Local potential profiles that can be created by embedding quantum dots can simulate the effect of negative capacitance, thereby maximizing device switching efficiency and reducing leakage currents. Furthermore, the quantum dots' composition and size can also be tuned to optimize their transport properties to improve performance when employed in NC-FETs without compromising compatibility with existing fabrication techniques. Through the use of such quantum confinement properties, researchers are able to improve the electrostatic control of NC-FETs, leading to performance that is essential in low-power applications. This potential inspires ongoing research on the fabrication and use of quantum dot-integrated NC-FETs for future electronic devices. The study of vertical transport devices also offers another widely acknowledged opportunity for quantum dot integration, particularly those with regards to optimizing performance at small device structures. Vertical transport transistors take advantage of the spatial distribution of materials to reduce the effective channel length, allowing electron transportation at high speed and minimizing parasitic capacitance. Quantum dots within such architectures can result in increased carrier mobility of charges and controlled injection of electrons, which will enhance the efficiency of the device. Scalability of such vertical architectures is a high-stakes proposition for the use of quantum dots. By tuning carrier dynamics through the controlled synthesis and location of quantum dots, researchers not only enhance device performance but also enable seamless integration with complementary metal-oxide-semiconductor (CMOS) technology. This symbiotic relationship between quantum dots and vertical transport devices underscores the critical development underway in sophisticated transistor architectures, reaffirming the relevance of quantum materials to the future of electronics.

5.5.2. Use of Two-Dimensional (2D) Materials

In cutting-edge transistor architectures, utilization of 2D materials has turned out to be a groundbreaking action towards the accomplishment of bypassing the confines established by conventional silicon-based technologies. Specifically, the properties of 2D materials such as graphene and transition metal dichalcogenides (TMDCs) enable the design of new devices such as Tunnel FETs (TFETs). TFETs leverage the unique band structure of 2D materials to offer lower subthreshold swings compared to traditional MOSFETs, thus enhancing energy efficiency. Moreover, their capability to operate at reduced voltage levels presents significant potential for reduced power consumption, consistent with the growing emphasis of the industry on sustainability. As demand for low-power electronic devices grows, 2D material use provides a compelling path to satisfy these demands through more economical transistor technology (Lee et al., 2025). Another major application of 2D materials in future transistor devices is the creation of negative capacitance field-effect transistors (NC-FETs). These new devices utilize ferroelectric materials along with 2D layers to achieve negative capacitance effects that can indeed decrease the gate voltage needed to deliver a certain level of transistor action. This impact not only causes significant reductions in power loss but also allows transistors to be scaled down further without loss of performance. The use of 2D materials in NC-FETs enhances their performance through reduced electrostatic shielding of the channel and consequently preventing short-channel effects. This integration is particularly important with the sector moving towards the research of devices that address the limitations of conventional scaling methods, presenting a competitive edge in efficiency and performance (Arsid, 2025). Last but not least, the research of vertical transport devices made of 2D materials is revolutionizing the landscape of semiconductor technology. The vertical architectures leverage the higher mobility of 2D materials to provide efficient charge transport through reduced channel lengths. The combination of verticality with the intrinsic nature of 2D materials facilitates the formation of strong transistors with the capability for high current density support and low power dissipation. Moreover, devices with vertical transport can facilitate the realization of compact multi-tier architecture comprising more than one transistor in a confined area, essential to meet the increasing requests for high-performance computing. The developments in the application of 2D materials pave the path to such future innovations in semiconductor technology in favor of the growing trend towards novel architectures rather than traditional scaling solutions (Lee et al., 2025).

5.5.3. Advances in Nanotechnology

Transistor design progress has been significantly aided by nanotechnology advances, specifically towards devices that can operate at lower voltages without reducing performance. A good example of this technology is the Tunnel Field-Effect Transistors (TFETs), which leverage quantum tunneling to maximize device efficiency. In contrast to traditional MOSFETs driven by thermionic emission, TFETs can switch on at lower threshold voltages with significantly less power consumption. Nanotechnology incorporation in TFETs entails extremely precise control of material dimensions and properties to enable the fabrication of devices with nanoscale dimensions to maximize performance parameters. As the requirement for scalable and energy-efficient devices is on the rise, the adoption of TFETs is a breakthrough to achieve the necessary performance in future computational architectures, eventually leading to applications ranging from mobile electronics to large data centers. Based on the platform provided by TFETs, the development of negative capacitance FETs (NCFETs) is emerging with rising prominence in cutting-edge transistor technology. Using ferroelectric material, NCFETs achieve significant reduction in subthreshold swing by using negative capacitance, hence saving energy. This approach addresses critical scaling issues for traditional FETs while minimizing static power loss. The miniaturization facilitated by nanotechnology enables precise stacking of material to create NCFETs that are not only more efficient but also compatible with existing semiconductor manufacturing techniques. Recent studies focus on improving ferroelectric materials and combining them, to develop more competitive NCFET implementations that will revolutionize power efficiency in electronic devices. Therefore, NCFETs are a transistor technology revolution, filling the gap between energy limitations and performance demands. Vertical transport devices, and particularly the vertical channel transistors (VCTs), are capitalizing on advancements in nanotechnology to redesign transistor structure extensively. In contrast with conventional planar architectures, VCTs take advantage of the three-dimensional nature to enhance the flow of current and reduce the channel length while maintaining high scalability. Latest advancements indicate that VCTs not only possess the ability to increase the cell area efficiency but also the potential to significantly enhance write speed and data retention characteristics. Notably, VCTs have been integrated into hybrid gain cell architectures that are optimized for state-of-the-art technology nodes, offering a new paradigm for access to embedded dynamic random-access memory solutions. These implementations demonstrate the ways nanotechnology facilitates the implementation of efficient, miniaturized devices central to contemporary computing demands. The interoperability of these developments in vertical transportation and innovative design emphasizes the relentless quest for maximization of performance in the field of advanced transistor architectures, opening the door to new innovation.

5.6. CHALLENGES AND LIMITATIONS

The need for more energy-efficient transistors has ignited research into alternative architectures, including Tunnel Field-Effect Transistors (TFETs) and negative capacitance FETs. Yet, these innovations are beset by large challenges that can slow their practical adoption in advanced electronics. TFETs, for example, hold the promise of smaller subthreshold swing and lower operation voltage than traditional MOSFETs, but incorporating these devices into current fabrication technology is problematic. Among the issues are the lack of enough drive current, increased off-state leakage, and compatibility with CMOS processes, all representing daunting obstacles. Also, negative capacitance FETs, although theoretically hold the potential for better gate control and lesser power loss, must battle the wars of material stability and optimal biasing conditions, essential for realizing maximum performance. Hence, realization of theoretical promise to applied benefit is still plagued by these problems, and the potential of such promising device structure is limited. Vertical transport devices became a solution to overcome the limitations of traditional planar transistors but suffer from their own issues (Figure 5.13).

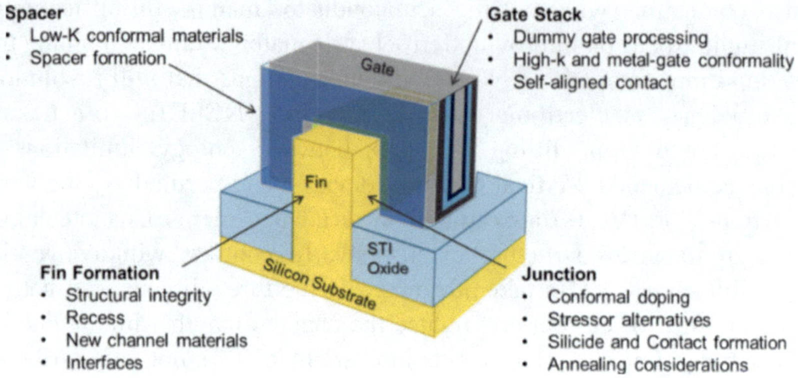

Figure 5.13. *Specific advanced transistor challenges in the formation of fin, spacer, gate stack and junction.*

Source: https://www.researchgate.net/profile/Klaus-Schuegraf/publication/260589813/figure/fig4/AS:305554317692934@1449861234985/Specific-advanced-transistor-challenges-in-the-formation-of-fin-spacer-gate-stack-and.png.

The unique structural arrangement of vertical devices allows greater current flow, hence potentially improving performance parameters. Nevertheless, their development encompasses very high technical difficulties, particularly as regards exact layer deposition and interface state control with a tendency to deteriorate electrical parameters. Moreover, scaling of the vertical transistors to nanometer dimensions aggravates short-channel effect problems, making transistor operation even more unpredictable. Predictability is also aggravated by the lack of holistic benchmarking methodologies capable of properly quantifying performance under various conditions of operation. As the industry pushes device miniaturization to its extremes, extending beyond these limitations is essential for the successful integration of vertical transport architectures into future semiconductor technologies. In parallel with these innovations, integration of two-dimensional (2D) transition metal dichalcogenide (TMD) materials into transistor technology presents opportunities as well as challenges. Despite the remarkable electronic characteristics of 2D TMDs, which have the potential to increase the scalability of the device, their utility is extremely impeded by material quality and characterization approaches. Some of the challenges that hinder the efficiency of high-performance TMD-based transistors include high contact resistance and the need for efficient dielectrics (Zheng et al., 2025). The efforts to improve fabrication techniques, as well as the development of scalable growth approaches, remain ongoing but complicated. Furthermore, incorporating TMDs into current semiconductor structures entails addressing challenges of compatibility with established technology. Through these diverse challenges and limitations, it is clear that innovation in transistor structure relies on overcoming such inherent bottlenecks to fully leverage the promise offered by new designs and materials for future electronic devices (Pandey, 2025).

5.6.1. Manufacturing Complexities

With the increasing need for reduced power consumption and increased performance in the semiconducting world, there has been a push to come up with new transistor structures like Tunnel FETs (TFETs) and Negative Capacitance FETs. These are incorporated into current devices through a high manufacturing barrier, particularly with the transition of the industry away from traditional MOSFET architecture to vertical transport devices with better characteristics. The complex structure of TFETs, grounded in quantum mechanical tunneling, requires precision in the process of fabrication to ensure proper material alignment at atomic scale. Additionally, fabrication of Negative Capacitance FETs introduces additional complexity since achieving stable negative capacitance effects with reliability within operational parameters requires advanced materials and techniques. As identified in recent reports, Desai (2025) presents the major innovations in physical design flows using AI-based optimizations to address these manufacturing issues. Additionally, the use of advanced materials and new architectures necessitates re-examination of existing manufacturing environments. As the designs increasingly scale the nano-barriers, shrinking transistors below the 28nm node involves significant risks with variability and defects at manufacturing. The application of novel materials, such as ferroelectrics for Negative Capacitance FETs, also makes production more complex due to their unique interaction with substrate materials and environmental conditions. Even though new developments hold the promise of better device performance, they also involve strict testing and validation procedures to ensure compatibility with existing semiconductor fabrication lines. Better Process Design Kits are at the forefront of managing such complexities, allowing for different scenarios to be simulated for pre-empting potential manufacturing faults. With developments in semiconductor technology, François Andrieu (2025) comments that integration of state-of-the-art memory solutions with transistor structures can potentially transform the embedded systems landscape. Lastly, balancing manufacturing efficiency and enhanced performance is an imperative within the semiconductor industry. Ongoing development towards cutting-edge transistor geometries has to weigh the potential of novel designs against the needs of fabrication processes. For instance, while vertical transport devices offer mind-boggling electrical characteristics, implementation reduces manufacturing tolerance so that there must be tighter regulation over environmental conditions and critical dimensions. The need to integrate new memory technologies, including MRAM and ReRAM, introduces additional complexity to the manufacturing issue through supply chain and manufacturing process adjustments. Solutions therefore must overcome not only the transistor level changes but also the overall system integration complexities. By overcoming these issues with creative design methodologies, the semiconductor industry is able to continue moving forward as it navigates the challenging landscape of manufacturing complexities so that future innovation is delivered to market and satisfies both market demand and reliability needs.

5.6.2. Thermal Management Issues

Since performance in transistor designs is being pushed increasingly higher by consumer demand, good thermal management becomes increasingly important. Highly advanced devices, particularly Tunnel Field-Effect Transistors (TFETs) and Negative Capacitance FETs, have unique thermal characteristics that need sophisticated cooling schemes. These transistors operate at reduced voltage levels, which can lead to greater on-chip temperature at high-frequency operation. Moreover, when transistors are packed higher in density on integrated circuits, the heat dissipation capability of the material becomes highly critical in order to prevent thermal runaway situations. Fixing these thermal issues not only prolongs device life but also optimizes operation efficiency for reliable performance across a broad spectrum of applications, ranging from telecommunications to high-performance computing. The addition of new cooling technologies can significantly boost the thermal effectiveness of future transistor structures. For instance, utilizing efficient heat sinks or phase change materials directly contributes to better temperature control in integrated circuits. Besides, recent studies emphasize the importance of waste heat recovery systems, which are utilized to harness otherwise lost thermal energy. Thermal management for the case of proton exchange membrane fuel cells (PEMFCs) presented in Wu et al. (2025) comprises both efficient cooling techniques and waste heat recovery. Growing interest in such techniques is a consequence of the broader field of electronics in which similar principles are envisioned to be used in optimizing next-generation transistor thermal management. In addition to improved protocols of cooling, the development of high-temperature-resistant materials plays a significant role in avoiding thermal management issues in transistor designs. Advanced thermal films, illustrated in Song et al. (2025), possess a good future, combining solar heating and radiative cooling properties. Such films possess broad potential for year-round thermal regulation, allowing devices to better manage heat. The two-way capacity of such films could bring innovation in using transistors, which could provide better thermal stability and efficiency. As new transistor architectures are being developed, controlling both passive and active thermal management will be key to going beyond current technology boundaries and finding commercial success.

5.6.3. Scalability Concerns

An important aspect of advancing transistor architecture is in addressing the inherent scalability challenges that come with continuously shrinking device sizes. As the requirement for increased transistor densities grows, standard scaling methods are constrained by power consumption and heat dissipation problems. New technologies such as Tunnel FETs (TFETs) offer improved electrostatics and can operate at reduced voltages, and hence solutions to transcend such limitations. With the utilization of new conduction mechanisms that enable steep subthreshold swings, TFETs can shrink while maintaining performance, representing a revolutionary shift in the ways in which scalability can be employed without compromising energy efficiency. With this, TFET research releases new opportunities for system design for high-density integration use in high-compute scenarios where lower power consumption and enhanced thermal design are a priority. Furthermore, the advent of Negative Capacitance FETs (NC-FETs) has been of interest as a promising route to circumvent traditional scaling limitations. By employing the exploitation of negative capacitance effects, NC-FETs can be designed to reduce switching voltages and yet preserve excellent on/off current ratios. The more devices are scaled down, the more difficult it becomes to manage short-channel effects; NC-FETs remedy this through enhanced electrostatic gate control over the channel. This improvement not only enhances on-state current drive but also minimizes leakage current, ultimately promising enhanced scalability in device performance. The implications of integrating NC-FETs into traditional transistor layouts have much potential to address the growing demand for high-performance applications throughout computing platforms based on innovations that optimize both energy efficiency and scaling potential. At the same time, the study of vertical transport devices, including Vertical Tunnel FETs and transistors, underscores the need for innovative design strategies to achieve scalable solutions. By providing support for vertical current transport, such geometries make it possible to achieve higher packing densities and efficient use of space, which is critical as devices increasingly trend toward atomic dimensions. The architecture transition from lateral to vertical geometries has the potential to reduce parasitic capacitances significantly and improve electrical behavior, in turn, overall performance. Ironically, these advances address the pressing need for advanced high-performance computing and sophisticated data-intensive applications. However, manufacturing complexity and interconnect scaling challenges in the emerging transistor structures must be overcome for realization of reliable and scalable integration into advanced systems. Such research discovers contributions notably towards enhancing functionalities and performance while maintaining uniformity within the paradigm of frontier transistor technologies.

SUMMARY

As semiconductor technology continues to advance, the development of novel transistor structures like TFETs, negative capacitance FETs, and vertical transport devices becomes an imperative reaction to the demand for better device performance. Low power consumption combined with high-speed performance in integrated circuits is driving researchers to push the boundaries to research these new designs. TFETs, for instance, are promising a great deal in ultra-low-power uses due to the fact that they possess a potential to create steep subthreshold slopes, thereby achieving greater switching efficiency. This renders TFETs trendsetters in new technology, particularly in uses where power efficiency is important. As such architectures evolve, regular performance testing will be required to place them within the context of current technologies. Moreover, the concurrent presence of multiple strategies to improve performance, as in the Dual Metal Gate-Split Source-Bilateral Tunnel Field-Effect Transistor (DMG-SS-BTFET), implies the potential for revolutionary transistor efficiency and effectiveness enhancements. By combining dual-metal gates with a split-source structure, this transistor maximizes tunneling efficiency with optimal electrostatic control. These advancements are necessary, especially as conventional transistors are constrained in scaling to sub-nanometer nodes. The DMG-SS-BTFET not only outperforms conventional TFETs in conventional specifications like ION/IOFF ratios but also promises high-speed operation, demonstrating its usability in various applications. Thus, research into these novel structures is necessary for meeting future semiconductor requirements. In conclusion, the resolution of the highly complex power management, performance, and miniaturization challenges for advanced transistor structures is crucial if semiconductor technology evolution is to continue. The experience gained from the understanding of power consumption phenomena, as outlined in current studies, indicates that there is a need for a global approach. Techniques such as Multi-Threshold CMOS and other new power-reduction approaches must be synergistically put together to leverage the complexities of modern-day chip design. As the industry continues to advance into new territories of transistor architecture, for example, the implications of thermal management, the task of achieving maximum performance and reliability in advanced devices will require continuous research and innovation. The complex structures discussed demonstrate the paradigm-shifting potential and necessity of continued research within this continually advancing technology space (Desai, 2025; Hamza et al., 2025).

The evolution of next-generation transistor structures has introduced new concepts that have the potential to revolutionize the landscape of electronic devices. Of them, TFETs are significant due to their ability to achieve subthreshold swing abilities less than the traditional boundaries set by mainstream MOSFETs. The higher performance therefore translates to lower power consumption and greater efficiency, which are critical for modern electronic devices. The tunneling mechanism used in TFETs not only makes them faster but also enables them to work effectively at lower voltage levels, fulfilling the rising demand for energy-saving technologies in various domains. Overall, TFETs have a significant role to play in enhancing the functionality of transistors in future circuits to pave the way for highly efficient and powerful electronic systems. At the same time, NC-FETs have also drawn interest based on the potential of further decreasing power dissipation in semiconductor technology. With ferroelectric materials that allow negative capacitance, NC-FETs have the capability to reduce the voltage required for operation without compromising performance. This unique feature makes NC-FETs operable with lower subthreshold slopes that equate to increased drive currents and reduced leakage currents. Thus, the integration of NC-FETs into circuit design might unveil significant enhancements in operational performance across various applications, ranging from portable electronics to high-end computing devices. Through continued research, the transistors can redefine energy conservation in integrated circuits, similar to the call for sustainability while advancing technology. Vertical Transport Devices studies constitute another monumental achievement across advanced transistor designs. This design offers individual advantages of space usage and thermal control, reasonably compensating for some of the disadvantages observed in traditional planar-based devices. With vertically stacked layers of different transistors, designers are able to achieve greater current densities and improved thermal dissipation, a vital necessity for high-performance applications. As has been seen in recent studies, vertical devices, when combined with organic electrochemical transistors (OECTs), have superior performance characteristics such as ultra-stable switching and quick transient times, as depicted in research (Bauer et al., 2023) and (Huang et al., 2023). These breakthroughs are crucial for the development of next-generation flexible electronics, including wearable and implantable devices, and point to the revolutionary impact of vertical transport technologies on the future of electronics.

REFERENCES

1. Agopian, P., Martino, J. A., Rooyackers, R., Simoen, E., & Claeys, C. (2025). (Invited) Basic analog blocks using advanced devices. *ECS Meeting Abstracts.* https://www.semanticscholar.org/paper/3f15999064fa30c313faf4fac9632094b5dff262 (accessed on 27th September 2025).
2. Andrieu, F. (2025). (Invited) Advanced non-volatile-memory embedded with CMOS: Performance and challenges. *ECS Meeting Abstracts.* https://www.semanticscholar.org/paper/4130f88acfc6ec33c0d1c28ad254137ef4fa8628 (accessed on 27th September 2025).
3. Arsid, R. (2025). Evolution of transistor architecture: From FinFET to gate-all-around technology. *International Journal of Advanced Research in Science, Communication and Technology.* https://www.semanticscholar.org/paper/ab9c1c1ad70fe536eff213e3955611a13e60f984 (accessed on 27th September 2025).
4. Bauer, D., Khazdozian, H., Mehta, J., Nguyen, R. T., Severson, M., Vaagensmith, B., Toba, A. L., et al. (2023). *2023 Critical Materials Strategy.* https://doi.org/10.2172/1998242 (accessed on 27th September 2025).
5. Burns, G. T., & Joubert, D. P. (2024). Running shoes of the postmodern footwear era: A narrative overview of advanced footwear technology. *International Journal of Sports Physiology and Performance.* https://www.semanticscholar.org/paper/3efc45922e58e3f79c97a927a36c92aecd2c12fb (accessed on 27th September 2025).
6. Choi, W., Choudhary, N., Han, G., Park, J., Akinwande, D., & Lee, Y. H. (2017). Recent development of two-dimensional transition metal dichalcogenides and their applications. *Materials Today, 20*, 116–130. https://doi.org/10.1016/j.mattod.2016.10.002 (accessed on 27th September 2025).
7. Chowdhury, P. (2025). Power optimization in VLSI circuits: A comparative study of flip-flop architectures with NOR clock gating. *International Journal of Scientific Research in Engineering and Management.* https://www.semanticscholar.org/paper/acdd4f2213771b6fa317093dd7bcaec66fde3983 (accessed on 27th September 2025).
8. Chuang, Y. H., & Hu, V. P. H. (2025). Vertical channel transistor-based OS-Si hybrid gain cell for high-density and high-retention embedded DRAM. *2025 Silicon Nanoelectronics Workshop (SNW).* https://www.semanticscholar.org/paper/3b1db53a00fc1bca4feb3df0b754b6cdc9bb918d (accessed on 27th September 2025).

9. de Lima, C. T., Belot, D., Berkvens, R., Bourdoux, A., Dardari, D., Guillaud, M., Isomursu, M., et al. (2021). Convergent communication, sensing and localization in 6G systems: An overview of technologies, opportunities and challenges. *IEEE Access, 9*, 26902–26925. https://doi.org/10.1109/access.2021.3053486 (accessed on 27th September 2025).
10. Desai, N. S. (2025). Power management challenges and solutions in advanced technology nodes: A comprehensive analysis. *International Journal of Computing and Engineering.* https://www.semanticscholar.org/paper/1784c73af3df4f35004e29665c4dd557017d416a (accessed on 27th September 2025).
11. Georgieva, M., Petkov, N., & Duffy, R. (2023). 3D to 2D perspectives–Traditional and new doping and metrology challenges at the nanoscale. *Materials Science in Semiconductor Processing, 163*, 107584. https://doi.org/10.1016/j.mssp.2023.107584 (accessed on 27th September 2025).
12. Ghose, S. (2024). General-purpose multicore architectures. https://www.semanticscholar.org/paper/e723e1dcdafbf9c9944ce4977a95b29059cd0077 (accessed on 27th September 2025).
13. Hamza, M., Goel, R., Verma, M., & Agrawal, S. (2025). Comparative analysis of dual metal gate-split source and bilateral tunnel field effect transistor. *2025 Devices for Integrated Circuit (DevIC).* https://www.semanticscholar.org/paper/4a6f4412097f2796a5dd09c580d121165dcf6228 (accessed on 27th September 2025).
14. Haq, S. U., & Sharma, V. K. (2022). Review of the nanoscale FinFET device for the applications in nano-regime. *Current Nanoscience, 19*, 651–662. https://doi.org/10.2174/1573413719666221206122301 (accessed on 27th September 2025).
15. Huang, W., Chen, J., Yao, Y., Zheng, D., Ji, X., Feng, L. W., Moore, D. C., et al. (2023). Vertical organic electrochemical transistors for complementary circuits. *Nature, 613*, 496–502. https://doi.org/10.1038/s41586-022-05592-2 (accessed on 27th September 2025).
16. Hueting, R. J. E. (2018). The balancing act in ferroelectric transistors: How hard can it be? *Micromachines, 9*, 582–582. https://doi.org/10.3390/mi9110582 (accessed on 27th September 2025).
17. Ihsan, R. (2025). Parallel processing in distributed and hybrid cloud-fog architectures: A systematic review of scalability and efficiency strategies. *The Indonesian Journal of Computer Science.* https://www.semanticscholar.org/paper/c6ced9c33ce9b99af5ab989ab7dd2d00819a8ee4 (accessed on 27th September 2025).

18. Karaosmanoğlu, M. F. (2024). Design of energy-efficient CMOS circuits for AI applications. *Human Computer Interaction.* https://www.semanticscholar.org/paper/83289fba72a3ed462bd63c8daa728cf98ea8c16a (accessed on 27th September 2025).

19. Kim, Y., Han, J., Lee, J. S., Jin, T., Jang, P., Shin, H., Lee, J., et al. (2020). Power-efficient CMOS cellular RF receivers for carrier aggregation according to RF front-end configuration. *IEEE Transactions on Microwave Theory and Techniques, 69*, 452–468. https://doi.org/10.1109/tmtt.2020.3038775 (accessed on 27th September 2025).

20. Kumar, P., & Sivani, K. (2023). Double gate TFET simulation with different gate oxides and thickness. *2023 14th International Conference on Computing Communication and Networking Technologies (ICCCNT).* https://www.semanticscholar.org/paper/ee8e78fa3987c3b67326d2841f5351b994bde644 (accessed on 27th September 2025).

21. Lee, G., Oh, Y., Hwang, J., Hong, S. W., Song, S., Shin, K. W., Lee, J., et al. (2025). Scalable van der Waals integration of III-N devices over 2D materials for CMOS-compatible architectures. *Advanced Materials, 37.* https://www.semanticscholar.org/paper/b9bb3b93312a5b74681afdb3f023b21aa3e6ddfe (accessed on 27th September 2025).

22. Luo, Y., Abidian, M. R., Ahn, J. H., Akinwande, D., Andrews, A. M., Antonietti, M., Bao, Z., et al. (2023). Technology roadmap for flexible sensors. *ACS Nano, 17*, 5211–5295. https://doi.org/10.1021/acsnano.2c12606 (accessed on 27th September 2025).

23. Malinowski, P., Pejović, V., Lieberman, I., Kim, J., Siddik, A. B., Georgitzikis, E., Lim, M., et al. (2023). Image sensors using thin-film absorbers. *Applied Optics, 62*(17), F21–F30. https://www.semanticscholar.org/paper/400bf137f98e3ae0c83bfdd2baa2f85cc8ee85c2 (accessed on 27th September 2025).

24. Narayanan, A. (2025). Scalability and security in cloud-native financial systems: A dual-pillar approach to modern fintech architectures. *World Journal of Advanced Research and Reviews.* https://www.semanticscholar.org/paper/700ae6b273b2b45e2ba21198c370fc2df23c8002 (accessed on 27th September 2025).

25. Pandey, J. K. (2025). Leveraging artificial intelligence in in-system test: A new paradigm for predictive and adaptive chip validation. *World Journal of Advanced Engineering Technology and Sciences.* https://www.semanticscholar.org/paper/

f2885ffc3846eec60cdcd54d33557db994db9b62 (accessed on 27th September 2025).

26. Pandey, M., Bhaiyya, M., Rewatkar, P., Zalke, J. B., Narkhede, N., & Haick, H. (2025). Advanced materials for biological field-effect transistors (Bio-FETs) in precision healthcare and biosensing. *Advanced Healthcare Materials, 14.* https://www.semanticscholar.org/paper/be0ad0a42fb8ac32d78ee375411f4d546cb29064 (accessed on 27th September 2025).

27. Pathak, A. (2025). Advanced ternary addition circuits leveraging carbon nanotube field-effect transistors. *International Journal for Research in Applied Science and Engineering Technology.* https://www.semanticscholar.org/paper/6ecba7590c8874baf744726a17bae83a4fe766fe (accessed on 27th September 2025).

28. Pelloquin, S., Muller, J., Mannaa, S., O'Connor, I., & Larrieu, G. (2025). Scalable 3D logic cells using vertical sub-20 nm gate-all-around nanowire FETs for high-density integration. *ECS Meeting Abstracts.* https://www.semanticscholar.org/paper/6750e5cdc5b0fec83f1a2556015c7bdaf41bc436 (accessed on 27th September 2025).

29. Ramakrishnan, K., & Vidhya, K. (2025). Development and optimization of frequency divider in integrated circuits leveraging Transvidkan gate architecture. *2025 5th International Conference on Trends in Material Science and Inventive Materials (ICTMIM).* https://www.semanticscholar.org/paper/ee80627fe7486ff758db7086188a444ccf42bc1f (accessed on 27th September 2025).

30. Rana, A., Gautam, D., Kumar, P., & Das, A. K. (2025). Architectures, benefits, security, and privacy issues of Internet of Nano Things: A comprehensive survey, opportunities, and research challenges. *IEEE Communications Surveys & Tutorials, 27*, 1152–1190. https://www.semanticscholar.org/paper/3c12d24af7fe3b7acb717056bed9bb17c28e0e94 (accessed on 27th September 2025).

31. Rasheed, A., San, O., & Kvamsdal, T. (2020). Digital twin: Values, challenges and enablers from a modeling perspective. *IEEE Access, 8*, 21980–22012. https://doi.org/10.1109/access.2020.2970143 (accessed on 27th September 2025).

32. Singh, S., Devi, P., & Dhar, R. (2025). Development and analysis of gate underlap GaN gate-all-around FETs. *2025 3rd International Conference on Smart Systems for Applications in Electrical Sciences (ICSSES).* https://www.semanticscholar.org/paper/

ce1c225ba025bb9e85a4436a4e6c35ceaa92edeb (accessed on 27th September 2025).

33. Song, Q., Rettinger, S., Xu, C., Mansfeld, U., & Retsch, M. (2025). Dual-mode film based on highly scattering nanofibers and upcycled chips-bags for year-round thermal management. *Advanced Functional Materials.* https://www.semanticscholar.org/paper/f4fd08e38472e14232b8c500636feaa133c7df35 (accessed on 27th September 2025).

34. Sze, V., Chen, Y. H., Yang, T. J., & Emer, J. (2017). Efficient processing of deep neural networks: A tutorial and survey. *Proceedings of the IEEE, 105*, 2295–2329. https://doi.org/10.1109/jproc.2017.2761740 (accessed on 27th September 2025).

35. Tamersit, K., Kouzou, A., Rodríguez, J., & Abdelrahem, M. (2024). New label-free DNA nano sensor based on top-gated metal–ferroelectric–metal graphene nanoribbon on insulator field-effect transistor: A quantum simulation study. *Nanomaterials, 14.* https://www.semanticscholar.org/paper/c91433e0a78675b9315ff6e2c0c004abafc41b13 (accessed on 27th September 2025).

36. Tataria, H., Shafi, M., Molisch, A. F., Döhler, M., Sjöland, H., & Tufvesson, F. (2021). 6G wireless systems: Vision, requirements, challenges, insights, and opportunities. *Proceedings of the IEEE, 109*, 1166–1199. https://doi.org/10.1109/jproc.2021.3061701 (accessed on 27th September 2025).

37. Wu, F., Chiu, C. Y., Lin, T. Y., Wu, C. H., Hu, P. H., Su, P., & Su, C. J. (2025). Enabling broader memory windows by double-gate nanosheet ferroelectric FETs for next-generation non-volatile memory storage. *2025 9th IEEE Electron Devices Technology & Manufacturing Conference (EDTM).* https://www.semanticscholar.org/paper/b49c982398c0de9003c4fdc0d8dc442f1bc60aa4 (accessed on 27th September 2025).

38. Wu, Q., Dong, Z., Zhang, X., Zhang, C., Iqbal, A., & Chen, J. (2025). Towards more efficient PEM fuel cells through advanced thermal management: From mechanisms to applications. *Sustainability.* https://www.semanticscholar.org/paper/6398400e84ea1fd59fe97a6d80bb93816b4095fc (accessed on 27th September 2025).

39. Wu, Q., Zhang, S., Zheng, B., You, C., & Zhang, R. (2021). Intelligent reflecting surface-aided wireless communications: A tutorial. *IEEE Transactions on Communications, 69*, 3313–3351. https://doi.org/10.1109/tcomm.2021.3051897 (accessed on 27th September 2025).

40. Yu, J., Cho, H., Kim, Y., & Kim, G. T. (2024). Tunable band-to-band tunneling and conducting path transition in local-control-gate heterostructure transistor. *Advanced Functional Materials, 34.* https://www.semanticscholar.org/paper/67172bfd38ccd5184f149e5d8cf71601400fd93b (accessed on 27th September 2025).
41. Yumigeta, K., Yusufoglu, M., Sarker, M., Raj, R., Daluisio, F., Holloway, R., Yawit, H., et al. (2025). Scalable etch-free transfer of low-dimensional materials from metal films to diverse substrates. https://www.semanticscholar.org/paper/df8d48f2c9fb7589732d541404190976983f6873 (accessed on 27th September 2025).
42. Zhang, Q., Hu, W., Sirringhaus, H., & Müllen, K. (2022). Recent progress in emerging organic semiconductors. *Advanced Materials, 34.* https://doi.org/10.1002/adma.202108701 (accessed on 27th September 2025).
43. Zheng, F., Meng, W., & Li, L. J. (2025). Continue the scaling of electronic devices with transition metal dichalcogenide semiconductors. *Nano Letters, 25*, 3683–3691. https://www.semanticscholar.org/paper/58fa49433e90d086fdda93869356072f4a70d80f (accessed on 27th September 2025).
44. Zhu, H., Yang, Y., Zhu, X., Raju, P., Ioannou, D., & Li, Q. (2023). Graphene-integrated negative quantum capacitance field-effect transistor with sub-60-mV/dec switching. *IEEE Transactions on Electron Devices, 70*, 4899–4904. https://www.semanticscholar.org/paper/e7b764e656287c81469b2949060e4f07c691d849 (accessed on 27th September 2025).

CHAPTER 6

HIGH-POWER AND HIGH-FREQUENCY DEVICES

Contents

Introduction 221

6.1. Types of High-Power Devices..226

6.2. Materials Used in High-Frequency Devices..232

6.3. Applications of High-Power Devices...236

6.4. Thermal Management Strategies 241

6.5. Challenges in High-Power and High-Frequency Device Design....247

6.6. Future Trends in High-Power and High-Frequency Devices 252

Summary...256

References..259

INTRODUCTION

Progress in semiconductor materials has had tremendous contributions towards the development of high-power and high-frequency devices. Among these, the innovation of wide band gap (WBG) semiconductors such as gallium nitride (GaN) and silicon carbide (SiC) has propelled the performances of modern electronic systems. These materials not only optimize devices to be more efficient but also enhance their ranges in a big way, making them worth considering for applications previously impossible with standard silicon-based technology. High-κ dielectric with its better insulation properties and larger permittivity than silicon dioxide has become a mainstay component of future transistors. Their applications as gate dielectric layers in GaN and SiC transistors indicate the ongoing drift from silicon technology, spurring innovation in high-frequency and high-power electronics (Nigro et al., 2022). High-frequency device design, particularly resonant converters such as LLC converters, is challenge and opportunity. Application of these converters leverages the soft-switching benefit of WBG devices to attain unrivaled switching frequencies in the megahertz range. This innovation results in drastic minimization in converter size and maximization in power density, addressing increasing demands for space-saving and power-efficient energy solutions. These are at the expense of increased magnetic design and control method complexity. High-frequency magnetic design is typified by certain challenges, specifically to achieve stable and optimized operation under rapid transient response requirements prevalent in many applications (Lee et al., 2021). Consequently, novel magnetic design and control strategies have been created to make LLC converters resilient, a reflection of the necessity of continuous improvement and innovation in the field. Moreover, the advent of high-power and high-frequency devices is a strong indication of the necessity of suitable thermal management practices to achieve optimal performance criteria. As these devices operate at higher levels of power and frequency, thermal dissipation takes on a vital function to support longevity and reliability. Advanced cooling techniques, such as liquid cooling and phase change materials, have begun to play integral roles in solving thermal issues, representing a total device design methodology. Extensive analysis of such thermal management systems is warranted as they have direct impacts on the performance and efficiency of high-power and high-frequency devices. By incorporating new materials and new styles of architecture, the area of electronics is poised to meet the pressing challenges of thermal management and therefore facilitate the creation of solid, reliable, and efficient high-power and high-frequency devices appropriate for today's requirements.

Definition of High-Power and High-Frequency Devices

High-power and high-frequency devices are necessary to meet today's electronic applications that need to be efficient at a high power level and high frequency. Among the many technologies that have been developed, High Electron Mobility Transistors (HEMTs), Lateral Diffused Metal Oxide Semiconductors (LDMOS), and Insulated Gate Bipolar Transistors (IGBTs) are prominent based on their unique properties. HEMTs are particularly loved in radio frequency applications for their efficiency and high electron mobility. LDMOS devices integrate high frequency with increased power handling and are thus specially employed for use in communication system base stations. IGBTs, though, integrate the strength of bipolar transistors and MOSFETs to present rugged solutions for high-voltage and high-current applications. The devices have the performance enhancement needed by industries like telecommunications and power generation, presenting the strategic place they hold in high-power and high-frequency applications. Advances in GaN and SiC materials have been a key contributor to propelling the development of high-power and high-frequency devices. GaN, being highly thermally conductive and having a wide bandgap, enables it to support higher power densities and frequencies than silicon-based counterparts. GaN technology has thus accelerated the pace of development of RF power amplifiers and high-frequency communication equipment. SiC is, however, characterized by higher voltage and thermal resistances and is hence widely used in power electronics for numerous industrial applications. The innovation of these materials has enabled devices to operate best under stressful conditions, ultimately leading to enhanced performance levels in commercial and military use (Figure 6.1).

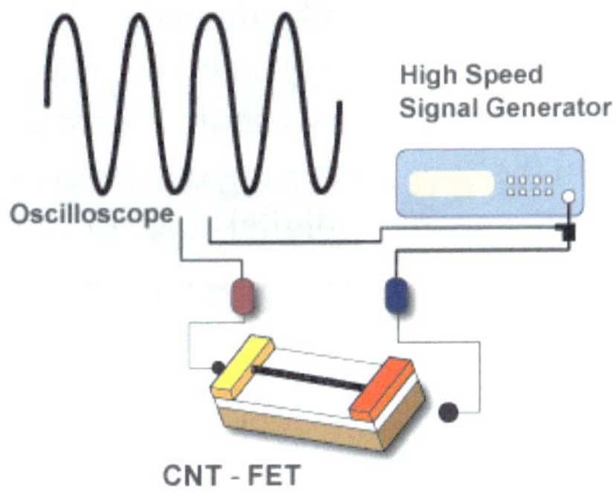

Figure 6.1. *The modern landscape of high-frequency devices.*

Source: https://nanointegris.com/wp-content/uploads/2017/10/Img-High-Frequency-Devices.png.

This advancement is a major achievement towards meeting the world's needs for efficient energy conversion and signal processing. Heat management is a critical issue in the operation and design of high-frequency and high-power devices. Because these devices are driven at increased power levels, heat generation is more significant, compromising performance and reliability unless managed properly. Advanced heat management methods, such as heat sinks, thermal interface materials, and tailored packaging, are employed to spread heat and maintain operational integrity. Along with this, other innovative strategies like micro-channel cooling systems are also being studied for enhancing thermal performance while minimizing space requirements. All these strategies need to be integrated to make high-power and high-frequency devices operational and reliable at high-stress levels. Such pre-emptive thermal management not only enhances device longevity but also significantly boosts systemwide performance as a whole, thus abiding by industry requirements for reliability and efficacy in high-use applications.

Importance in Modern Technology

High-power and high-frequency device development has transformed modern technology significantly, particularly in telecommunications and power management technologies. This development is spearheaded initially by advancements in materials such as GaN and SiC that provide greater efficiency and thermal response than devices based on silicon. They enable greater electron mobility and increased breakdown voltage, making them suitable for effectively being used in applications involving fast switching and high power processing. Moreover, the utilization of HEMTs has facilitated radio frequency (RF) devices to be optimized, leading to better signal processing and communication speed. The importance of these advancements cannot be overstated, as they are the building blocks to sustain the ever-increasing needs for speed and performance of electronic devices in the modern digital age. In RF applications of high frequency, the use of advanced semiconductor technologies has vast promise in various industries like renewable energy and automotive. Devices like LDMOS and IGBTs are vital in revolutionizing the efficiency and performance of electrical systems to unprecedented levels. LDMOS, for instance, is widely used in RF power amplifiers and is thus a major device for 4G and 5G telecom networks. On the other hand, IGBTs play a central role in power inversion for clean energy sources, such as wind and solar, to expand the ability to convert and deliver energy efficiently. The interaction of such devices is a good example of the critical need for high-performance devices to facilitate sustainable development as well as respond to increasing power demands of the modern world. With progress in technology further in recent years, concerns regarding high-voltage as well as high-frequency operation cannot be overlooked. The increasing reliance upon quick processing is matched by the pressing demand for reliable insulation systems, particularly for electrical machinery. Research has shown that high-frequency square wave pulses can considerably affect the lifespan of motor winding insulation, emphasizing the imperative relationship between switching frequencies and electrical component integrity. For instance, insulation reliability is a critical concern that must be addressed in the context of modern power electronics and inverter-based systems, where lifetime tests reveal the implications of high rise times on insulation breakdown characteristics (Islam et al., 2025). The integration of robust designs founded on high-power technologies not only enhances performance but also the sustainability and safety of advanced electrical devices (Fan et al., 2024). With these innovations, the technology is capable of addressing the needs of modern electronics through ensuring the durability and efficiency of high-power and high-frequency systems.

Overview of Applications

High-power and high-frequency devices have become more of a focal point in many applications, which make possible progress that makes them more efficient and productive. Among such devices, HEMTs have garnered immense attention due to their ability to sustain high temperatures and voltages. This makes them particularly suited for application in telecommunication devices where their rapid switching ability comes in handy when it comes to efficient signal amplification and transmission. Both LDMOS and IGBTs are widely used in power electronics, especially in applications that demand high switching strength such as converters and inverters. With the advance of these technologies, their uses in core fields such as renewable energy and electric vehicles indicate how they are the basis of developments in system reliability and energy conversion efficiency. The growing use of GaN and SiC materials has also revolutionized the design of high-frequency and high-power devices. GaNs favorable properties like strong breakdown field strength and electron mobility make it particularly attractive for application in RF amplifier and power management system functionalities, as evident with current research on GaN HEMT that has shown promising performance in radio frequency amplifiers and charger circuits. On the contrary, SiC excels at withstanding high power and high temperature and is therefore ideally suited for use in electric vehicle charging stations and industrial power systems. Both materials are enabling new devices and architectures that push performance limits while simultaneously reducing energy losses, more vital than ever in our power-hungry world. Effective thermal management methods are essential to maximize the potential performance of high-power and high-frequency devices. Because these components are exposed to extreme thermal stress, inadequate heat dissipation can lead to diminished performance and shorter lifetimes. Breakthrough thermal management technologies, such as new cooling mechanisms and heat sinks, are essential to device reliability and performance. Furthermore, the inclusion of thermal management systems at and near GaN and SiC technologies can significantly minimize thermal concerns and facilitate continuous operation at maximum performance. Accordingly, the emphasis on effective thermal management not only maximizes the capability of high-power and high-frequency devices, but also makes industries more ecological by encouraging the optimal usage of energy resources in applications ranging from telecommunications to electric vehicle infrastructures.

6.1. TYPES OF HIGH-POWER DEVICES

High-power devices are the imperative in contemporary advanced electronic systems, with technologies being different for various purposes in industries. Among these technologies, High Electron Mobility Transistors (HEMTs), Lateral Diffusion Metal-Oxide-Semiconductor (LDMOS) transistors, and Insulated Gate Bipolar Transistors (IGBTs) are notable due to their specific characteristics. HEMTs, which are of high speed and efficiency, are mostly utilized in radio frequency (RF) applications where the switching has to be rapid. LDMOS devices, however, are good at handling high-voltage applications, and therefore they find place in power amplifiers in telecommunications. Similarly, IGBTs combine the virtues of both MOSFETs and bipolar transistors, making them suitable for high-power applications such as motor drives and renewable energy systems.

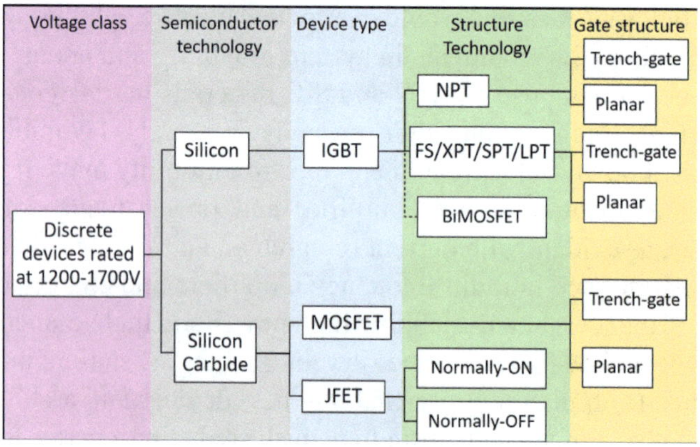

Figure 6.2. *Commercially available high-power semiconductor devices.*

Source: https://www.researchgate.net/publication/363205663/figure/fig1/AS:11431281082613472@1662082142381/Commercially-available-high-power-semiconductor-devices-suitable-for-solid-state-DC.ppm.

The choice between the two technologies usually falls back on specific performance parameters such as power density, switching time, and thermal management aspects. The emergence of wide bandgap semiconductors, i.e., Gallium Nitride (GaN) and Silicon Carbide (SiC), has revolutionized high-power electronics with the prospect of more efficiency and thermal management. GaN is particularly favored for RF applications due to their superior efficiency and frequency response, allowing thin and lightweight devices to be developed without compromising performance. SiC, on the other hand, has better thermal stability and is consequently aptly suited for applications in high-temperature environments and high-voltage use. An example of this application is the utilization of these materials in DC/DC converters like Dual-Active-Bridge (DAB) and Active-Bridge-Active-Clamp (ABAC) topologies, which are commonly utilized in More Electric Aircraft (MEA) systems. The transition to these high-end materials is not entirely driven by theoretical benefit; empirical research underscores their efficiency in limiting energy losses and enhancing operation ranges of high-frequency devices (Fernandez-Hernandez et al., 2024) (Figure 6.2).

Thermal management methods are of utmost significance when dealing with high-power devices since effective heat dissipation is directly related to the reliability and performance of devices. The incorporation of new materials and geometries improves heat transfer, maintaining these devices in a safe temperature. Heat sinks, thermal interface materials, and the phase change cooling mechanism are imperative in retaining optimal operating conditions. Besides, the development of computational models ensures that thermal performance can be analyzed and predicted for high-power applications. For instance, using models when simulating power devices such as HEMTs or IGBTs allows engineers to contrast thermal effects of varying configurations and improve designs. Efficiency and thermal control-focused new strategies are vital to the continued success of high-power devices, especially as market demand for higher performance in smaller packages continues to intensify (Tellefsen et al., 2024).

6.1.1. High Electron Mobility Transistors (HEMTs)

The evolution of power electronic devices has witnessed an unprecedented shift from traditional silicon-based technologies to newer, more advanced alternatives that provide higher performance and efficiency. In particular, HEMTs, especially those on the basis of gallium nitride (GaN), have emerged as important devices for high-power and high-frequency applications. The unique material characteristics of GaN, including a wide bandgap and high thermal conductivity, enable HEMTs to be operated at high voltages and temperatures with low on-resistance. This technology addresses the growing demand for high efficiency power conversion systems in many applications, such as renewable energy technologies and electric vehicles, that increasingly rely on high efficiency transistors to reduce energy wastage and enhance overall performance (Adekunle, 2025). In addition to their superior electrical behavior, HEMTs possess huge advantages over some of the alternative technologies, such as laterally-diffused metal-oxide-semiconductor (LDMOS) and insulated gate bipolar transistors (IGBTs). GaN-based HEMTs' ability to deliver high-frequency operation means smaller circuit size and higher power density, which are of paramount importance in current RF communications and high-power applications (Figure 6.3).

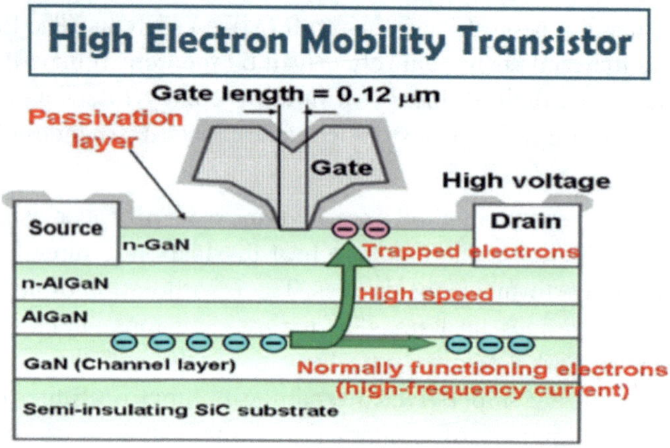

Figure 6.3. *High electron mobility transistor (HEMT).*

Source:https://www.elprocus.com/wp-content/uploads/2017/05/High-Electron-Mobility-Transistor.jpg.

Also, the advances in device structures have facilitated other innovations in HEMT design for thermal and reliability improvements. As scientists continue to work towards solving material defect and manufacturing costs issues, the integration of GaN HEMTs into electronics signifies a paradigm shift in power electronics and high-frequency designs (Islam et al., 2024). Thermal management is a crucial problem for the widespread use of HEMTs in high-power applications due to their operating conditions. Effective thermal management methods like newer heat sinks, thermal interface materials, and substrate engineering must be utilized to ensure optimal performance and long-term reliability of HEMTs. With the devices being compelled to operate in severe conditions or at high frequencies, proper heat dissipation is increasingly important in order to prevent performance degradation and device failure. It will become achievable to employ them in high power and efficient applications with the improvement of packaging and cooling techniques coupled with GaN-based HEMTs' thermal dynamics understanding (Adekunle, 2025).

6.1.2. Laterally Diffused Metal-Oxide-Semiconductor (LDMOS)

High-power and high-frequency device development has witnessed significant focus on LDMOS technology development. The revolutionary design of LDMOS enables it to support high breakdown voltages with low on-resistance, which makes it well adapted to utilize in applications where efficiency as well as reliability is required. This aspect comes in handy in power amplification, where energy loss must be minimized to the barest level.

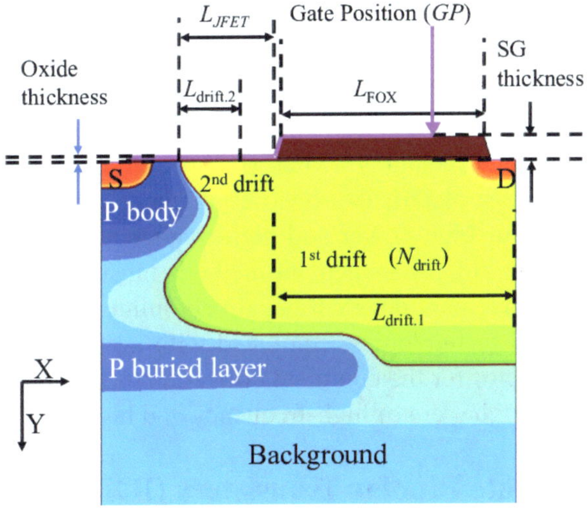

Figure 6.4. *Schematic cross-section of a laterally-diffused metal-oxide-semiconductor (LDMOS).*

Source:https://www.researchgate.net/publication/371708762/figure/fig1/AS:11431281169243595@1687266184290/Schematic-cross-section-of-a-laterally-diffused-metal-oxide-semiconductor-LDMOS-with.png.

The integration of new materials, such as gallium oxide (Ga_2O_3), also enhances the performance of LDMOS devices through the development of high-voltage capabilities (Bian et al., 2024). Through optimization of parameters such as gate oxide thickness and drift zone concentration by structuring devices through simulation tools like Sentaurus TCAD, scientists can achieve improved voltage resistance, satisfying the needs of future power systems with high efficiency and thermal stability. The relevance of LDMOS devices is best utilized in radio frequency (RF) applications, where their compatibility with integrated circuit processes allows them to be miniaturized. One of the most notable achievements in this area is the combination of new device structures with field plates that enhance electric field control. By employing a hybrid field plate method, LDMOS devices have exhibited remarkable improvements in breakdown voltage and on-resistance specific values, showing their potential for high-frequency operational control (Yu et al., 2024) (Figure 6.4).

This strategic device design suppresses hot carrier injection degradation, enabling reliable long-term operation under stressful electrical conditions. These advances not only improve device performance but also shrink the size and thermal management needs of RF systems, making them more easily integrated into state-of-the-art communication networks. This development of LDMOS technology exemplifies an overall trend toward the use of wide-bandgap semiconductors in high-power applications beyond the confines of the conventional silicon material. With the development of silicon carbide (SiC) and gallium nitride (GaN) materials, LDMOS devices are being utilized in power electronics that need higher switching rates, efficiency, and operating temperatures. The better electrical properties of these materials make them attractive candidates for high-power and high-frequency applications. As the capability increases, thermal management methods must also develop so that the performance improvement does not negatively influence device reliability. Efficient heat removal technologies, combined with novel device topologies, will be the decisive factor for the complete utilization of LDMOS technologies within the rapidly changing era of high-frequency and high-power devices.

6.1.3. Insulated Gate Bipolar Transistors (IGBTs)

The field of power electronics has witnessed remarkable advancements with the application of advanced devices capable of supporting high-power and high-frequency applications. These include the IGBTs, which stand out due to their superior combination of the high input impedance of the MOSFET with bipolar transistor low conduction losses. The hybrid configuration allows IGBTs to efficiently deal with significant voltage and current levels while giving them perfect places in motor drives, renewable energy systems, and power converters. However, it is a compounded requirement to utilize their best with their potential when applied in series topologies because these create voltage balancing issues and performance variation under loading conditions, which are crucial in high-voltage applications (Hamidi et al., 2025). Fixes for these issues are essential for optimizing IGBT performance in actual applications. Another complication to the operation efficiency of IGBTs is the effect of parasitic inductances that has a tendency to be highest in high-frequency applications. The complication lies in the fact that they create problems of switching losses and electromagnetic interference that are on the switching transitions. Conventional approaches that involve snubbers minimize some of these problems but add extra complexities, especially with increasing operating speeds. Recent technology has witnessed the production of designs that incorporate passive components, e.g., coupled inductors together with snubber capacitors, to enable soft-switching conditions. These technological developments in IGBT are aimed at minimizing switching losses, thus enhancing performance under stressful operating conditions (Xiao et al., 2025). Satisfactory thermal management measures also accompany these advancements by protecting the IGBT within safe thermal margins with sustained reliability and efficiency. As demand for efficient power conversion systems is increasing, the evolution of IGBT technology will be a crucial factor in power electronics' future (Figure 6.5).

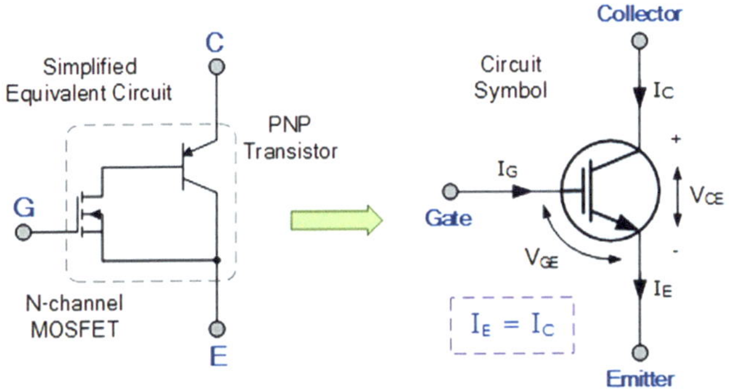

Figure 6.5. *The advancements in IGBT technology.*

Source:https://www.electronics-tutorials.ws/wp-content/uploads/2013/08/igbt1.gif.

Future designs need to maximize not only electrical but also thermal performance to prevent degradation and failure when being used. Study of such materials as Gallium Nitride (GaN) and Silicon Carbide (SiC) in conjunction with IGBTs can offer additional paths toward high-frequency capability enhancement. In combination with wisely planned thermal management strategies, the effective integration of IGBT technology into the next generation of power systems can lead to more consolidated, energy-efficient designs. Lastly, continuous research on the use of IGBT in the overall context of high-power and high-frequency devices must highlight breakthrough solutions in an effort to meet the evolving demands of modern energy systems.

6.2. MATERIALS USED IN HIGH-FREQUENCY DEVICES

Development of high-frequency devices is highly dependent on the materials employed in their production, with gallium nitride (GaN) and silicon carbide (SiC) being strong rivals in RF and power electronics applications. The materials have improved electrical properties that enable efficiency and performance at high frequencies to be increased. GaN, due to its wide bandgap and high breakdown, allows devices to operate at high power densities without inducing thermal issues, and therefore it is optimally suited for application in telecommunications and space satellite systems. Similarly, the strength of SiCs under harsh conditions improves thermal management, an important area of application in high-power usage where excessive heat dissipation can lower the performance of the devices.

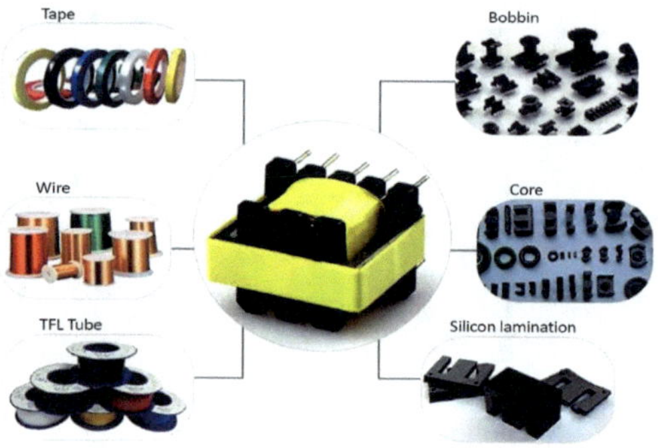

Figure 6.6. *Types of materials used in high-frequency devices.*

Source: https://cdn-diaea.nitrocdn.com/hbfYEphstQZnIIJKnyibxbXZNQoUxubD/assets/images/optimized/rev-4daeffe/unicreed-transformer.com/wp-content/uploads/2024/09/What-Materials-Are-The-High-Frequency-Transformers-Made-Of.jpg.

This focus on material selection is immediately connected with the sustainability of high-power device performance, and therefore an ongoing process of research and development must be employed to optimize these materials for specific applications within the high-frequency spectrum. High-frequency operation challenges are closely tied with the need for effective thermal management practices. High-power devices, particularly those constructed from GaN and SiC, possess an intensive generation of heat that can inhibit their operating capability and lifespan. In order to neutralize this effect, new cooling methods such as liquid cooling and the incorporation of heat sinks are being utilized in order to sustain the most favorable temperature levels. High-tech packaging technologies are also under development in order to increase thermal dissipation and thus secure the device integrity in operation. It is indispensable in developing the design and performance of high-frequency devices to comprehend the complicated interaction between heat handling and material properties (Figure 6.6).

The interaction between these variables ultimately contributes to the overall efficiency, reliability, and scalability of future power electronics. The application of high-frequency devices makes its way into a wide variety of industries, with significant implications for system performance and energy efficiency. For instance, the high-electron-mobility transistors (HEMTs), laterally-diffused metal-oxide-semiconductor (LDMOS), and insulated-gate bipolar transistors (IGBT) devices show how material properties can lead to increased capability within high-power applications. The devices exploit the corresponding material advantages in order to achieve increased switching speeds and enhanced energy handling capabilities, critical for power electronics in today's world. In line with current studies, integrating advanced composite materials with standard substrates holds very promising avenues for enhancing device performance further. Well-designed materials not only improve performance in existing applications but also enable new applications in renewable energy devices, electric vehicles, and smart grids, and therefore underpin the growing importance of material science in the high-frequency and high-power world.

6.2.1. Gallium Nitride (GaN)

The rapid emergence of GaN as a high-power and high-frequency device is indicative of its expanding area of application across a series of applications. Being a multifunctional wide bandgap semiconductor material, GaN has improved electron mobility and high breakdown voltage characteristics that have a dramatic effect on switch-mode operating efficiency. This enhanced efficiency has particular expression in power electronic converters, where GaN-based products will reduce switching losses significantly. For instance, recent developments have enabled the fabrication of GaN DC-DC converters that have both more power density and conversion efficiency compared to current silicon products. The application of GaN in devices such as photovoltaic systems showcases the possibility of improved energy harvesting because two-input converters based on GaN technology not only possess an efficiency rate that is more than 98% but are also operable at high switching rates, for instance, 600 kHz, significantly enhancing overall system performance (Feng et al., 2024). In wireless power transfer applications, GaN devices consistently come out on top, particularly in inverter design applications intended for multi-megahertz (MHz) operation. Thermal management and overheating become the primary concerns in such high operation frequencies in which GaNs have a definite edge. GaN FETs have much better thermal performance compared to silicon carbide (SiC) devices, which is a prerequisite in high-frequency operations. For example, inverters using GaN have, during high-power testing, achieved up to 3.8 kW conversion capacities at 500V with still maintaining peak efficiencies of over 97%. Such performance not only facilitates easier development of power systems that are lighter and smaller in size, but also ensures that devices will be efficient even at higher temperatures, an issue that would otherwise be faced with traditional materials (Wang et al., 2024). Integrating GaN technology into high-frequency and high-power devices needs creative thermal management technologies to enhance performance even further. Applying sophisticated cooling techniques such as liquid cooling or phase change materials can mitigate the thermal challenges accompanying high switching frequencies in GaN devices. Additionally, taking advantage of the unique property of GaN to have high-temperature operation allows creative design methodologies that minimize size but optimize efficiency. As the technology advances towards miniaturization and increased power densities, dependable thermal management increasingly comes to the fore. The combination of GaNs thermal properties with sophisticated cooling techniques can lead to groundbreaking advancements in the efficiency and reliability of next-generation power electronics. Through precise optimization of the synergy between material properties and thermal methods, engineers can push the boundaries of what is feasible in high-power and high-frequency devices (Feng et al., 2024).

6.2.2. Silicon Carbide (SiC)

Growing demands for high-performance power electronics necessitate the research and exploitation of new materials such as SiC. The compound semiconductor delivers unique characteristics that significantly enhance the performance of high-frequency and high-power devices, thereby addressing the silicon-based traditional systems' limitations. SiC devices, including High Electron Mobility Transistors (HEMTs) and Insulated Gate Bipolar Transistors (IGBTs), have been found to be thermally conductive and possess a wide bandgap, thus being efficient at high temperatures and voltages. They are fast-switched, thereby making them more efficient in power converters, which is essential in applications such as electric vehicles and renewable energy. With the market performance towards more energy-efficient power devices, SiC technology adoption becomes all the more significant with immense promises of energy savings and device efficiency in general (Nielsen et al., 2025). Along with this, advancements in compact modeling methods for SiC devices have further cemented its position in high-power electronics. Fourierier modeling is necessary to predict device performance and optimize it, especially under varying operational conditions. Recent work has shown that applying non-linear bias-dependent drift resistance allows accurate modeling of SiC-based Field Effect Transistors (FETs). The given BSIM-BULK HV compact model replicates dominant features from weak inversion to strong inversion regimes, allowing accurate circuit design and simulation (Tripathi et al., 2025). As the players in the industry go on to further develop these modeling techniques, the reliability and efficiency of SiC devices in practical applications will continue to enhance, paving the way for more efficient and powerful electronic solutions in industrial and commercial contexts. Thermal management remains a substantial issue when incorporating SiC devices into high-power and high-frequency applications. The excellent thermal conductivity of SiC facilitates improved heat dissipation, but increasing power densities pose challenges that must be addressed by innovative thermal management strategies. Effective cooling devices, including advanced heat sinks and thermal interface materials, are necessary to maintain junction temperatures at optimal values and device reliability. Furthermore, as SiC technology becomes widely used in industrial applications such as automotive and renewable energy systems, thermal factors will be one of the most influential aspects of the reliability and overall efficiency of high-power electronic modules. By harnessing SiCs inherent capabilities using optimized thermal solutions, the next generation of power systems for high-performance applications can fulfill the escalated demands of modern-day electrical infrastructure, driving the green energy revolution (Nielsen et al., 2025).

6.2.3. Comparison of Gallium Nitride (GaN), Silicon Carbide (SiC), Properties

Power electronics development has been significantly influenced by the properties of wide bandgap (WBG) semiconductors, specifically GaN and SiC. Their characteristics enable the production of high-power and high-frequency devices necessary for the application in electric vehicles and renewable energy systems. GaN is valued for its high saturation velocity and electron mobility, enabling high-frequency efficient operation. Its suitability renders it best suited for high-speed switching applications, such as RF amplifiers and communication systems. SiC, on the other hand, has high thermal conductivity and breakdown strength, enabling suitability for stable performance under high-temperature conditions. As a result, technology like high electron mobility transistors (HEMTs) in GaN and insulated gate bipolar transistors (IGBTs) in SiC have become crucial devices in current power electronics, with each of them serving different operation conditions and demands. In comparing GaN and SiC, it is necessary to take into account their thermal management characteristics, which determine device reliability and operation. GaN typically has lower thermal conductivity than SiC, which results in a greater heat dissipation problem when it is in operation. This necessitates the implementation of advanced thermal management schemes in devices based on GaN to enable efficient operation. However, its ability to work at higher frequencies compensates for this disadvantage for most applications at high frequencies. In contrast, SiC's improved thermal capability allows devices to be as efficient at high temperatures as at normal temperatures. The combination of high breakdown voltage and resistance to thermal degradation renders SiC a solid contender for high-power applications with a requirement of reliability under extreme conditions, which can be observed in automotive devices and industrial devices (Mitrovic et al., 2022). The exclusive strengths of both GaN and SiC are not only exemplified by their respective individual performance profiles but also by their combined complementary positions within the group of high-power and high-frequency devices. GaN excels in applications that require compact and efficient performance, such as in RF amplification and high-frequency power conversion. Its low parasitic capacitance allows it to reach rapid switching rates necessary in modern electronic systems. On the other hand, SiCs properties are appropriate for those power electronics requiring superb durability and thermal stability and therefore are appropriate for use in the automotive and aerospace industries. With advances in semiconductor processing technology ongoing, further exploitation of these materials in hybrid systems would lead to better performance characteristics and further widen the scope of applications in consumer and industrial markets (Liu, 2022).

6.3. APPLICATIONS OF HIGH-POWER DEVICES

Power electronic devices' history has significantly impacted industrial application, particularly with high-power devices. A particularly notable category here is the insulated gate bipolar transistor (IGBT) used extensively in numerous adjustable speed drives (ASDs) for induction motors. The products allow improved efficiency and control in high power applications, especially when paired with devices such as pulse width modulation voltage source inverters (PWM-VSI). Where cable lengths of many meters are inevitable, as in heavy industry, concessions like locating the inverter bridge and DC bus capacitors closer to the motor can prove useful. It converts the long cables to DC transmission lines, effectively eliminating high-frequency issues that arise in the inverter-cable-motor interface, but creating new problems with device reliability and performance under harsh environments (Silva et al., 2025). As the demand for greater efficiency and performance quickened, the development of devices with high frequency such as High Electron Mobility Transistors (HEMTs) and laterally diffused MOSFETs (LDMOS) has become increasingly significant. HEMTs in particular possess outstanding capabilities for applications involving high frequency, suitable for telecommunications and RF systems. Their design enhances higher mobility of electrons, which significantly increases efficiency parameters such as transconductance and cut-off frequency. As illustration, composite channel HEMT technology demonstrates a device parameter optimization for a lattice-matched structure that offers performance and energy efficiency in perfect balance. Drain resistance enhancement as well as capacitance effects are brought about by precise tuning of parameters like gate length and barrier thickness to result in outstanding maximum frequency of oscillation suitable for cutting-edge applications (Krishna et al., 2021) (Figure 6.7).

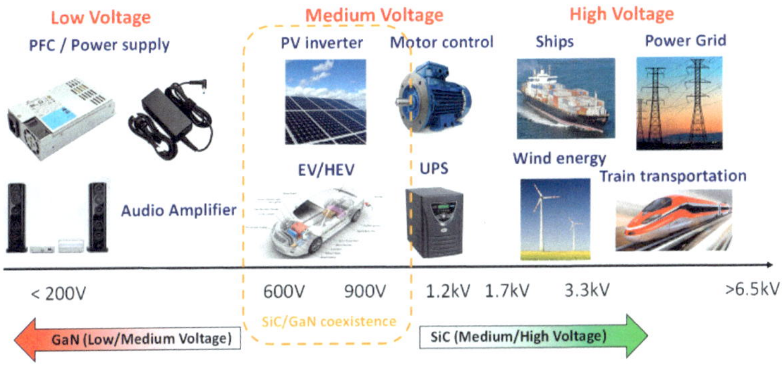

Figure 6.7. *Possible applications of gallium nitride (GaN) power devices as a function of the voltage. The range of application of silicon carbide (SiC) devices is also shown for comparison.*

Source: https://www.mdpi.com/materials/materials-12-01599/article_deploy/html/images/materials-12-01599-g001.png.

Materials integration, such as gallium nitride (GaN) and silicon carbide (SiC), still further supports the area of high-power applications. These wide bandgap semiconductors enable the devices to operate under far greater voltages, temperatures, and frequencies, positioning them as highly capable contenders for power electronics use in renewable energy technologies, electric vehicles, and other high-usage industries. Their ability to withstand more stringent conditions enables improved thermal management strategies, which remain paramount to maintaining device integrity and lifetime. Thus, synergy between modern materials and next-generation cooling technologies not only enhances performance but also lessens operation failures, resulting in more efficient and more reliable high-power applications. Growing synergy between state-of-the-art semiconductor technology and good design practices holds out a promising future for high-power devices in various industries.

6.3.1. Power Electronics in Renewable Energy Systems

The integration of renewable resources into power grids demands state-of-the-art power electronic technologies that can efficiently handle the conversion and storage of energy. High Electron Mobility Transistors (HEMTs), laterally-diffused Metal Oxide Semiconductors (LDMOS), and Insulated Gate Bipolar Transistors (IGBTs) are archetypal components in the development of renewable energy systems.

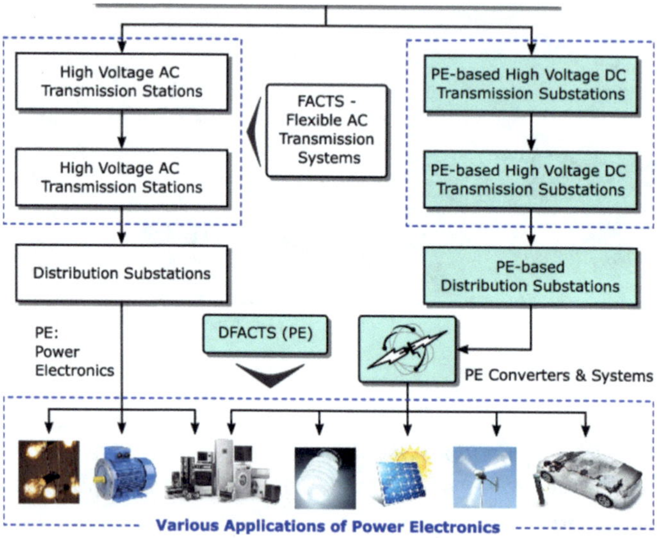

Figure 6.8. *Power electronics in modern power transmission systems and its increasing applications in future energy systems.*

Source: https://www.researchgate.net/publication/369020956/figure/fig3/AS:11431281139172232@1680829843495/Power-electronics-in-modern-power-transmission-systems-and-its-increasing-applications-in.png.

These devices have certain features that qualify them for numerous applications, particularly for high-frequency and high-voltage applications. For instance, HEMTs exhibit low on-resistance and fast switching capability, qualifying them for efficient renewable energy conversion. IGBTs, however, excel where high voltage and high current are required, which are usually found in grid-connected systems. Choice of these devices straight away influences the operation of renewable energy systems in terms of their ability to seamlessly integrate into the power grid and achieve high levels of renewable penetration. As the demand for renewable energy continues to rise, the development of wide-bandgap materials such as Gallium Nitride (GaN) and Silicon Carbide (SiC) has been revolutionary when it comes to power electronics. These materials possess greater electric field strength and thermal conductivity compared to traditional silicon-based devices, enabling efficient and compact power conversion system designs. GaN devices, in particular, can handle greater frequencies, which is beneficial for high-efficiency RF power and energy harvesting applications. In the context of renewable energy systems, the enhanced efficiency provided by GaN and SiC materials translates to reduced energy losses and increased overall system performance (Figure 6.8).

Thus, the use of these materials not only facilitates the integration of renewable sources into the grid but also provides a boost to the technical infrastructure necessary for future energy solutions. Effective thermal management practices are essential to the effective functioning of power electronic components in renewable energy systems. High-power and high-frequency electronic components such as those involving GaN and SiC technology create significant amounts of heat when in operation, and this heat impacts efficiency and reliability if not managed effectively. Sophisticated cooling technologies such as liquid cooling and heat sinks enhance the thermal dissipation and therefore the longevity and performance of such devices. Moreover, integrating thermal management with control techniques allows for the dynamic regulation of operational parameters based on temperature profiles, thereby optimizing energy conversion and storage processes in real-time. Thermal concerns are addressed not as a reactive means but as a proactive solution that assures the reliability of renewable energy systems and, by implication, stability in the context of variable power loads and the intrinsic intermittence of renewable resources.

6.3.2. Telecommunications and Radio Frequency (RF) Applications

Device development of high-frequency and high-power has relied extensively on semiconductor material and technology developments. Amongst the materials, Silicon Carbide (SiC) and Gallium Nitride (GaN) have been the most prominent materials for RF applications due to their superior thermal management as well as wide bandgap characteristics. These characteristics not only improve power electronics efficiency but also facilitate higher breakdown voltages and thermal stability, which can be employed in harsh telecommunications environments.

With the increasing need for high-performance devices in commercial and military devices, the addition of GaN and SiC to RF technology is a promising area for creating and optimizing systems for high-power applications. Notably, the development of High Electron Mobility Transistors (HEMTs), Lateral Double-Diffused MOSFETs (LDMOS), and Insulated Gate Bipolar Transistors (IGBTs) accentuates the critical role of cutting-edge transistor design to telecommunications. These transistors have a key role in managing the increasing power requirements with efficiency and linearity in signal processing. HEMTs, particularly GaN-based ones, possess a high transconductance and are microwave frequency operable, which is needed for applications demanding precise signal amplification. LDMOS devices, however, are rugged and often used in base stations, enabling dramatic improvement in overall system performance. Increased awareness of these technologies informs the broader debate on the future of high-frequency device innovation in telecommunications. The latest advancements in RF photonics have also attracted significant interest as the field approaches the integration of breakthroughs such as optical frequency combs developed with micro-resonators. These micro-combs represent a new paradigm for RF photonics with the potential to provide miniaturized footprints and low power usage but with enhanced functionalities in various applications including telecommunications, spectroscopy, and sensing. The integration of these optical technologies into one will transform traditional RF systems dramatically, facilitating higher efficiencies and allowing for more adaptable telecommunications platforms. However, as the extent of micro-combs unfolds, it is imperative to address current shortcomings and challenges in their real-world application, so they can smoothly complement existing high-power and high-frequency devices for applications in real-life technology, maximizing performance and reliability in telecommunication (Chu et al., 2018).

6.3.3. Industrial Motor Drives and Automation

Power electronics has immensely contributed to the advancement of industrial motor drives through higher efficiency, performance, and application flexibility. Modern designs incorporate devices such as High Electron Mobility Transistors (HEMTs), Lateral Double-diffused Metal-Oxide-Semiconductor (LDMOS) transistors, and Insulated Gate Bipolar Transistors (IGBTs) for facilitating precise control of motor performance. These machines enable variable torque and speed control, thereby providing maximum energy utilization and operational reliability across various industrial processes. Application of high-power and high-frequency devices in motor control is critical to preventing energy wastage and environmental preservation. Above all, the successful functioning of such drives is more or less tied to the functioning of the power devices utilized in enabling smooth and reliable operation, which has become fundamental in the case of emerging industrial automation. The introduction of wide-bandgap materials like Gallium Nitride (GaN) and Silicon Carbide (SiC) has also revolutionized industrial motor drives, particularly for high-power and high-frequency applications. Such materials have better thermal conductivity and higher breakdown voltage, allowing for more efficient and compact designs to work at elevated temperatures. GaN and SiC application has resulted in power converters with much improved performance, making them longer-lasting and more reliable and reducing the physical dimensions of devices. As such, the integration of new materials and innovative electronic design forms a new generation of motor drives with the ability to meet the demands of contemporary industrial automation. As such, research into wide-bandgap technologies is at the forefront of research activities with the aim of further improving the performance of motor drive systems towards seeking sustainable industrial solutions. Thermal management methods are critical to ensuring optimal performance and lifespan of industrial motor drives, particularly through the utilization of high-power apparatus. The increased heat generated due to high-frequency operation can lead to inefficiency and probable malfunction of the components. Effective thermal management systems incorporate advanced cooling methods that include liquid cooling, heat sinks, and thermal interface materials that offer high efficiency in heat dispersion. In brushless DC (BLDC) motors, temperature control is required not only to maintain performance but also to minimize the incidence of issues such as torque ripple, which is possible due to overheating. Emerging trends in thermal management can play a major role in increasing the reliability of motor drives and address the challenge of high-frequency operation. By giving high priority to effective thermal solutions in motor drive systems, industries can make use of improved performance along with the operation efficiency of high-power and high-frequency devices (Darba et al., 2016; Guarnieri et al., 2018).

6.4. THERMAL MANAGEMENT STRATEGIES

The increasing demand for high-power and high-frequency devices necessitates the need for aggressive thermal management solutions in order to ensure reliability and optimal performance. With advancing technology in electronic devices, materials such as Gallium Nitride (GaN) and Silicon Carbide (SiC) have emerged as go-to semiconductors due to their higher thermal conductivity and high operating temperatures. Such materials are of maximum benefit in uses with High Electron Mobility Transistors (HEMTs) and Lateral Diffused Metal-Oxide-Semiconductor (LDMOS) technologies, in which cooling the heat dissipation is an important consideration. Conventional cooling schemes, like using heat sinks or forced convection, can be insufficient through the use of the small form factor of many contemporary devices. Therefore, new fasteners, including the inclusion of phase change materials and advanced thermal interface materials, are being designed to help counter overheating and foster device longevity and consistent performance under rigorous conditions (Albisinni et al., 2015). As a factor of thermal management, predictive software embedded in the system are becoming increasingly important. Using machine learning to forecast thermal performance in high-power devices can enable anticipatory modulation and optimize thermal efficiency. Through intelligence that is data-driven, dynamic voltage and frequency scaling systems can better control energy usage while, concurrently, decreasing thermal output. The ability to forecast thermal events converts conventional reactive cooling methods into proactive thermal solutions, allowing for more control of thermal profiles. These innovations not only improve the efficiency of devices but also increase battery lifespan in devices like smartphones and drones, where thermal stability is paramount. This new predictive approach is a step towards smarter thermal management approaches, with a potential solution to fight the issues of ever-growing demands of high-frequency operation (Ababei et al., 2019) (Figure 6.9).

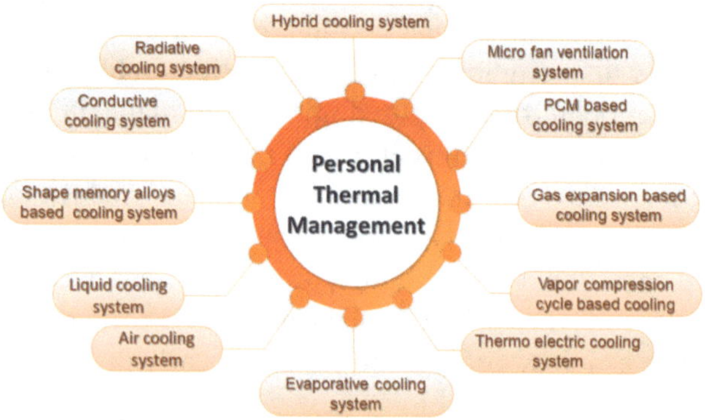

Figure 6.9. *Personal thermal management.*

Source: https://ars.els-cdn.com/content/image/1-s2.0-S0735193321006321-gr2.jpg.

Applications like MRI-guided focused ultrasound surgery show how advanced thermal management can be applied from traditional electronic systems to medical equipment. This approach utilizes the precision of ultrasound for non-invasive treatment while ensuring that tissues are not harmed by excessive heat. The interaction between imaging and therapeutic modalities focuses on controlled thermal environments, echoing the importance of effective thermal management in high-power electronics. From medicine to power electronics, methods employed to control heat transfer are universal; both aim to optimize device performance while preventing damage or failure. With more clinical applications integrating cutting-edge imaging technologies and materials science, inter-disciplinary thermal management technology innovation opportunities will expand even more, accelerating the development of durable, high-performance systems (Albisinni et al., 2015).

6.4.1. Importance of Thermal Management in Device Performance

Increasing growth in high-power and high-frequency devices necessitates an inquiry into advanced thermal management methods to guarantee optimum performance. Adequate thermal management is significant owing to the introduction of devices like Gallium Nitride (GaN) and Silicon Carbide (SiC) for radio frequency (RF) and power electronic applications.

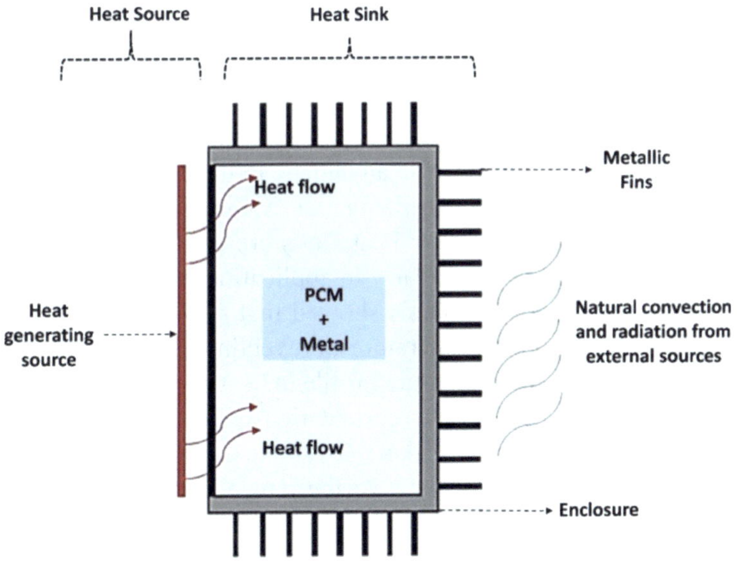

Figure 6.10. *A comprehensive evaluation on thermal management of electronic devices* |.

Source:https://media.springernature.com/lw1200/springer-static/image/art%3A10.1186%2Fs44147-023-00309-2/MediaObjects/44147_2023_309_Fig1_HTML.png.

High electron mobility transistors (HEMTs), lateral-diffused metal-oxide-semiconductor (LDMOS) transistors, and insulated gate bipolar transistors

(IGBTs) are increasingly put under their thermal limits. With these devices handling increased frequencies and power levels, excessive heat generation can erode their reliability and overall functioning. Thus, embracing innovative thermal management techniques is necessary, particularly with challenges introduced by close packaging and miniaturizing electronic components. One of the most compelling thermal management techniques is incorporating advanced cooling systems that effectively dissipate heat generated by high-powered devices (Figure 6.10).

For example, the incorporation of microchannel cooling designs facilitates improved thermal dissipation due to their lower thermal resistance in comparison to standard cooling schemes. With the help of direct bonding technologies, such as nanosilver sintering coupled with microchannel architectures, researchers have significantly improved the optimization of heat transfer. The most recent research takes note of the potential of integrated-cooling technologies, where SiC power modules showed ability to control the heat flux levels over 1000 W/cm^2 without increasing the thermal resistance heavily (Tang et al., 2025). This technology not only provides more stable devices but also the potential for compact system designs that enhance overall efficiency for high-power applications. Additionally, advances over the past several years in vapor chamber technology have shown considerable advantage in the thermal management of high-power electronic devices. Utilizing the evaporation-condensation cycle of a working fluid, vapor chamber heat sinks are capable of providing very good cooling performance, particularly in applications such as electric vehicle (EV) chargers. Comparative analysis showed that vapor chamber geometries significantly reduce junction temperatures, preventing hotspot overheating and achieving a more uniform temperature profile in heat sinks (Awais et al., 2025). These benefits enable heat sink size reorientation and overall cooling efficiency enhancement, enabling device makers to create smaller, lighter, and more efficient thermal management systems. As the demand for high-power electronic solutions keeps rising, these sophisticated thermal management techniques will play an even more important role in maximizing the performance and reliability of high-power and high-frequency devices.

6.4.2. Cooling Techniques for High-Power Devices

Sophisticated cooling techniques are essential for effective thermal management of high-power devices such as High-Electron-Mobility Transistors (HEMTs), Lateral Double-Diffused Metal-Oxide-Semiconductor (LDMOS) transistors, and Insulated Gate Bipolar Transistors (IGBTs). The high power densities associated with these devices generate heat that leads to performance and life degradation. Standard cooling measures, such as passive heat sinks, are largely insufficient to meet the cooling requirements of these high-power devices. More sophisticated solutions are therefore beginning to step in and include active cooling systems using liquid coolants, phase change materials, and thermoelectric coolers. These methods effectively enhance thermal dissipation for the purpose of device reliability and performance. The adoption of these advanced technologies becomes more critical as semiconductor materials like Gallium Nitride (GaN) and Silicon Carbide (SiC) increasingly dominate RF and power electronics, where thermal management is a critical factor. The integration of new materials with higher thermal conductivity is crucial to optimize cooling technology performance in high-power devices. GaN and SiC, being high-performance at high temperature conditions, enable designs that require less cooling effort than conventional silicon-based designs. This feature, not only enables improved thermal management strategies but also allows for higher operating frequencies and power levels. Technologies such as liquid cooling, which passes the coolant directly over the chips, and microchannel cooling designs bring the coolant and device surface into close contact, enhancing thermal performance significantly (Figure 6.11).

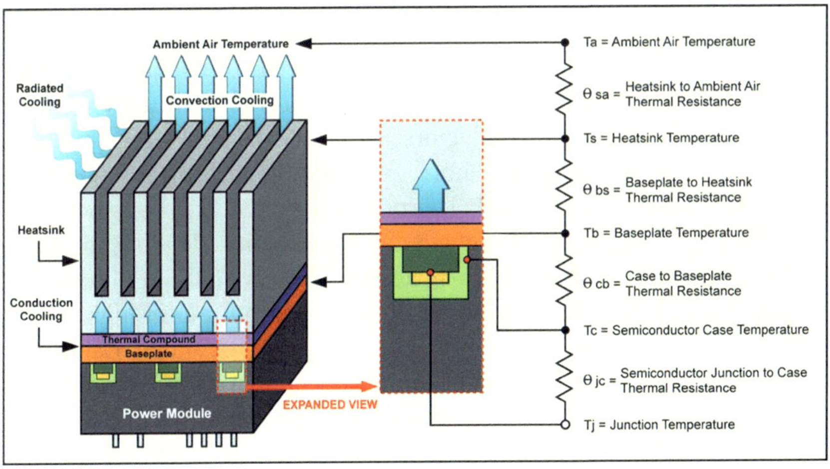

Figure 6.11. *Techniques for cooling power and other electronic devices -.*

Source: https://res.cloudinary.com/tbmg/c_scale,w_400,f_auto,q_auto/v1549183885/sites/tb/articles/sup/ET/2008/features/15130-400-fig2.png.

This combination of state-of-the-art cooling with new materials addresses the unique challenges posed in controlling heat in high-frequency applications. Under such environments, advanced techniques of cooling need to be understood and implemented so that it is possible to use high-power and high-frequency devices effectively in extreme conditions. Energy-efficient thermal management techniques become vital for the functioning and sustainability of massive data centers, where excessive power consumption is a prevalent issue. As data center sizes grow, so does server heat output, and effective cooling solutions become necessary to supply ideal working conditions. New cooling-related technology innovations, for example, the use of dynamic thermal management systems, have emerged to respond to escalated power dissipation challenges. Examples include techniques such as data-driven dynamic voltage and frequency scaling (DVFS) where power consumption is scaled according to real-time demand, thus avoiding unnecessary heat generation. Moreover, the classification of these architectural approaches facilitates exploring different means of energy efficiency in data centers with emphasis on the pivotal role of advanced cooling solutions for ensuring the performance and lifetime of high-power devices (Mittal et al., 2014). The ongoing advancement in cooling technologies will ultimately lead to more reliable and efficient high-power applications across industries.

6.4.3. Materials and Technologies for Heat Dissipation

The advent of advanced high-power and high-frequency devices has necessitated the redressal of materials and technologies devoted to efficient heat dissipation. As the likes of High Electron Mobility Transistors (HEMTs), Lateral Double-Diffused Metal-Oxide-Semiconductors (LDMOS), and Insulated Gate Bipolar Transistors (IGBTs) gain more visibility, the task of efficient thermal management cannot be overstated. Thermal energy generated by such sources can significantly impair performance and reliability, leading to thermal runaway or premature failure. New materials such as Gallium Nitride (GaN) and Silicon Carbide (SiC) are therefore under research for their higher thermal conductivity and high breakdown voltage capability. These materials not only facilitate better thermal management but also accommodate higher efficiency in power electronics, and hence they are the best candidates to be used in aerospace and automotive sectors where reliability under extreme conditions is of utmost importance. Transistor structures and packaging materials also play a significant part in thermal dissipation. Rising interconnects, such as carbon nanotube interconnects, are a revolution in overcoming challenges pertaining to traditional copper interconnects, which typically withstand greater resistance and heat dissipation. Literature today points towards carbon nanotubes in overcoming traditional materials in terms of conductivity as well as thermal dissipation, paving the way for better use of on-chip interconnects without impacting device performance. Amoroso et al. (2017) presents the detailed considerations, including the growth processes and theoretical modeling needed for the integration of such materials into systems. With the use of such high-level materials, the efficiency of heat dissipation can be improved considerably, thus improving the reliability and life of high-frequency devices operating in severe conditions. For streamlined energy consumption, integrated systems are now resorting to advanced voltage tuning methods that actively manage power dissipation. A novel technique employing multi-functional Ring Oscillators (ROs) shows dynamic voltage tuning enhancements, indicating a correlation between RO observed frequency and the high-power device minimum safe operating voltage. Such a technique assures an adaptive process variation compensation, thereby optimal energy utilization without compromising the performance. When used on chips such as the XMOS XS1-U8A-64 embedded microprocessors, the strategy achieved substantial power reductions of as much as 25% while maintaining real-time performance levels. The strategy emphasizes the essential necessity of integrating advanced thermal management with voltage control to ensure that high-power and high-frequency devices operate effectively and stably in a wide variety of applications, which ultimately prolongs their service life in mission-critical applications (Eder et al., 2015).

6.5. CHALLENGES IN HIGH-POWER AND HIGH-FREQUENCY DEVICE DESIGN

High-power and high-frequency device design is subject to numerous technical challenges, particularly in material selection and thermal management. Wide bandgap semiconductors such as Gallium Nitride (GaN) and Silicon Carbide (SiC) are gaining attention among engineers as these semiconductors have more favorable performance metrics for applications requiring high efficiency and thermal management. Interestingly, devices containing High Electron Mobility Transistors (HEMTs), Lateral Double-diffused Metal-Oxide-Semiconductor (LDMOS) devices, and Insulated Gate Bipolar Transistors (IGBTs) tend to need added focus on thermal management strategies. As operating frequencies rise, normal cooling schemes become insufficient, ultimately resulting in substantial performance degradation or device failure. The optimal control of the thermal dynamics is therefore of paramount significance, not just dictating the life expectancy of the devices but also their functioning integrity in high-stress applications (Fu et al., 2017). Aside from thermal management, the intricate design requirements to support fault conditions present substantial challenges for high-power and high-frequency device applications. For instance, dc microgrid topologies exhibit unusual fault behavior that makes protection systems challenging, and this takes ingenuity to achieve consistent load operation under such conditions. Non-unit protection schemes have been researched, but they prove inadequate when applied in more complex network topologies. Thus, the designers are forced to put keen design considerations into action, particularly while integrating existing differential schemes, more involved but addressing amply available resources and planning to address intricate fault behaviors (Burt et al., 2012) (Figure 6.12).

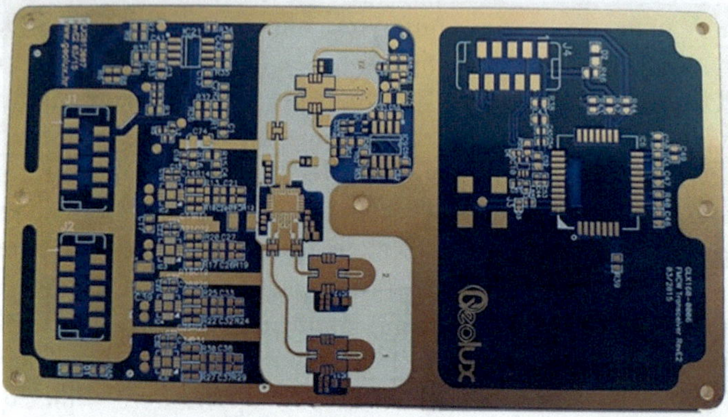

Figure 6.12. *The design of high-frequency devices.*

Source:https://www.rigaopcb.com/uploads/img1/20231207/657194f55d6bb.png.

This is not only emphasizing the crucial requirement for robust protection schemes but also mounting pressure on design teams to shake possible threats through work stages. Another key issue in high-power and high-frequency device design is achieving a balance between device performance and system efficiency and manufacturing cost. With the market trend toward miniaturization and increased functionality, there is an imperative need to enhance device capability without incurring a proportional rise in cost. Consequently, deviations from typical manufacturing processes to enable the ADAS or IoT markets can come at either sacrifices of efficiency or material quality. Further, achieving scalability of manufacturing remains a significant factor, particularly for technologies using specialty materials like GaN or SiC, where the supply chain may still be in the process of creating broadly applicable manufacturing techniques. Thus, where design innovation and economic viability remain an ongoing discussion among engineers as they strive to meet new emerging market needs under stringent performance constraints.

6.5.1. Reliability and Lifespan Issues

As a result of the incorporation of high-power and high-frequency devices in today's electronic applications, these new developments come with some level of reliability and lifespan problems. Specifically, High Electron Mobility Transistors (HEMTs), Lateral Diffused Metal-Oxide-Semiconductor transistors

(LDMOS), and Insulated Gate Bipolar Transistors (IGBTs) are inevitable in high-performance applications. However, their operations are normally hindered by extreme thermal stress and power density, leading to premature failure if not constrained. As these components support enormous electrical loads, dissipation of heat is a matter of extreme concern. Application of heat management methods, such as augmented heat sinks or active cooling systems, becomes essential to enhance the reliability of these devices. Failure to maintain thermal-related problems can result in decreased lifespan, affecting not only individual devices but can also destroy entire systems that are based on their performance. Furthermore, the development of wide-bandgap materials like Gallium Nitride (GaN) and Silicon Carbide (SiC) also presents opportunity as well as reliability challenges for high-power and high-frequency applications. Even though the GaN and SiC devices possess superior thermal efficiency and performance to conventional silicon-based devices, their relatively recent introduction has raised concerns regarding long-term reliability under aggressive application conditions. Power cycling, thermal fatigue, and electromagnetic compatibility effects are vital variables for lifespan in applications where stability and performance at high levels over long durations are necessary. Their resolution is necessary; otherwise, manufacturers would need to endure large operation downtimes and increased maintenance costs. Improved understanding of failure modes in such advanced materials is necessary to offer better designs and predict longevity accurately. Finally, the significance of power quality and EMC issues cannot be overstated in evaluating the reliability of high-power and high-frequency devices. Power Quality (PQ) disturbances like voltage transients and harmonics can adversely affect the operation and lifespan of sensitive components. A Power Quality session highlights critical details such as voltage continuity and the effect of electric disturbances on device operation. Sustained focus on safety and reliability guidelines continues to ensure that an orderly approach to PQ and EMC management is imperative to device integrity and operational excellence. Developing strategies that value power quality, in addition to device performance, enables manufacturers to significantly minimize risks from electric outages and extend the lifespan of such critical technologies. Mitigation of thermal and electromagnetic problems will ultimately enhance the lifetime and efficiency of high-performance electronic systems.

6.5.2. Cost Considerations and Manufacturing Challenges

The fast development of high-frequency and high-power devices necessitates strict attention to cost factors and manufacturing issues of such technologies. With producers seeking to use such materials as Gallium Nitride (GaN) and Silicon Carbide (SiC) because of their better RF and power electronic performance, they are confronted with significant economic factors. Supplying and processing these materials are major challenges that can lead to a higher cost of their production mainly due to sophisticated fabrication methods needed to realize their unique properties. Joined with the great degree of capital required to finance specialty equipment, manufacturers must balance cost and innovation, so efficiency becomes a paramount concern in production. Lastly, in a competitive marketplace, seeking profitability can facilitate a focus on cost-reduction measures that unwittingly compromise performance or reliability. High-power and high-frequency device manufacturing also creates a host of technical issues. The fragility of the design and production procedures of technologies such as High Electron Mobility Transistors (HEMTs) and Lateral Double-Diffused Metal-Oxide-Semiconductor (LDMOS) structures demand attention to precision and detail. Energy-efficient production techniques are becoming more vital in light of these issues. For instance, application of energy management devices directed by model predictive control (MPC) strategies can optimize the operation of machinery, reducing energy consumption and cost significantly. These innovations are significant as they lead to a greater comprehension of energy models and allow for the simulation of various manufacturing behavior, thus making things easy and increasing overall productivity. These innovations demonstrate how manufacturing problem solutions can cross with cost management actions to ensure more sustainable manufacturing techniques for high-power electronics. Environmental considerations make manufacturing problems and related costs more complex when working with high-power and high-frequency devices. The burden of reducing waste and power consumption in the midst of industry outrage places additional weights on manufacturers. The need to adopt greener practices cannot be overstated, as seen in the complexities of the life cycle analysis (LCA) of semiconductor technology. Lacking in putting much effort into the environmental footprint of manufacturing can lead to penalties way bigger than brief money loss. For example, strategic planning of resource utilization, like the use of energy and water, can inform more sustainable practices and innovation at the level of manufacturing. The considerations are most important when products experience positive impacts through intelligent design and operating efficiencies. Having an eco-friendly approach in dealing with cost and production concerns is what producers must have in place to thrive within the high-power and high-frequency device manufacturing competitive landscape (Ayerbe et al., 2019; Cheriet et al., 2014).

6.5.3. Integration with Existing Technologies

Emergence of high-power and high-frequency devices has required intense emphasis towards hassle-free integration with existing technologies. For this purpose, high-electron mobility transistors (HEMTs) have emerged as a top choice due to their better performance capabilities, particularly when developed from gallium nitride (GaN). The adaptability of HEMTs enables them to integrate into conventional electronic structures, such as radio-frequency (RF) amplifiers and power conversion equipment. Specifically, the use of AlGaN/GaN HEMTs that take advantage of the unique features of two-dimensional electron gases has been shown to result in a drastic enhancement of the operating efficiency of these devices. The proposed AlGaN/AlN/GaN HEMT structure encourages device design innovations facilitating improved heat dissipation and device reliability in high power and frequency operation applications when employed in systems requiring such operations, emphasizing their applicability in contemporary technological foundations (Hidayat et al., 2024). For silicon-germanium (SiGe) semiconductors, compatibility with prevailing silicon manufacturing processes remains a major advantage towards integration into prevalent technologies.

The enhanced electronic performance of SiGe, such as improved electron mobility and improved high-frequency characteristics, make the semiconductors a perfect selection for the production of future devices. By employing SiGe in heterojunction bipolar transistors (HBTs) and radio-frequency integrated circuits (RFICs), engineers can be creative without compromising the reliability that accompanies traditional silicon devices. By incorporating the new technology in established fabrication facilities, the latter may implement the new technology without incurring significant disruption. In addition, the tunability of germanium composition to achieve fine control over thermal and bandgap characteristics allows for the possibility of creating devices tailor-made for specific applications using current fabrication methods (de Azevedo et al., 2024). Thermal management methods are also important to the successful integration of high-power and high-frequency devices into existing technologies. With devices such as IGBTs and LDMOS transistors being pushed to their maximum in high-standard applications, more sophisticated thermal solutions are needed to combat performance degradation. Techniques such as new heat sinking, phase change materials, and active cooling systems act as powerful methods of combating heat dissipation during use. The development of thermally conducting materials, coupled with designs that consider thermal dissipation in the system design, can help improve reliability and endurance of systems. When these thermal management techniques are integrated into HEMTs, SiGe, and other high-performance devices, device manufacturers can optimize their performance so that high-frequency and high-power operation becomes possible in the long term in a broad spectrum of technology applications.

6.6. FUTURE TRENDS IN HIGH-POWER AND HIGH-FREQUENCY DEVICES

Emergence of semiconductor technologies is all set to change the paradigm of high-power and high-frequency devices over the next few years. The utilization of Gallium Nitride (GaN) and Silicon Carbide (SiC) in power electronics and Radio Frequency (RF) applications is a milestone that increases efficiency and performance across various applications. GaN and SiC devices are superior to traditional silicon semiconductors with enhanced thermal conductivity, higher breakdown voltages, and faster switching speeds. As a result, they are increasingly employed in High Electron Mobility Transistors (HEMTs) and laterally-diffused Metal-Oxide-Semiconductor (LDMOS) devices that enable enhanced telecommunications and renewable energy systems. This is likely to drive future innovation as there is growing demand for high-performance devices at harsh conditions. Development and research on these technologies are continuing, and the future looks bright for SiC and GaN in high-power applications. In addition to semiconductor technology advancement, novel thermal management methods are essential to the sustainable operation of high-frequency devices. With increasing power density and higher operational frequencies, heat management is becoming a growing issue. Emergent cooling technologies, such as future phase-change materials and microchannel heat sinks, are gathering momentum in solving this challenge effectively. These systems are capable of optimizing the feedback loop among heat creation and dissipation, ensuring maximum device reliability and lifetime. Furthermore, designing high-power converters with thermal management solutions, such as in inverters, minimizes the dangers associated with overheating and offers system robustness (Figure 6.13).

Figure 6.13. *The future demand of high power devices.*

Source: https://eepower.com/uploads/articles/2024-power-electronics-a-look-at-the-future-fig0.jpg.

Effective heat dissipation not only enables better performance but also miniaturization and better efficiency of device arrangements, particularly for applications that demand high operating stability. The mounting need for energy harvesting technologies is emerging as a trend-enabling factor in high-power and high-frequency devices, largely propelled by sustainability efforts. With increasing awareness regarding saving energy, piezoelectric energy harvesting developments are being considered to tap the ambient energy provided by various sources such as vehicular transportation and fluid flows. The technology can operate effectively in the mode of high frequencies so that more energy can be harvested compared to low frequencies. Recent advancements in the piezoelectric materials and design of piezoelectric energy harvesters (PEHs) indicate a trend towards the maximization of these devices for higher currents and efficiency. With new energy-conversion mechanisms, like high-speed rotation and frequency up-conversion, being found by research, the application of novel piezoelectric materials is bound to be a dominant factor. Continuous research in this field is essential to integrating energy harvesting solutions into high-power and high-frequency applications so that future development is sustainable.

6.6.1. Advancements in semiconductor Materials

Recent technological advancements have realistically transformed the realm of high-power and high-frequency devices, mostly due to the availability of wide-bandgap (WBG) semiconductors such as gallium nitride (GaN) and silicon carbide (SiC). They possess improved electrical characteristics compared to traditional silicon-based materials, enabling the devices to operate under higher voltages, frequencies, and temperatures. This translates to higher power density and efficiency, and they emerge as top candidates for applications like radio frequency (RF) amplifiers in 5G technology and electric vehicle (EV) power converters. The ability of these materials to effectively dissipate heat enhances the reliability and performance of the device, opening up new applications and improving the performance of existing technology. The use of these developments represents a revolutionary shift in power electronics design paradigms, highlighting the pivotal role semiconductor materials play in modern electronic systems (Liu, 2025). Additionally, the development of high electron mobility transistors (HEMTs), laterally-diffused metal-oxide semiconductor (LDMOS) transistors, and insulated gate bipolar transistors (IGBTs) has also helped improve semiconductor technology. In particular, HEMTs using GaN are noteworthy for their extremely high efficiency in high-frequency applications and enabling greater miniaturization and integration of RF systems. By minimizing power loss and improving linearity, such devices are particularly useful for application in accurate signal amplification and power distribution. LDMOS devices, also noted for thermal stability, serve applications in high-voltage applications, while IGBTs are optimally suited for dealing with higher amounts of power in industrial applications. All these technologies are symptomatic of ongoing progress in semiconductor materials that have made different aspects of performance better (Srivastava et al., 2025). With the growing semiconductor frontiers, thermal management methods are now a necessity in maintaining functionality and life of high-power semiconductor devices. Effective heat dissipation is crucial, particularly in WBG applications where high operating temperatures are the norm. Techniques such as new cooling systems, thermal interface materials, and heat spreaders are increasingly being implemented to provide optimum operating temperatures. Synergy between next-generation semiconductor materials and next-generation thermal management systems ensures the reliability and efficiency of future electronic devices. While work continues to optimize these parameters, more emphasis is now placed on creating materials that not only offer better electrical performance but also thermal stability. This comprehensive approach positions the semiconductor industry in a good position to meet the high-power and high-frequency applications' strict demands optimally by emphasizing the necessity of integrating material advancements as well as thermal technologies (Liu, 2025).

6.6.2. Innovations in Device Architecture

The rapid advancement of device architecture has been pivotal in fulfilling the increasing demands for efficiency and performance in high-power and high-frequency applications. The most significant developments of the high electron mobility transistors (HEMTs) and laterally diffused metal-oxide-semiconductor (LDMOS) devices have enormously enhanced the systems' performance to work at microwave frequencies. The improvement is due to the use of wide bandgap materials like gallium nitride (GaN) and silicon carbide (SiC), which exhibit considerably superior thermal conductivity and electric field breakdown characteristics. When the devices are employed at higher frequencies, efficient heat management becomes an imperative requirement. This is where innovations in new thermal techniques, such as enhanced heat sinks and inbuilt cooling mechanisms, ensure devices are able to continue functioning at their optimal capacity without heating up. Consequently, these technologies not only increase the reliability of devices but also increase the lifespan of high-power apps. In the area of radio frequency (RF) technologies, the architectural advances in the development of filters have transformed the ability to treat signals. SDRs' focus today shows the importance of flexibility and adaptability in contemporary communication systems. Advancements such as the application of extreme throw MEMS actuators have enabled the development of tunable RF filters with the potential to operate optimally throughout a wide frequency, specifically between 4 and 15 GHz (Beacken et al., 2018). The challenge remains that of detecting weak signals against multiple strong transmitters. With these new MEMS-based filter designs, scientists are constructing devices that support dynamic manipulation of the RF environment, enabling real-time frequency agility and increased signal integrity. This enhances the capabilities of existing devices while preparing for next-generation high-frequency applications. Wireless network architecture is experiencing a renaissance, as well, with the requirement for programmability across high-frequency domains. Traditional network structures based largely on fixed settings are being reshaped to encompass software-defined networking (SDN) and cognitive radio structures. These technologies allow for a highly malleable communications framework capable of addressing diverse demands on many platforms. Programmable wireless networks center on the concept of providing seamless operation through intelligent management and configuration capabilities (Ahad et al., 2014). Such adaptability enables the merging of disjointed technologies into a cohesive system, eventually producing more efficient, higher-performing high-power wireless communications. As researchers venture into these technologies, the vision for scalable, robust, and dynamically reconfigurable network structures becomes a clearer reality, opening doors to future solutions for the high-frequency device world.

6.6.3. Emerging Applications and Markets

The increasing demand for high-speed and high-power-efficient power electronics is driving the development of next-generation semiconductor devices such as Gallium Nitride (GaN) and Silicon Carbide (SiC). They are particularly suitable materials for the design of high-power and high-frequency devices due to their superior thermal conductivity, enhanced breakdown voltages, and higher electron mobility than the traditional silicon-based devices. As a result, GaN and SiC are destined to be at the forefront of future markets in applications requiring advanced RF (radio frequency) applications and power electronics. Applications involving high electron mobility transistors (HEMTs), laterally-diffused metal-oxide-semiconductor (LDMOS) devices, and insulated-gate bipolar transistors (IGBTs) are revolutionizing circuit designs, promoting efficiency and reliability in applications ranging from telecommunications to electric vehicles. The wave created by these technologies represents a paradigm shift in power electronics integration in modern systems to address the needs of various industries in the pursuit of enhanced performance and reduced energy usage. Advances in the performance characteristics of GaN and SiC devices are attracting significant investments from key industrial players, as evidenced in the emerging mobile networking technology market. As companies become aware of the need to adapt in response to fast-evolving consumer demands, the challenge is how to integrate these technological innovations into market-driven business strategies. High-frequency devices offer a robust solution to applications requiring strong and fast data delivery, such as 5G wireless communications and advanced IoT systems. The intricacies of managing market volatility in such high-risk environments necessitate the use of a strategic business planning model, one that can help companies effectively assess the viability of wireless networking use. Employing this tailored model, businesses can better be able to manage the challenges that come with introducing new technologies into the marketplace, offering a smoother ride from research and development through profitable market offerings (Coster et al., 2009). In the future, the integrated strategy of thermal management techniques will be the most important factor in attaining high performance and reliability of high-power devices made from advanced materials. In the context of increasing operating frequencies and power, special consideration to heat dissipation is demanded, affecting the reliability and efficiency of the device. New cooling techniques and packaging strategies are under research to counter such thermal issues, thereby enhancing the operational potential of GaN and SiC devices. The convergence of these technologies offers new opportunities for application in renewable energy, electric vehicles, and smart grid technology, which effectively contribute to supporting the goals of sustainable development. Moreover, future research and collaboration between academic and industrial partners are targeted at maximizing these thermodynamic models and material properties, which ultimately makes it a robust platform to support the development of high-power and high-frequency devices in the growth markets (Amano et al., 2018).

SUMMARY

The investigations into high-power and high-frequency devices have revealed unprecedented innovation in semiconductor materials and device structure. As the applications in telecommunications, renewable energy, and automotive electronics grow, the demand for compact, efficient, and reliable power devices has never been more imperative. HEMTs, LDMOS, and IGBTs are some of the technologies that have emerged as the frontrunners in delivering the needed performance specifications. Materials selection has also evolved with GaN and SiC becoming more preferred based on their improved thermal and electrical properties. This development reflects a shift towards an elemental change in the landscape of power electronics, where capability and efficiency during operation increasingly hinge upon the device structures and materials used. Thus, the need for sustained innovation in such areas will continue to serve as the backbone of research and development of high-power applications. Apart from this, the application of advanced thermal management has now become vital in the delivery of the maximum performance of such high-power devices. While there is higher operation frequency and power, heat generation also increases, which can lead to reduced efficiency and life of the device. Innovative cooling techniques, such as liquid cooling systems and novel heat sinks, play a highly critical role in the thermal performance control. Dynamic thermal cycling management methods are also gaining steam to enable devices to more effectively handle changing thermal conditions. This move towards improved thermal management will ultimately make high-frequency devices efficient and reliable, which are the most critical in high-performance applications. Research that focuses on these advancements will enable high-power and high-frequency devices to meet the strict demands of future technology applications. By way of illustration, the development of high-power and high-frequency devices is marked by a convergence of device engineering, material science, and thermal optimization. Projects such as the development of a single-phase DC-AC converter for use in wind generator systems illustrate the practical significance of these developments and the ways in which mathematical models and artificial intelligence are being utilized to fine-tune device control and output stability (Buswig et al., 2011). The results of simulation as well as experimental validation validate the applicability of these innovations in real-world uses. Additionally, since industries continue to leverage GaN and SiC technologies with their improved performance, the broader implications for energy efficiency and sustainability emerge. Thus, the ongoing research in high-power and high-frequency devices is not only essential to technological progress but also to addressing actual global concerns such as energy efficiency and environmental sustainability (Green et al., 2012).

The development and utilization of high-power and high-frequency devices have revolutionized the area of RF as well as power electronics. Among these innovations, the LDMOS transistors and HEMTs are most prominent based on power efficiency and operating frequency performance. GaN and SiC materials are currently strong contenders for these applications with improved thermal ability and the capability to enable devices to operate at higher voltage and temperature levels than silicon-based devices. As industries make the move towards progressively smaller and more efficient designs, these semiconductor materials make the basic building block on which to build robust, high-frequency signal switching and amplification possible. Their presence in next-generation devices highlights the imperative for future breakthroughs in materials science to support even higher efficiencies and operating performance.

Moreover, thermal management of high-power devices cannot be discounted as playing a pivotal role in ensuring device reliability and performance. High power-level high-frequency operations generate a great deal of heat, which, without adequate dissipation, causes the devices to fail catastrophically. The use of advanced thermal management schemes including the application of phase change materials and new heat sink designs is necessary in order to extend device life and ensure operational stability. Scientists are increasingly interested in enhancing thermal conductivity with new materials and engineered structures, which can also assist in heat spreading and minimizing junction temperatures. Furthermore, an understanding of thermal dynamics of such devices is crucial to develop more durable microgrid structures, particularly for high-risk applications, such as renewable energy systems, where reliability is of paramount importance. The recent research in high-power and high-frequency devices also aims at the emerging role of security in the physical layer, especially in wireless communication systems. Physical layer security exploits the natural randomness of the transmission channels for data security, which is an emerging direction in the light of the growing challenge due to the advance in 5G technologies. Technologies within physical layer security, combined with vast multiple-input multiple-output systems and millimeter-wave communication, fulfill the necessity of secure data transfer in high-power environments. As industries increasingly reliant on networked systems exist, the need for robust security measures that are complementary to high-frequency, high-power applications cannot be overemphasized. To sum up, the convergence of new semiconductor materials, effective thermal management methods, and enhanced security functionalities embodies the different modes of advances in high-frequency and high-power devices.

REFERENCES

1. Ababei, C., & Moghaddam, M. G. (2019). A survey of prediction and classification techniques in multicore processor systems. https://core.ac.uk/download/322825320.pdf (accessed on 27th September 2025).
2. Adekunle, O. I. (2025). Review of ultra-wide bandgap (UWBG) GaN based HEMTs: Opportunities for high efficiency power conversion in renewable energy and electric vehicle applications. *International Journal of Multidisciplinary Research and Growth Evaluation*. https://www.semanticscholar.org/paper/19acbf865ab415c63b8b3a6fc221955d87f90df1 (accessed on 27th September 2025).
3. Ahad, N., Ahmed, N., & Qadir, J. (2014). Building programmable wireless networks: An architectural survey. http://arxiv.org/abs/1310.0251 (accessed on 27th September 2025).
4. Albisinni, U., Battista, G., Bazzocchi, A., Catalano, C., et al. (2015). MRI-guided focused ultrasound surgery in musculoskeletal diseases: The hot topics. https://core.ac.uk/download/98341478.pdf (accessed on 27th September 2025).
5. Amano, H., Baines, Y., Borga, M., Bouchet, T., et al. (2018). The 2018 GaN power electronics roadmap. https://core.ac.uk/download/154431388.pdf (accessed on 27th September 2025).
6. Amoroso, S., Asenov, A., Brown, A., Dhavamani, P., et al. (2017). A survey of carbon nanotube interconnects for energy efficient integrated circuits. https://core.ac.uk/download/82958608.pdf (accessed on 27th September 2025).
7. Awais, M., Amin, S., Brito, F. P., Afonso, J., Costa, N., & Monteiro, V. (2025). Enhanced and uniform thermal management system based on an integrated vapor chamber and air-cooled heat sink for high-power EV chargers. *2025 IEEE 19th International Conference on Compatibility, Power Electronics and Power Engineering (CPE-POWERENG)*. https://www.semanticscholar.org/paper/287866eca989f3871ddf3a9c81f0b8bafe11f474 (accessed on 27th September 2025).
8. Barraj, I., Mestiri, H., & Masmoudi, M. (2024). Overview of memristor-based design for analog applications. *Micromachines, 15*. https://www.semanticscholar.org/paper/9428170c27f71a60b0afec26b30ce86f010efde2 (accessed on 27th September 2025).
9. Beacken, M., Bishop, D. J., Chang, J., Holyoak, M., et al. (2018). High performance, continuously tunable microwave filters using MEMS devices with very large, controlled, out-of-plane actuation. http://arxiv.org/abs/1804.08658 (accessed on 27th September 2025).

10. Bermeo Ayerbe, M. Á., & Ocampo-Martínez, C. (2019). Energy efficiency improvement through MPC-based peripherals management for an industrial process test-bench. https://core.ac.uk/download/294829585.pdf (accessed on 27th September 2025).

11. Bian, D., & Luan, S. (2024). Design and research of high voltage *β-*Ga_2O_3*/4H-SiC* heterojunction LDMOS. *Engineering Research Express, 6.* https://www.semanticscholar.org/paper/9d543048c62c5d81fb9e0f5b9b2496a9ffcac80b (accessed on 27th September 2025).

12. Burt, G., Crolla, P., Fletcher, S., Galloway, S., et al. (2012). Optimizing the roles of unit and non-unit protection methods within DC microgrids. https://core.ac.uk/download/9040571.pdf (accessed on 27th September 2025).

13. Buswig, Y., & Mohammed, Y. (2011). Development of a DC-AC power conditioner for wind generator by using neural network. https://core.ac.uk/download/12007862.pdf (accessed on 27th September 2025).

14. Caire, G., Gao, X., Khisti, A., Wong, K., et al. (2018). A survey of physical layer security techniques for 5G wireless networks and challenges ahead. https://core.ac.uk/download/158170707.pdf (accessed on 27th September 2025).

15. Charalambous, M. C., Mavromoustakis, C. X., & Yassein, M. B. (2012). A resource intensive traffic-aware scheme for cluster-based energy conservation in wireless devices. http://arxiv.org/abs/1209.3721 (accessed on 27th September 2025).

16. Cheriet, M., Dandres, T., Lemieux, Y., Moghaddam, M., et al. (2014). Challenges and complexities in application of LCA approaches in the case of ICT for a sustainable future. https://core.ac.uk/download/556226324.pdf (accessed on 27th September 2025).

17. Chu, S. T., Little, B. E., Mitchell, A., Morandotti, R., et al. (2018). Harnessing optical micro-combs for microwave photonics. http://arxiv.org/abs/1710.08611 (accessed on 27th September 2025).

18. Darba, A., De Belie, F., & Manzolini, V. (2016). Improving the torque generation in self-sensing BLDC drives by shaping the current waveform. https://core.ac.uk/download/55822928.pdf (accessed on 27th September 2025).

19. De Coster, R. (2009). A business planning framework for WiMAX applications. https://core.ac.uk/download/335901.pdf (accessed on 27th September 2025).

20. Desmet, J., Heimbach, B., & Renner, H. (2015). Power quality and electromagnetic compatibility: Special report, session 2. https://core.ac.uk/download/55822716.pdf (accessed on 27th September 2025).

21. Eder, K., & Kerrison, S. (2015). A software-controlled voltage tuning system using multi-purpose ring oscillators. http://arxiv.org/abs/1503.05733 (accessed on 27th September 2025).

22. Fan, Y., Liu, X., Zhang, W., Wu, Y., Liu, Z., Zhang, C., Jiang, Y., et al. (2024). Monolithic heterogeneous integration of Si (100)/GaN CMOS inverters and normally-off GaN power devices for high switching frequency and high-power applications. *2024 36th International Symposium on Power Semiconductor Devices and ICs (ISPSD)*. https://www.semanticscholar.org/paper/9a8a7e7aff39a70525482feef734282dd9b7df53 (accessed on 27th September 2025).

23. Feng, J., Yuan, X., Liu, Q., Zhao, C., Zhang, Y., Wang, K., & Ma, Y. (2024). A 5-kW gallium nitride (GaN) photovoltaic (PV) DC-DC converter with a power density of 4.73 W/cm³ and efficiency of 98%. *2024 IEEE 9th Southern Power Electronics Conference (SPEC)*. https://www.semanticscholar.org/paper/60db638238891146f73d2c8b0af9ff17f5445dad (accessed on 27th September 2025).

24. Fernandez-Hernandez, A., Avila, A., González-Hernando, F., & Villar, I. (2024). Minimum dead time for complete ZVS in bidirectional isolated DC/DC converters for more electric aircraft. *2024 Energy Conversion Congress & Expo Europe (ECCE Europe)*. https://www.semanticscholar.org/paper/792f9127077961ca564c8fe84550b261040554bf (accessed on 27th September 2025).

25. Fu, H., & Yeatman, E. M. (2017). A methodology for low-speed broadband rotational energy harvesting using piezoelectric transduction and frequency up-conversion.

26. Gill, E., Monna, G. L. E., Scherpen, M., et al. (2007). MISAT: Designing a series of powerful small satellites based upon micro systems technology. https://core.ac.uk/download/pdf/12941826.pdf (accessed on 27th September 2025).

27. Green, T. C., & Mitcheson, P. D. (2012). Maximum effectiveness of electrostatic energy harvesters when coupled to interface circuits.

28. Guarnieri, M. (2018). Solidifying power electronics [Historical]. https://core.ac.uk/download/158825404.pdf (accessed on 27th September 2025).

29. Hamidi, N., & Mohsenzade, S. (2025). Determination of centralized/decentralized snubbers power capacity for series connected devices in inductive loads. *2025 16th Power Electronics, Drive Systems, and Technologies Conference (PEDSTC)*. https://www.semanticscholar.org/paper/4ceb02d821d248f6749de8b7ed56336bc00e4622 (accessed on 27th September 2025).

30. Hidayat, W., Usman, M., Shakir, S. W., Anum, Anjum, I., Ali, S., & Mustafa, L. (2024). Breaking performance barriers: AlN spacer integration boosts GaN HEMTs to higher drive drain current for HEMT-LED. *13440*, 1344003–1344003-6. https://www.semanticscholar.org/paper/90852f4e59bbefe0156a0325f6c0f8bd1daf2adc (accessed on 27th September 2025).
31. Islam, A., Sepelak, N., Miesle, A., Lee, H., Snure, M., Nikodemski, S., Walker, D. E., et al. (2024). Effect of high temperature on the performance of AlGaN/GaN T-gate high-electron mobility transistors with ~140-nm gate length. *IEEE Transactions on Electron Devices, 71*, 1805–1811. https://www.semanticscholar.org/paper/43ae5a35e2e5fef583e4d7d58596213a18d71997 (accessed on 27th September 2025).
32. Islam, F., Arafat, E., & Ghassemi, M. (2025). Impact of rise time and frequency on motor winding insulation lifetime under high frequency, high slew rate bipolar square wave voltages. *2025 IEEE 18th Dallas Circuits and Systems Conference (DCAS)*. https://www.semanticscholar.org/paper/8308bf1beee18a743400498b52c4269391672075 (accessed on 27th September 2025).
33. Karthik, M., Hemalatha, G., Mathankumar, M., Sreevidya, L., Deepalakshmi, R., & Kandasamy, V. (2025). Design a high-frequency dual active bridge converter for discrete time frequency. *2025 3rd International Conference on Advancements in Electrical, Electronics, Communication, Computing and Automation (ICAECA)*. https://www.semanticscholar.org/paper/58c52e36764beb092b109528524dfae9489dcd52 (accessed on 27th September 2025).
34. Krishna, M., Rahool, M. M., Kumar, K. D., Raj, A., & Kumar, R. S. (2021). Performance analysis of InP-based composite channel e-mode HEMT device for high frequency applications. *Journal of Physics: Conference Series, 1917*. https://www.semanticscholar.org/paper/717d8b42f2b835b10ffd93738149689798e5ee98 (accessed on 27th September 2025).
35. Lee, F., Li, Q., & Nabih, A. (2021). High frequency resonant converters: An overview on the magnetic design and control methods. *IEEE Journal of Emerging and Selected Topics in Power Electronics, 9*, 11–23. https://www.semanticscholar.org/paper/50fe677161fd6439b376282b11d0574aaa628282 (accessed on 27th September 2025).
36. Li, Z. (2024). Advancements in GaN HEMT structures and applications: A comprehensive overview. *Journal of Physics: Conference Series, 2786*. https://www.semanticscholar.org/paper/24b54c94808835505ecf1fcee6fca86b9007f299 (accessed on 27th September 2025).

37. Lima-López, D., & Gómez-Gil, P. (2024). Butterworth CNN: An improvement on memory use for Fourier convolutional neural networks. *2024 IEEE International Autumn Meeting on Power, Electronics and Computing (ROPEC), 8*, 1–6. https://www.semanticscholar.org/paper/a0b235314dcceb0db4ae2978602047325e0866c1 (accessed on 27th September 2025).

38. Liu, J. (2025). From silicon to wide bandgap semiconductors: Exploration of compound semiconductors in high frequency and high-power devices. *Applied and Computational Engineering.* https://www.semanticscholar.org/paper/8c0107e04e68e99a109d79657463335c13afb144 (accessed on 27th September 2025).

39. Liu, K., Feng, Y., Han, C., Chang, B., Chen, Z., Xu, Z., Li, L., et al. (2024). High-speed 0.22 THz communication system with 84 Gbps for real-time uncompressed 8K video transmission of live events. *Nature Communications, 15.* https://www.semanticscholar.org/paper/534fd2dce7e08ad72458da53ccf23590b145b476 (accessed on 27th September 2025).

40. Liu, Y. (2022). 2: Porous electronics in an era of metaverse. *SID Symposium Digest of Technical Papers, 53.* https://www.semanticscholar.org/paper/444013b67149817bb3a454bc0db5700e54446b26 (accessed on 27th September 2025).

41. Lo Nigro, R., Fiorenza, P., Greco, G., Schilirò, E., & Roccaforte, F. (2022). Structural and insulating behaviour of high-permittivity binary oxide thin films for silicon carbide and gallium nitride electronic devices. *Materials, 15.* https://www.semanticscholar.org/paper/8463a2b84e3c65d3ee578599192f3e0864e0d383 (accessed on 27th September 2025).

42. Machado de Azevedo, A., Lopes, T. J., D'Oliveira Cardoso, D., Monterio, S. N., Rocha Silveira, P. C., & da Silva Figueiredo, A. B. H. (2024). SiGe semiconductor electronic component: A review on fundamentals and applications. *Observatório de la Economía Latinoamericana.* https://www.semanticscholar.org/paper/2a66365d863863fe46091580cfb6b047743ef7ba (accessed on 27th September 2025).

43. Mitrovic, I., Finch, H., Jones, L., Dhanak, V., Hannah, A., Valizadeh, R., Renz, A., et al. (2022). (Invited) Rare earth oxides on wide band gap semiconductors. *ECS Meeting Abstracts.* https://www.semanticscholar.org/paper/d87121a7f62eb18423b6b621d82f81b902d076d8 (accessed on 27th September 2025).

44. Mittal, S. (2014). Power management techniques for data centers: A survey. http://arxiv.org/abs/1404.6681 (accessed on 27th September 2025).

45. Nagesh, S., & Manasa. (2025). An analytical model of gallium nitride MESFET for parasitic capacitance for improvement of frequency response. https://core.ac.uk/download/648635413.pdf (accessed on 27th September 2025).
46. Nielsen, M. R., Deng, S., Mirza, A., Kjærsgaard, B. F., Jørgensen, A. B., Zhao, H., Li, Y., et al. (2025). High-power electronic applications enabled by medium voltage silicon-carbide technology: An overview. *IEEE Transactions on Power Electronics, 40*, 987–1011. https://www.semanticscholar.org/paper/0aee86c56d36fbd468b1cac0f93a261cc84cbfce (accessed on 27th September 2025).
47. Silva, C., Reis, G. L., Alzamora, A., & de Paula, H. (2025). Reliability analysis of IGBT modules of multilevel inverters operating in hostile mining environments. *IEEE Transactions on Industry Applications, 61*, 6798–6809. https://www.semanticscholar.org/paper/8ca36eebc766b88e26448370f5c63b9a33b3736c (accessed on 27th September 2025).
48. Srivastava, P., Upadhyaya, A., Yadav, S., Negi, C., & Singh, A. K. (2025). Analysis of nanoscale short channel effects in cylindrical gate-all-around junction less FETs and performance enhancement with GaAs and III–V materials for low-power, high frequency applications. *Electronics*. https://www.semanticscholar.org/paper/c52469e84b8dbb36ac474aff2d318e7f3113ca3c (accessed on 27th September 2025).
49. Sun, G., Li, J., Hu, S., Zhang, Q., & Ma, X. (2025). Research on grid-forming flywheel energy storage-supported frequency regulation strategy for renewable energy integration. *2025 5th International Conference on Advances in Electrical, Electronics and Computing Technology (EECT)*. https://www.semanticscholar.org/paper/41f609e8012efed38927b2ec3aef8d11ec17a358 (accessed on 27th September 2025).
50. Tang, W., Huang, X., Chen, Z., Sheng, K., & Wu, Z. (2025). Integrated thermal management for a high-power-density silicon carbide power module with die-level heat flux over 1000 W/cm². *IEEE Journal of Emerging and Selected Topics in Power Electronics, 13*, 2125–2137. https://www.semanticscholar.org/paper/bd1260bc325b5d6d21f6c4f49dbab1b6b422cf37 (accessed on 27th September 2025).
51. Tellefsen, A., Yakkati, R. R., & Cenkeramaddi, L. R. (2024). Vehicle type classification using lightweight CNN from aerial images for traffic management applications. *2024 IEEE 19th Conference on Industrial Electronics and Applications (ICIEA)*. https://www.semanticscholar.org/paper/2ac6684c7f18d261757e9292918481ccabf22ca7 (accessed on 27th September 2025).

52. Tierens, W., Kumar, A., Lore, J., Urbanczyk, G., Diab, R., et al. (2025). Gyrotropy-induced symmetry breaking of erosion hot spots on antenna limiters during high-power tokamak RF operation: Mechanisms and mitigation strategies. https://www.semanticscholar.org/paper/c8b3a6e226bb4f86456046b4a443f64e6e06ca88 (accessed on 27th September 2025).
53. Triana-Camacho, D. A., Meoni, A., Quintero-Orozco, J. H., D'Alessandro, A., & Ubertini, F. (2025). Low-cost electrometer for self-sensing construction materials: Design and measurement insights for smart concrete and bricks. *Measurement Science and Technology, 36*. https://www.semanticscholar.org/paper/55999388a66d2cf9eeb4de674105f73217c7dfb5 (accessed on 27th September 2025).
54. Tripathi, K. K., Machhiwar, Y., Raja, D., & Agarwal, H. (2025). Compact modeling of silicon carbide (SiC) power FETs. *2025 9th IEEE Electron Devices Technology & Manufacturing Conference (EDTM)*. https://www.semanticscholar.org/paper/278f384e5fb8f26e94e6f88b4da1633e353f5ec7 (accessed on 27th September 2025).
55. Upadhyay, D., Deliri, S., Mattsson, A., Peltoniemi, P., & Aarniovuori, L. (2025). Review on silicon carbide-based high-fundamental frequency inverters for high-speed drives. *IEEE Access, 13*, 106055–106076. https://www.semanticscholar.org/paper/27a64adbdc55a37ff331ef92e3dff88cbd7021c1 (accessed on 27th September 2025).
56. Wang, Y., Wang, K., Yang, Y., & Hui, S. Y. R. (2024). Design, implementation, and comparison of multi-MHz multi-kW H-bridge inverters based on 650V SiC and GaN devices for high-frequency WPT. *2024 IEEE Energy Conversion Congress and Exposition (ECCE)*. https://www.semanticscholar.org/paper/52418972770e9ce81f24ba62f873ae507e7bf704 (accessed on 27th September 2025).
57. Xiao, Y., Han, Q., & Wu, N. (2025). Piezoelectric energy harvesting: A review of energy sources, structures, and working mechanisms in high-frequency excitations and operations. *Smart Materials and Structures, 34*. https://www.semanticscholar.org/paper/1415aa57eb428c3d55ed455d5de482784a098ee9 (accessed on 27th September 2025).
58. Xiao, Y., Ji, S., Yang, M., & Zhao, Z. (2025). An easy-to-implement flexible commutation design with reduced switching losses, *di/dt* and *dv/dt* for high-speed and high-power IGBTs. *IEEE Transactions on Power Electronics, 40*, 12123–12142. https://www.semanticscholar.org/paper/eec1075bd99bbe8f0939dcc1793a521c47a0075f (accessed on 27th September 2025).

59. Yu, S., Chen, R. H., Shao, W., Yu, W., Zhao, X., Chen, Z., Shan, W., et al. (2024). A high-performance and low HCI degradation LDMOS device with a hybrid field plate. *IEEE Journal of the Electron Devices Society, 12*, 605–612. https://www.semanticscholar.org/paper/801758fb0e6deee314d1a27fa7a0c94fc1504d09 (accessed on 27th September 2025).

CHAPTER 7

SPINTRONICS AND MAGNETIC DEVICES

Contents

Introduction 267
7.1. Fundamentals of Spintronics..272
7.2. Spin Transport and Spin Torque Transfer..278
7.3. Magnetic Random Access Memory (MRAM) 282
7.4. Spin Field-Effect Transistor (Spin-FET) Design...................................288
7.5. Prospects for Neuromorphic Computing..293
7.6. Current Research and Innovations 298
Summary..302
References...305

INTRODUCTION

As the rate of electronics innovation continues to rise, research in spintronics and magnetic devices has emerged as an important area of study. Spintronics takes advantage of the intrinsic spin of electrons, besides their charge, to create more efficient and useful devices. The technology promises a revolution in data processing and storage, as seen in innovations such as magnetic random-access memory (MRAM). MRAM takes advantage of the magnetic state of matter to store data, allowing for faster, more energy-efficient non-volatile memory than conventional electronics. Furthermore, spin torque transfer, wherein spin-polarized currents cause a torque on magnetization, has significant implications for the optimization of the performance and speed of spintronic devices. As researchers are unfolding the promise of spintronics, applications are increasing, making way for a paradigm shift in modern computing and information technology. Breakthrough materials, such as two-dimensional (2D) van der Waals magnetic materials, are also under the spotlight of spintronic investigation. Such materials, such as few-layer $Cr_2Ge_2Te_6$ and monolayer CrI_3, among others, not only possess unique magnetic properties but also allow devices to be achieved at the nanoscale. The conflict between low dimensionality and ferromagnetism is problematic; however, anisotropy within these materials can form more effective magnetic ordering, which is critical for efficient spintronic devices. This modulation potential assists in countering the inherent shortcomings of 2D systems. Newer advancements are shedding light on intrinsic and induced ferromagnetism, which in turn inform device fabrication techniques. By using such new materials in spintronic devices, researchers can potentially achieve higher operation temperatures, paving the way for practical application in widely used technology (He et al., 2022). Besides, spintronics is available not just for traditional data storage but also for emerging computing paradigms such as neuromorphic computing. Neuromorphic systems designed to replicate the neural structure and function of the brain can leverage spintronic principles to provide efficiency and speed unavailable using classical von Neumann architectures. As one example, ongoing investigation of doped monolayer GaS has appeared to demonstrate how judicious material modifications can induce magnetic moments and enhance electronic properties, moving these materials toward being ideal for neuromorphic use. With the aid of ferromagnetism and half-metallicity, researchers are exploring architectures capable of processing information in a more biologically compatible manner. The fusion of spintronics and neuromorphic computing is not just a push toward more efficient processing but also beckons an era of greater computational capabilities that can redefine artificial intelligence and machine learning (Khan et al., 2021).

Definition of Spintronics

Spintronics emerged because there is a requirement to use not only the electrons' charge but also their intrinsic spin, a type of intrinsic angular momentum. Spintronics is distinct from traditional electronics because it exploits spin transport and spin torque transfer processes to enable manipulation of data at practically reduced energy levels. The potential of controlling and manipulating spin states offers a number of advantages such as enhanced speed and efficiency in information processing. The feature is of great utility in applications like Magnetoresistive Random Access Memory (MRAM), which uses the magnetoresistive effect to enable nonvolatile memory storage. MRAMs reliability and scalability are two essential properties that position it as a possible replacement for conventional memory technologies, drawing interest from industry and academia to develop next-generation computing architectures. As the technology matures over time, spin field-effect transistors (spin-FETs) have emerged as a vital component in spintronic devices. Spin-FETs utilize the spin property of electrons in order to offer control and manipulation beyond charge-based transistors (Figure 7.1).

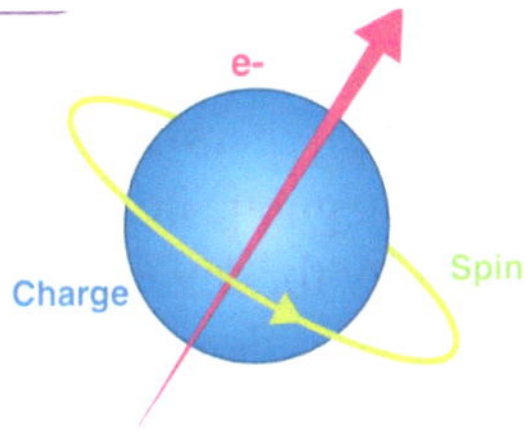

Figure 7.1. *The base of semiconductor architecture.*

Source:https://cdn1.byjus.com/wp-content/uploads/2023/04/Spintronics.png.

Through the insertion of magnetic components in semiconductor structure, researchers aim to construct devices that draw lower levels of power consumption, which could significantly shake up the electronics sector. Additionally, spin-FET technology development may assist in bridging the transition gap between existing electronic devices and next-generation quantum technologies, enabling a continuity of improvements. Through the facilitation of more profound manipulation of electron states with spin effects, these devices are at the cutting-edge of new, multi-functional applications that can interface easily with existing circuits and open the way to future innovations in information technology. Merging spintronics and neuromorphic computing is an area of study that is just waiting to be researched and innovated in the future. Neuromorphic computers that replicate the way human brains process information would be significantly boosted by what spintronics could offer. Magnons' unique characteristics, as outlined in the introduction to magnonics, allow for information to be conveyed as spin waves, and this would lead to the creation of energy-efficient, brain-inspired computing architectures that can process large amounts of information at once. Moreover, research into spin-based neural networks may provide new perspectives in machine learning and artificial intelligence. The revolutionary implications of the combination of these two ideas may revolutionize computational paradigms and render spintronics an inevitable part in the future of high-level computational devices, thereby solidifying its top role in the timeline of technology (Barman et al., 2021; Yuan et al., 2022).

Historical Context and Development

The progress of spintronics can be predominantly attributed to its leading research in spin transport and spin torque effects. These principles allowed researchers to manipulate electron spins with electric fields and later spawned technology reliant on spin rather than charge. The invention of spin-orbit torque (SOT), a phenomenon that combines electric fields with magnetism, has been a breakthrough for the discipline. Researchers have found a variety of materials with higher SOT efficiency, including magnetic insulators and topological insulators, that extend the range of applicability for devices to be used in logic and memory. Historical device structure analysis demonstrates transition from classical magnetic devices towards sophisticated spintronic devices, demonstrating tremendous growth in theoretical research and experimental application over recent years (Shao et al., 2021). As the technology evolved, applications such as magnetoresistive random access memory (MRAM) emerged, allowing data to be stored using non-volatile configuration. MRAM uses spin-polarized currents to change the magnetic bits' state, achieving permanent power-free data retention. Among the outstanding landmarks in the past is combining spin field-effect transistors (spin-FETs), which leverage carrier spin to achieve increased performance in computer operations. The promise provided by such hybrid devices has motivated continued work on other architectures that integrate semiconductor and magnetic technology. This thrilling history places in perspective the rapid progress and cross-disciplinary collaborations that form the basis for the emergence of spintronics and magnetic devices and chart the course for future research and applications (Shao et al., 2021). Future uses of this research indicate the extremely high potential of neuromorphic computing, a technology that imitates neural networks and operations. Using the inherent characteristics of spintronic devices, such as low power consumption and high speed, researchers hope to construct more efficient computational systems that are capable of executing complex computations like biological processes. As the history of spintronics progresses, new discoveries of 2D materials like phosphorene and derivatives are another breakthrough offering fresh hopes for memory and spin-based logic device integration. As pioneers in the field continue pushing the frontier of these technologies, neuromorphic computing prospects continue to evolve, promising an interesting intersection of practical applications and theoretical developments in spintronics and magnetic devices.

Importance of Magnetic Devices in Modern Technology

The evolving technology environment more and more relies on spin transport and spin torque transfer efficiencies, a paradigm shift that emphasizes the functionality of magnetic devices. These mechanisms serve as the foundation of spintronic device functionality and utilize the spin of electrons, besides its charge, to drive information processing and storage. The use of spintronics in MRAM and other devices enhances speed and efficiency, far outpacing conventional semiconductor technology. Further, new developments in spin-transfer torque techniques allow for faster switching times in MRAM, which provides room for more sophisticated memory designs required in modern computing environments. Therefore, spin transport optimization not only facilitates high-performing computing but also becomes the basis for designing devices that not only consume less energy but can even have greater data processing rates, which eventually transforms the digital world. Notably, spin-FET design becomes a revolutionary approach in integrating magnetic functionality into electronic circuits. Spin-FETs capitalize on the manipulation of electron spins to enhance logic functionality, which is fundamental in the quest to achieve power consumption reduction while maximizing computational density (Figure 7.2).

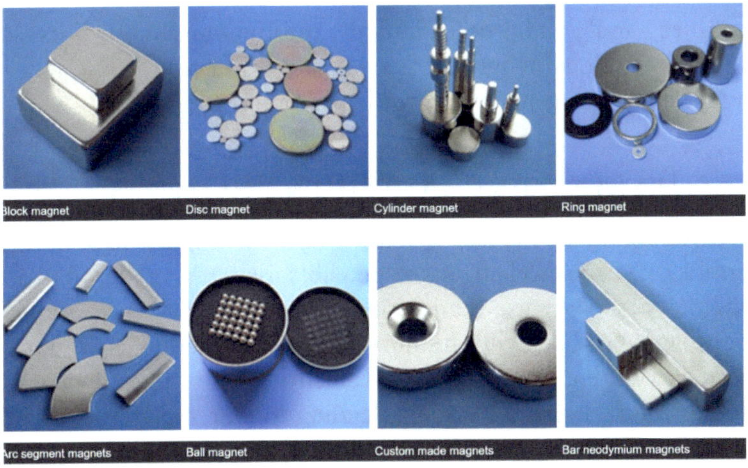

Figure 7.2. *Magnets and electronics: The close relationship driving modern devices.*

Source:https://media.licdn.com/dms/image/v2/D5612AQExXD6CxsQk9g/article-cover_image-shrink_720_1280/article-cover_image-shrink_720_1280/0/1700184647265?e=2147483647&v=beta&t=0kjGoGiSFf6Z-JKkMsHg_CM-1lFkPT-d8UPPdtrN70es.

Scientists were able to successfully demonstrate that employing spin current was able to switch states quite effectively in a manner that is quite similar to traditional transistors but with better advantages in terms of speed and fewer thermal issues. The study of these devices reveals a fascinating synergy of magnetic and electronic properties that allow hybrid technologies to have the potential of revolutionizing information processing. The adaptive quality of spin-FETs can also facilitate tailored solutions for advanced computing tasks, thus representing a chaotic move in the device design continuum that sets off on efficiency and performance. To the future, the horizon of magnetic devices in neuromorphic computing looks bright on promising cognitive computing and artificial intelligence research areas. This emerging direction is trying to emulate biological neural networks for the hope of developing machines that are as much capable of learning and adaptation as the human brain. Magnetic devices, particularly spin-based architectures, are especially well-positioned to fill a niche in this field with their intrinsic quality of speed and efficiency. Spin state control would potentially allow the high-speed and energy-efficient information processing that neuromorphic system performance demands. Concepts such as associative memory and synaptic connections in such systems can be maximized through the precision of spintronics, presenting new possibilities for machine learning applications—potentially reconceiving how machines handle and process intricate information. In this intersection of biology and technology, magnetic devices are soon going to form an integral part of redefining next-generation paradigms of computing.

7.1. FUNDAMENTALS OF SPINTRONICS

In the recent past, the two facets of spin transport and spin torque transfer have come forth as the prime mechanisms responsible for the drive towards developing spintronic devices. They use the internal spin of electrons, making it feasible to control magnetic states with minimal energy dissipation. Spin transport, in which there is a current of spin-polarized electrons, offers the potential for novel routes to data storage and processing, leading to extremely efficient magnetic random-access memory (MRAM). Similarly, spin torque transfer relies on employing spin currents to manipulate magnetic moments, thereby guaranteeing the potential for fast switching. Together, these impacts are revolutionizing the design of next-generation spintronic devices, particularly in data retention and speed of operation, attributes that are of supreme significance in the ever-evolving technological scenario of information processing and storage. The research into magnetoresistive random-access memory (MRAM) and spin-field-effect transistors (spin-FETs) demonstrates the tremendous potential of spintronics in practical applications. MRAM applies the advantages of spin-dependent electron tunneling in particular for utilizing non-volatile memory storage with high endurance and low access times. Spin-FETs, however, which integrate magnetic phenomena into field-effect transistor device structure, aim to harness both charge and spin towards a more flexible mode of electronics. They have the potential to reduce power usage by a factor of many orders while ensuring ultimate performance efficiency. As researchers discover new means of integrating these technologies, the border between traditional electronics and spintronics dissolves further, paving the way for new applications that tap into both spin and charge, transforming the circuitry world (Figure 7.3).

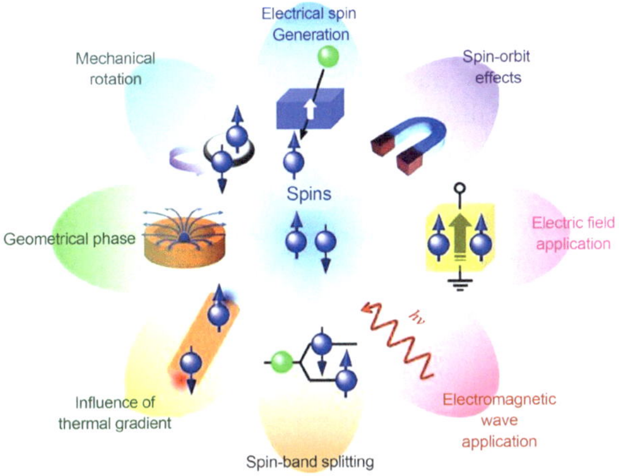

Figure 7.3. *Future developments in spintronics.*

Source:https://ars.els-cdn.com/content/image/1-s2.0-S0304885320302353-ga1.jpg.

Spintronics advancements of the future are becoming increasingly associated with the future of neuromorphic computing, which attempts to replicate human neural structures to process information. Such a paradigm change would anticipate spin-based synapses and neuron analogs, and thus potentially more efficient and adaptive computational paradigms. Spintronics devices, in the present instance spin-crossover material- and single-ion magnet-based devices, can potentially offer unique functionalities in the aspect of multifunctionality and responsiveness and can therefore be applied as reservoir computing units in neural networks. The role of molecular spintronics in this revolution can be seen from the use of advanced materials like CoII SCO-SIM complexes, which not just facilitate magnetic data storage but also function at the molecular level for improved computation. The use of spintronic concepts in neuromorphic architectures can foreseeably lead to innovations in artificial intelligence, machine learning, and related fields, propelling further study of their scalability and viability in future computing paradigms.

7.1.1. Spin and Charge of Electrons

The dance of the spin and charge of electrons has unlocked a universe of potential for spintronics, with a revolutionary shift of paradigm in data processing and storage. This breakthrough method utilizes spin transport and spin torque transfer to achieve the high efficiency which is impossible for traditional charge-based electronics. Spin transport processes include the transfer of spin-polarized electrons through various materials, with the electron's intrinsic angular momentum having the ability to exert torque onto neighboring magnetic layers in order to induce effects such as magnetization switching. Such a property is essential to the development of new memory devices like magnetoresistive random access memory (MRAM), based on the different resistance levels of magnetic states, in storing information. Thus, spintronics provides a potential alternative to conventional methods, addressing the acute necessity for increased speed and decreased power dissipation in present devices. Another groundbreaking application of electron spin dynamics is the application of spin to field-effect transistors (spin-FETs). Through the new properties of spin, spin-FETs attempt to improve semiconductor operation along with reducing energy consumption (Figure 7.4).

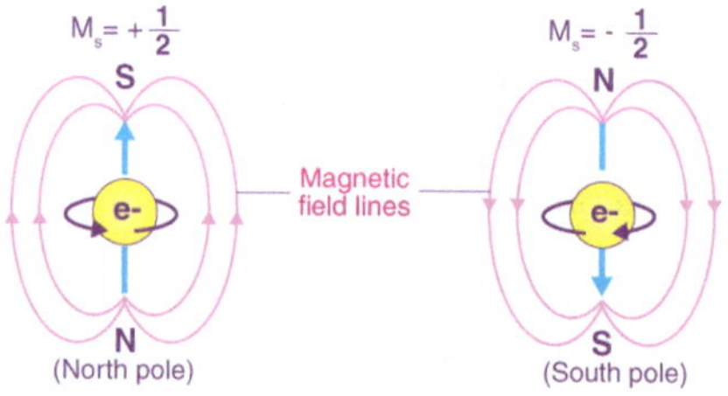

Figure 7.4. *The mechanism between spin and charge in electrons.*

Source: https://cdn1.byjus.com/wp-content/uploads/2023/03/Electron-Spin-Updated.png.

The underlying concept depends on manipulating the spin polarization of a semiconductor channel by a ferromagnetic source and drain, which can actually manipulate electrical currents. This integration has the potential for non-volatile operation such as MRAM but with the added advantage of increased integration density and scalability. The intricate designs of spin-FETs highlight the potential for the design of more effective computation structures. The unification of spin and charge has numerous opportunities for exploration, destined to transform not only electronic device operation but even the fundamental principles governing information processing. In the future, the potential of neuromorphic computing in the spintronics paradigm provides a novel and promising field of investigation. By duplicating the form and function of biological neural networks, neuromorphic devices present the possibility of revolutionizing the process of information, enabling computation that is faster and more efficient. The integration of spintronic devices can simplify artificial synapse realization with magnetic states as synaptic weights and effectively controllable by spin, allowing breakthroughs in machine learning and artificial intelligence by facilitating the possibility of real-time data processing and sophisticated pattern recognition. As scientists continue to explore the subtle interplay between spin and charge, the establishment of complete theoretical and experimental frameworks will become even more important. In general, spintronics development will not just improve current technologies but also set the stage for novel computing paradigms exploiting the intrinsic nature of electron spin.

7.1.2. Spin Transport Mechanisms

Behind the new science of spintronics lie the processes of spin transport, through which spin-polarized electrons are transmitted through various materials. Such processes are essential in determining the functioning of spintronic devices such as magnetic random-access memory (MRAM) and spin-field effect transistors (spin-FETs). Spin transport efficiency is characterized by properties like spin diffusion length and spin lifetime, and diametrically opposite conducting materials lead to different performance characteristics. With the progress of research, the understanding of how spin-polarized currents interact with adjacent materials—particularly in two-dimensional (2D) systems—has become important. Effects like spin-orbit coupling enable new functionalities, thereby enhancing device performance and operation reliability. This understanding not only improves existing applications but also drives research into novel structures that take advantage of proximity effects to include spintronics in conventional electronics (Roche et al., 2024). Examining the interaction between molecular materials and spin transport mechanisms holds immense potential for device design and functionality. Molecular spintronics exploits the unique character of organic semiconductors, which can offer better spin lifetimes and flexible electrical, optical, and magnetic characteristics (Figure 7.5).

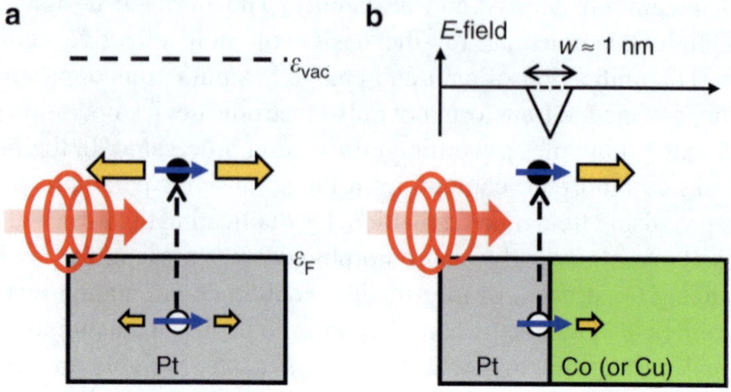

Figure 7.5. *Mechanisms for spin transport.*

Source:https://www.researchgate.net/publication/340060864/figure/fig5/AS:959161403723788@1605693307598/Mechanisms-for-spin-transport-a-The-mechanism-based-on-the-diffusion-of-electrons-and.png.

Researchers are investigating how molecular level modifications can enhance spin injection efficiency and transport characteristics in spintronic devices. Additionally, the spinterface effect, where spin-polarized tunneling is seen at interfaces between two unlike materials, is noted as a key area for improvements in performance. Future requirements are centered on developing reproducible ways for integrating molecular materials into device architecture that maintains high spin coherence. Beating these hurdles will prove important as the research strives to bridge molecular spintronics with practical application, driving memory technology innovation and beyond in the process (Gu et al., 2023). As spintronics and neuromorphic computing intersection advances, the utilization of spin transport mechanisms to replicate the behavior of neurons becomes increasingly sought after. The inherent ability of spintronic devices, such as spin-FETs, to be low in power dissipation without sacrificing their non-volatility makes it a perfect scenario for the use of neuromorphic systems that require effective data processing and storage. The use of spin-polarized currents to neural network implementation has the promise of significantly enhancing computational effectiveness and speed to open up new artificial intelligence applications. Besides, the spintronic devices' scalability positions them as the major players in the development of new models of computing derived from biological functions. The exploration of how spin transport functionalities can enable such capabilities has the potential to reshape our understanding of computation and memory and usher an intriguing future for spintronics as it merges with neuromorphic technologies (Roche et al., 2024).

7.1.3. Role of Magnetic Materials in Spintronics

Emerging advances in spintronics have emphasized the valuable role of magnetic materials in achieving efficient spin transport as well as spin torque transfer. The ability to manipulate electron spins as well as their charge provides a substantial advantage in the design of future devices. Magnetic materials, for instance, facilitate the design of spintronic devices wherein information processing and storage could be achieved through the spin state of electrons. This physical characteristic is crucial in enhancing performance in memory devices such as Magnetoresistive Random Access Memory (MRAM), which takes advantage of electrons' spin to achieve better read and write speeds. By using material with huge spin polarization, researchers can ensure the optimal efficiency of MRAM and, in the end, make a compelling case for its application in future computing systems, where speed of data access is of utmost concern. Spin-ferromagnetic field-effect transistor (spin-FET) structure is another where magnetic materials are essential. Incorporating ferromagnetic material into typical FET architectures, spin-FETs utilize the physics of spin transport to increase functionality. This allows magnetization to be reversed in magnetic layers to achieve functionality better than standard electronic devices. This also increases the scalability of the device and reduces power consumption by a significant amount. Furthermore, as the need for smarter, faster electronic devices continues to grow, the potential for utilizing multiferroic materials in spin-FET designs could usher in new applications. By leveraging the odd interaction between magnetic and electric states within the materials, researchers can create next-generation technologies with capabilities beyond their traditional counterparts without sacrificing energy efficiency (Ito et al., 2024). Neuromorphic computing opportunities are an amazing spintronic application area where magnetic materials have the potential to shape computational structures greatly. By replicating the neural pathway connectivity of the human brain, neuromorphic devices try to process data in a way similar to living organisms. Magnetic materials play a critical role in the development of the synaptic components that can dynamically adjust their connection strength depending on spin states, facilitating the simulation of learning and memory functions. Their incorporation with materials such as Fe_4N, with tunable magnetoelastic properties, may further advance this technology. Most importantly, their dynamic response to magnetic fields and strain renders them beneficial for spintronic strain sensor innovations, and this enhances the functionality of neuromorphic computing systems even further. Hence, the role of magnetic materials extends beyond the traditional applications in memory and situates them at the forefront in attempts to pursue brain-inspired computational architectures (Ito et al., 2024).

7.2. SPIN TRANSPORT AND SPIN TORQUE TRANSFER

The intricate interaction between spin transport and spin torque transfer is vital in achieving the capability of spintronic devices such as magnetic random-access memory (MRAM) and spin field-effect transistors (spin-FETs). Efficient spin transport is characterized by the flow of spin-polarized electrons in materials that enable controlling magnetic states. The efficacy of such a phenomenon depends on different factors including the magnetic and electronic nature of the materials. Spin current utilized in spin transport allows the utilization of spin torque, which can lead to magnetic layer switching at lower rates of power consumption. Such better control of magnetic states not only makes devices perform better but also emphasizes the possibility of further miniaturization of electronic devices, as smaller units may effectively make use of these quantum mechanical effects (Figure 7.6).

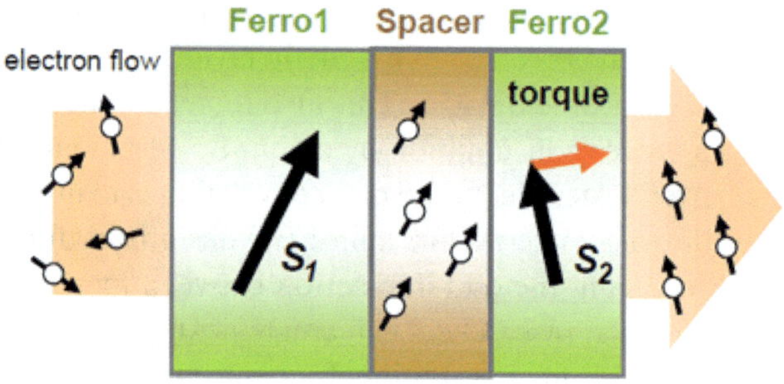

Figure 7.6. *Spin transfer torque.*

Source:https://docs.quantumatk.com/_images/torque.gif.

Recent developments in spin-transfer torque magnetic tunnel junctions (STT-MTJs) have greatly advanced the boundary of spintronics, as evidenced by their promising characteristics for future technologies. STT-MTJs' developments are most remarkable on account of their high tunnel magnetoresistance ratios, low critical currents, and efficiency in auto-oscillation behaviors that are essential for high-speed and dependable computing purposes. Research indicates that variations in the free layer composition of different types of STT-MTJs can significantly influence their performance, with certain configurations revealing improved characteristics. For instance, research has proven that B-MTJs have lower auto-oscillation critical currents than their A-MTJ equivalents, demonstrating the importance of material composition in the attainment of device efficiencies, a critical factor for future developments in data storage and processing technologies (Parvini et al., 2023). The possible use of spin transport and spin torque transfer extends beyond standard computing systems, and holds rich soil for neuromorphic computing. Neuromorphic computing seeks to copy the architecture and functionality of biological networks to enhance computational speed and efficiency. The spin control ability of precision enables the construction of devices that not only store and process information like synapses of the brain but do so with much reduced energy consumption. New developments in materials and structures, such as skyrmion devices, provide promising opportunities for the integration of spintronic effects into neuromorphic technologies, which could trigger revolutionary breakthroughs in artificial intelligence and machine learning. The ongoing exploration of the aforementioned principles promises a bright future for spintronics as a fundamental element of the models of next-generation computing (Liu et al., 2025).

7.2.1. Methods of Spin Transport

The underlying essence of the new field of spintronics is the complex relation between spin transport and spin torque transfer mechanisms. These properties are vital towards the realization of next-generation magnetic devices like Magnetoresistive Random Access Memory (MRAM) and spin Field-Effect Transistors (spin-FETs).

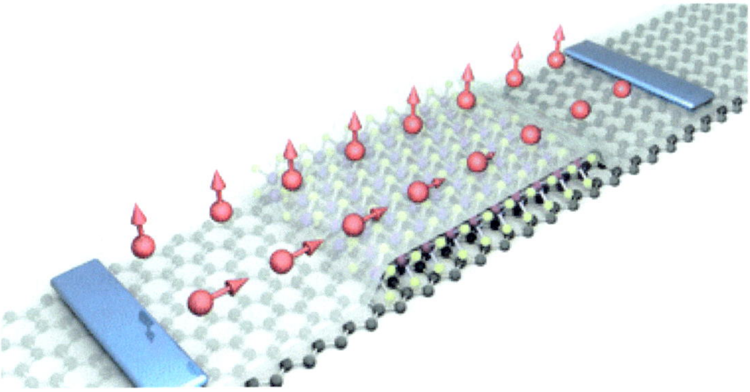

Figure 7.7. *Spin transport in graphene/transition.*

Source: https://pubs.rsc.org/en/Image/Get?imageInfo.ImageType=GA&imageInfo.ImageIdentifier.ManuscriptID=C7CS00864C&imageInfo.ImageIdentifier.Year=2018.

Spin transport refers to the transportation of electron spins in materials, while spin torque transfer refers to the manipulation of the spins by external magnetic fields or electric currents. The manipulation is required for writing information in MRAM, which utilizes the electron spin state to store data more efficiently and durably compared to conventional memory technologies. Continued optimization of these mechanisms also holds the promise to significantly enhance device performance, particularly in speed and energy efficiency. Breakthroughs in two-dimensional materials (2DM) have also opened up new avenues for spin transport mechanisms, further allowing deeper understanding of their functionalities within spintronic devices (Figure 7.7).

A decade and a half of investigation have clarified the underlying spin dynamics of the compounds, revealing new properties that can be leveraged by applications. Proximity effects can make materials like graphene spin-active, critical to the integration of these materials into useful spintronic devices. However, there is still a great deal of technological challenge that exists and which could hinder the widespread use of 2DM-based spintronics in areas such as spin-logic devices and non-volatile memory technologies. These are challenges that need to be overcome if the field is to be further pushed towards the development of ultracompact and energy-saving devices. The future of spin transport mechanisms also intersects with the newly developing field of molecular spintronics, which utilizes the unique characteristics of molecular semiconductors. These semiconductors are characterized by their extended room-temperature spin lifetimes and vast numbers of electrical, optical, and magnetic properties which can be exploited to create cutting-edge applications. Present research will investigate spin transport in molecular systems and the interface interaction between them, which critically affects spin injection and overall device performance. Addressing the existing issues with tailored molecular design and interface engineering, scientists hope to improve the efficiency of spin injection and bring different spin-related functions into play. This coordinated strategy not only vows to enhance the current backdrop of molecular spintronics but also paves the way for fascinating directions in neuromorphic computing, an area that requires sophisticated spin-induced dynamics to emulate brain-inspired processing capabilities.

7.2.2. Spin Torque Transfer Effects

Spin transport and spin torque transfer are at the heart of the advancements in spintronics and have been instrumental in defining the next generation magnetic devices. With spin torque effects controlling spin currents, angular momentum can be transferred directly to magnetic moments, allowing magnetic states to be controlled. This is particularly clear in the application of magnetoresistive random access memory (MRAM) and spin field-effect transistor (spin-FET) technologies where electron spin is employed for data storage and logic operations. Through an interfacing with the prowess of such processes, device architectures can be designed less power-hungry and more functionally efficient. The unrivaled scalability and pluripotency of spin torque-based methods allow engineers to be pivotal in the grand agenda of miniaturization and increased computing power within magnetic devices. Furthermore, spin torque transfer has important consequences in MRAM design and operation, a non-volatile memory technology with excellent speed and endurance. The mechanism underlying is magnetic moment switching via spin-polarized currents. Spin-FET device breakthroughs also depend on spin torque transfer mechanisms and are consequently key to the manufacture of devices that take advantage of the merits of spin and charge conduction. Notably, experimental investigation of spin-transfer torque effects on antiferromagnetically coupled ferrimagnets presents the prospect of enhanced domain wall motion, a critical element to facilitate efficient data manipulation for MRAM architectures (Okuno et al., 2019). The optimization and control of such transfer effects are towards better understanding and ability in the design of magnetic storage solutions that are high-performance and robust. As research extends to the domain of neuromorphic computing, the role of spin torque transfer effects grows increasingly significant. This emerging field is aimed at the replication of human cognitive processes through hardware, with an aim towards information-processing machines akin to the human brain. Utilizing the high-speed dynamics of antiferromagnetic materials, and spin transfer torque mechanisms, neuromorphic devices possess the potential for rapid state switching, hence accelerating synaptic processes. The use of spin-polarized currents in such computational frameworks also holds the potential to create effective, energy-saving systems capable of undertaking elaborate functions such as pattern recognition and learning algorithms. The ongoing work on the spin current-magnetic texture coupling and specifically skyrmions also offers grounds for broadening the frontiers of neuromorphic computing to ultimately lead to intelligence, adaptive machines (Saidi et al., 2024).

7.2.3. Applications in Memory and Logic Devices

The rapid development of spintronics has been driving the exploration of new applications in memory and logic devices through channels like spin transport and spin torque transfer. Using the inherent spin of the electrons rather than only their charge, devices become more efficient and faster, with greatly enhanced performance parameters compared to conventional electronics. A good example is Magnetoresistive Random Access Memory (MRAM), which uses the magnetoresistive effect for memory nonvolatile storage. MRAM retains data even without a power supply, and its potential of higher speeds in reading and writing with less energy consumption positions it as a good contender in the quest for next-generation memory technologies. Better features of MRAM are achieved with high-quality materials like Mn_4N, in configurations that optimize magnetic switching, thereby promoting more trustworthy MRAM platforms (Yasuda et al., 2023). Better features also exist with spin-FET devices, where magnetic features are integrated into field-effect transistors. Through spin being part of their operational paradigm, these devices have the potential to shatter the boundaries of traditional semiconductor technology, enabling smaller, faster, and power-efficient logic devices. The inclusion of ferrimagnetic materials provides a unique opportunity in this category, where they possess controllable magnetization that can be used in coordination with traditional charge-based signals. It is not just a method to enhance device operation but also a response to the need for compatible technologies that do not sacrifice the electrical signal processing speed while missing out on spin manipulation. With ongoing research into these spintronic applications, the possibility of the total optimization of spin-FETs to be utilized appears increasingly plausible, once more revolutionizing the area of electronic logic devices (Zhang et al., 2023). In addition to data storage and conventional logic calculations, neuromorphic computing potential starts to provide a fascinating prospect in the area of spintronic applications. Neuromorphic systems attempt to mimic the neural architectures of biological brains to facilitate energy-efficient handling of large amounts of data based on parallel computation. Spin-based devices can be particularly useful in this regard based on their inherent properties that permit rapid changes to magnetic states to be used for synaptic weights encoding. Ultrafast tunability of devices towards the realization of complex neural networks promises opportunities for constructing next-generation artificial intelligence systems. With continued research into magnetic materials and their unique characteristics, the contribution of spintronics to neuromorphic computing looks towards a convergence of energy efficiency, speed, and intelligent processing, highlighting the revolutionary potential of spin-based devices in the future of computing (Yasuda et al., 2023; Zhang et al., 2023).

7.3. MAGNETIC RANDOM ACCESS MEMORY (MRAM)

The integration of spin transport with spin torque transfer in the case of MRAM proves its innovativeness and efficiency in data storage technology. The use of the spin of electrons along with its charge by MRAM offers non-volatile memory devices for storing data even in the absence of constant power supply. This aspect not only enhances data holding capabilities but also paves the way for fast access times compared to traditional memory technology. At the heart of this process is the magnetic tunnel junction (MTJ), whose functionality is critically dependent on sophisticated fabrication technology and materials optimization. The interplay between spin transport mechanisms enables magnetization states to be controlled, which is essential for writing and reading operations in MRAM devices. Thus, the effective application of spin torque transfer is both power consumption cut and enhanced operating speed, which makes MRAM a viable candidate for modern computing applications (Thapa et al., 2025). Continuing the discussion on MRAM's architecture, spin-FET (spin field-effect transistor) designs prove highly promising for improving memory performance. This approach integrates spintronic principles into transistor structures directly, allowing for higher efficiency in storage operations. The concept of the spin-FET allows for manipulation of controlled spin states with low energy use, and this is a demonstration of the need for energy-saving data processing tools (Figure 7.8).

Figure 7.8. *Basic features of magnetic random access memory (MRAM).*

Source:https://media.licdn.com/dms/image/v2/C4D12AQFeDfFWUiMPxA/article-cover_image-shrink_600_2000/article-cover_image-shrink_600_2000/0/1649154641961?e=2147483647&v=beta&t=m3Vd_Oi-0evDfaC86A4b7LZAY9HM4s-4R9N88-Ws2CQ.

Moreover, the integration between spin-FET design and MRAM technology supports miniaturization and scalability to meet the demands of high-density and compact memory. The development of these architectures demonstrates a strategic shift toward integrating memory and processing units, and may result in the overhaul of computer design and capability. As advancing research continues in this direction, spin-FET uses in MRAM systems are likely to increase, with driving digital hardware innovations (Usami et al., 2024). Neuromorphic computing innovations also introduce promising intersection points with MRAM technology. The paradigm strives to replicate the neural architectures of biological networks, thereby enhancing computation efficiency through parallel processing and adaptive learning. The non-volatility of MRAM, coupled with its ability to perform fast read and write operations, makes it a strong contender for neuromorphic applications. The integration of synaptic functionality in MRAM devices has the potential to bring about the advent of memory systems with synaptic-like behavior, responding dynamically to input stimuli. Through the exploitation of spintronic material properties, researchers expect to create memory devices that are amenable to synaptic-like modifications, essential for algorithms of learning. These prospects not only unveil the adaptability of MRAM in various computing tasks but also show how it can transform data processing and storage, ushering in new avenues for intelligent systems (Thapa et al., 2025).

7.3.1. Structure and Operation of Magnetic Random-Access Memory (MRAM)

In spintronics, spin manipulation is at the heart of enhancing data storage solutions' functionality and efficiency, particularly in MRAM. The magnetic tunnel junction (MTJ), a fundamental building block of MRAM technology, typically consists of two ferromagnetic layers separated by an insulating thin barrier.

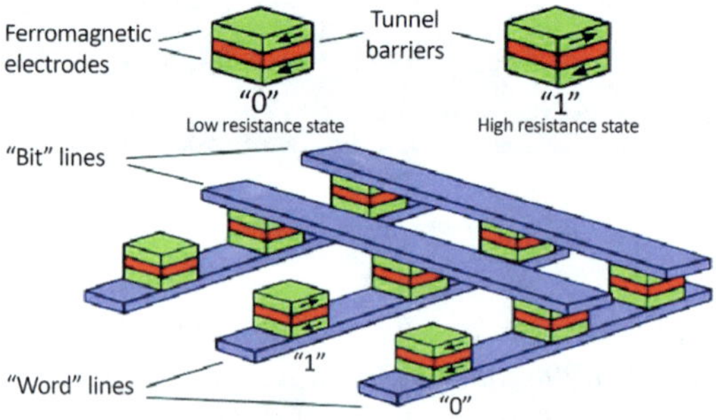

Figure 7.9. *The operational activity of MRAM (Magnetic random access memory).*

Source: https://www.researchgate.net/profile/Vijay-Kalyani/publication/281555469/figure/fig4/AS:669660693417984@1536670959017/Fig-3-MRAM-magnetic-random-access-memory-IV-SPINTRONICS-WITH-SEMI-CONDUCTORS-It-is.jpg.

The tunnel magnetoresistance (TMR) effect, which is critical to MRAM's function, happens when relative polarities between the magnetic layers are in alignment or opposition to one another, creating varying levels of resistance. This structural configuration allows MRAM to possess non-volatile memory, with data remaining unaffected by power. Emerging advancements, including the incorporation of low-dimensional materials (LDMs) in MTJ design, are anticipated to further improve TMR ratios and operation efficiency, and the study of such materials is required for future MRAM technology (Thapa et al., 2025). Spin transport and spin torque transfer processes are key to MRAM operation, enabling effective write and read operation of information. When a write current is provided, the spin-polarized electrons enter into the MTJ, leading to an angular momentum transfer, which governs the magnetization state of one of the ferromagnetic layers—a phenomenon known as spin transfer torque (STT). This process allows for rapid switching among magnetic states in parallel and anti-parallel configurations (Figure 7.9).

Additionally, spin field-effect transistors (spin-FETs) are emerging as a highly promising device that combines spintronic concepts with traditional electronics. In this device, spin polarization can be employed to control charge currents, with the possibility of lower power consumption and improved performance compared to existing transistor technologies. While the industry pursues innovation in microelectronics, understanding the function of MRAM and its interaction with spin-FETs will be critical to the design of more complex computational architectures (Thapa et al., 2025). The future for MRAM is thus bright, with prospects crossing into the new domain of neuromorphic computing, where simulation of cognitive function holds transformative potential for computation efficiency and pattern recognition. Unlike conventional computing architectures that rely solely on binary operations, neuromorphic systems leverage the intrinsic nature of MRAM, such as non-volatility and scalability, to create synaptic connections mimicking the natural neural network in biology. Through the exploitation of spintronic operations, MRAM devices can emulate the dynamical signaling in biological systems to enable energy-efficient learning algorithms and memory access. Furthermore, the marrying of MRAM with LDMs can enhance synaptic functionality to improve performance in neuromorphic computing. With continued progress, it will be important to understand the flexibility and robustness of MRAM in neuromorphic design to enable next-generation computing machines that are capable of resolving intricate, data-heavy issues.

7.3.2. Advantages Over Traditional Memory Technologies

New spintronics technologies offer a gigantic range of advantages over previous memory systems, particularly through mechanisms like spin transport and spin torque transfer. Both processes utilize the intrinsic magnetic behavior in nanoscale material to facilitate higher data processing rates with reduced energy consumption. An example includes magnetic random access memory (MRAM), which exploits the electron spin to store data, offering non-volatile memory that retains information when power is no longer available. This is as opposed to conventional volatile memory technologies such as DRAM, which must be constantly powered. Moreover, MRAM's endurance and speed present a good argument, and it is thus best suited for applications that require immediate access to gobs of data without the energy cost of conventional memory technology. These advancements significantly enhance the reliability and performance of data storage technology, positioning spintronics as a pivot for next-generation electronic products, especially for memory technologies (Sud et al., 2025). In addition to MRAM, the fabrication of spin field-effect transistors (spin-FETs) represents another frontier of spintronics beyond the limits of traditional semiconductor devices. Spin-FETs utilize the spin and charge of electrons rather than just their charge, not only enhancing the functionality of the device but reducing power consumption. The two uses of electrons enable the potential to undertake intricate computation processes at higher speed and more efficiently. Moreover, compared to the conventional transistors, which dissipate heat due to electron flow, spin-FETs have reduced thermal output, hence they will function optimally in extremely integrated chip designs. The integration of such devices into mainstream technology would not only be more efficient in energy but also bypass the miniaturization issues faced by conventional electronics, paving the way for next-generation computing architectures with the ability to satisfy the increasing demands of modern applications (Wang et al., 2025). With ongoing advancements in neuromorphic computing, spintronics has proven to be a highly promising avenue for simulating the dynamics of biological neural networks. Neuromorphic devices are characterized by low power consumption and effective handling of complex computations, thus offering advantages that are hard for traditional silicon-based systems to emulate. Based on magnetic tunnel junctions that act as artificial synapses and neurons, spintronic technology embodies the capability to emulate the brains structure effectively. Such spintronic devices not only possess high-speed operation and significant operational lifetimes but also the capability of operating without applying continuous power, thus such devices are more appropriate for long-term usage. Hence, the integration of spintronics in neuromorphic computing can transform algorithm execution such that solid architectures for artificial intelligence systems are implemented. This revolution to bio-inspired computing possesses the potential to achieve major improvements over the limitations of existing memory technologies, paving the way for a next-generation of intelligent systems (Sud et al., 2025).

7.3.3. Current Challenges and Limitations

Spintronics has introduced new technologies such as Magnetoresistive Random Access Memory (MRAM) and spin field-effect transistors (spin-FETs).

However, there exist various challenges towards the effective utilization of these devices at an expansive scale. One of the biggest challenges is how to realize efficient spin transport and spin torque transfer, two of the essential requirements for the successful operation of spintronic devices. Spintronic memory structures rely significantly on the degree of control over these events in the candidate materials.

Material optimization and interface engineering are the main research avenues to minimize spin scattering that degrades the spin polarization necessary for controlled switching. In addition, variation in operating speed and energy consumption compared to traditional electronic systems adds another factor of complexity, emphasizing the necessity of ongoing research and development to overcome these challenges for spin-based technologies. Aside from spin transport limitations, MRAM and spin-FETs are constrained in design by material compatibility and fabrication issues. For instance, although ferromagnets have been used more frequently in spintronics devices, antiferromagnets are increasingly being studied as new materials due to their unique attributes and benefits. Current studies show how antiferromagnets can be used to provide more efficient magnetic switching behavior, thereby enhancing the performance of devices.

Nonetheless, a transition from conventional ferromagnetic materials to antiferromagnetic systems is difficult to control and incorporate using existing technology, as such materials would preferentially have unique reading and writing mechanisms. Such inherent complexity is one of the factors why researchers and engineers find it hard to create useful and scalable spintronic systems because material properties vs. device efficiency is still a main issue. The potential of spintronics to assist in the realization of neuromorphic computing is an particularly intriguing field of study, but one that is filled with challenge. The building blocks of neuromorphic computing aim to emulate how the human brain functions, and that is decidedly not the traditional computational paradigm.

One of the new concepts that have been discussed is the use of spin waves as an information transport medium, which would improve information processing and information retrieving speed. Albeit, research by Bao et al. (2007) indicates that issues like raised signal attenuation and reduced signal propagation rates compared to conventional electronic transmission lines still exist, therefore inhibiting these spin-wave-based circuits from being integrated into existing technologies. With the advancement of the subject, overcoming these constraints will be necessary to realize the full capability of spintronics in traditional applications and new computing architectures such as neuromorphic systems.

7.4. SPIN FIELD-EFFECT TRANSISTOR (SPIN-FET) DESIGN

Spintronic devices came as a game-changer for paradigms in electronic engineering, particularly the Spin-FET. The Spin-FET utilizes the spin of electrons combined with their charge, leading to more efficient and stronger computer architectures. At the heart of Spin-FET operation is the principles of spin transport and spin torque transfer, through which manipulation of spin currents across various materials is possible. Specifically, the addition of magnetic material within these transistors enhances the potential for non-volatile use in memory, like Magnetoresistive Random Access Memory (MRAM). In addition to benefiting device operation, such technological advances also present paths to power reduction, scalability, and increased data storage, all of which can potentially be major factors in next-generation electronics. With the rising need for high-processing and memory systems, Spin-FETs are a critical breakthrough in addressing these needs. Besides, the integration of MRAM and Spin-FET design is a milestone enhancement of data retention and programming capabilities (Figure 7.10).

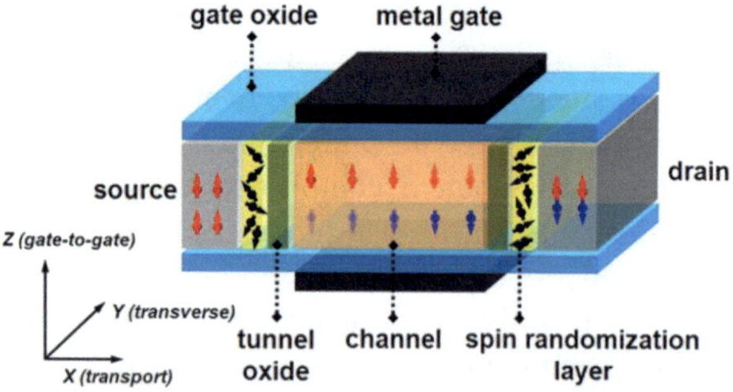

Figure 7.10. *The spin field-effect transistor (Spin-FET) schematic design.*

Source: https://www.researchgate.net/profile/Tony-Low/publication/260509232/figure/fig1/AS:297167504855055@1447861662880/The-spinFET-schematic-The-source-and-drain-are-HMF-The-magnetization-of-the.png.

The interaction among these technologies enables the use of magnetoresistive effects with the capability of maintaining swift switching tendencies, a significant quality for modern computing. Innovative architectures have focused on designing magnetic material characteristics as well as combining them into Spin-FETs. For instance, employing altermagnets with unique spin-splitting characteristics enhances the efficiency of spin injection and detection within the devices (Tian et al., 2024). In addition, spin-gapless semiconductor studies further unlock opportunities for the design of efficient spintronic devices. These advances suggest a trajectory where MRAM and Spin-FET integrate to produce powerful solutions addressing the demands of high-speed, low-power, and energy-efficient technologies in the digital era. With the advancement of spintronics, Spin-FET design implications spill over into revolutionary computing architectures, including neuromorphic computing. This paradigm change is inspired to mimic the operation of biological neural networks, eventually optimizing computational efficiency and processing power. Having spin-based devices brings unique benefits: better parallelism and reduced latency. Using materials that exhibit spin-filtering properties, such as Janus MXene material heterostructures, there can be a notable enhancement in the creation of brain-inspired systems that are energy-efficient but still very high performance (Zhang et al., 2023). With ongoing research aimed at incorporating these new materials, the future of Spin-FETs is poised to transform processing and transmission of information as it directly concerns creating robotics and artificial intelligence technology. The study of Spin-FETs thus not only addresses current electronic demand but anticipates emerging opportunities for the next generation of computational design.

7.4.1. Overview of Spin Field-Effect Transistor (Spin-FET)

Spin-FET development is a grand step forward in spintronics of interest to both processing and data storage. Spin-FETs, based on the intrinsic spin of electrons, hold out the prospect of better performance and less power usage than traditional charge-based transistors. Spin-FET devices take advantage of spin transport processes that enable the efficient control of spin currents as well as traditional electronic currents. One of the key features of spin-FETs is the application of spin torque transfer, offering precise control over magnetization states. This approach obviates the limitations of conventional magnetic devices and paves the way for enhanced switching speed and enhanced reliability during operation. The combination of spintronics principles with transistors is best exemplified by magnetic random access memory (MRAM) architecture, in which spin-FETs play a central role in redefining structure and function of memory and possessing potential for application in computing systems of the future.

Spin-FET architecture also significantly plays a role in developing MRAM technology that employs the management of magnetic states for storing information. Interestingly, the switching performance of spin-FETs is improved with the application of two-dimensional (2D) magnetic materials. The unique properties of these materials enhance spin detection and injection, which are critical for spin-FET operation. The application of the combination of spin-orbit torque (SOT) and tunneling magnetoresistance (TMR) in 2D structures has been found to provide better switching performance with reduced power requirements. Spin-FET technology advancements, driven by understanding the underlying mechanisms of magnetization dynamics, predict a futuristic avenue for MRAM applications within dense storage solutions. Spin-FET-based MRAM information retrieval and processing efficiency can be anticipated to surpass conventional memory technologies and open up vistas for electronic design. Exploration of the potential of spin-FETs extends beyond memory storage, i.e., to neuromorphic computing paradigms. The structure and operation of spin-FETs can be exploited to replicate the synaptic operations of biological systems, providing new options for neural network realization. The potential of spintronics, such as spin state manipulation at room temperature, is helping to achieve circuits that are more efficient than conventional electronic equivalents. Spin-FET advancements in interfacing with neural networks would be capable of enabling adaptive learning-capable systems, capable of carrying out sophisticated calculations at lower energy cost. As it goes on, understanding how to optimize the integration of spintronics in neuromorphic systems will be critical, as it would allow for artificial intelligence and machine learning technology breakthroughs, which would change computing environments and applications fundamentally.

7.4.2. Comparison with Conventional Field-Effect Transistors (FETs)

The development of spintronics has led to a re-evaluation of the traditional electronic device, particularly the FET. While traditional FETs are founded on the conduction of charge alone, spintronic devices capitalize on the inherent spin of electrons, furthering information processing capabilities. This union of spin transport and spin torque transfer provides an entire universe of advantages. For instance, spin-based devices can provide higher speeds of switching and require less power, qualities that are crucial for today's computing demands. Such devices include magnetic random-access memory (MRAM) and spin-FET design, which are actualizations of this technological shift, with MRAM providing non-volatility as well as high speed of access at reduced energy consumption compared to NAND flash memory. The transition away from conventional FETs and towards spintronic devices is not just a technological advancement but also one of paradigm in the way information can be stored and processed, as underlying principles like spin offer a new axis for the encoding of data. The tunability of materials such as molybdenum disulfide (MoS_2) also reflects the potential advantages of spintronic devices over their conventional counterparts. MoS_2 boasts excellent characteristics, including tunable band gap and superior magnetoresistance, which make it an attractive option for application in spintronics and a replacement for conventional semiconductors. Unlike standard FETs, which essentially operate within a restricted structural framework, MoS_2's layered structure allows for new interactions at the nanoscale level that can enhance the effectiveness of spin transport. This renders it more applicable to other applications such as biosensors and energy use, indicative of its versatility in applications that call for precision and reliability. As research continues to be conducted on MoS_2, the application of spintronics renders the materials used in electronic devices more versatile and pushes the development of future computer technologies, which conventional FETs cannot match (Figure 7.11).

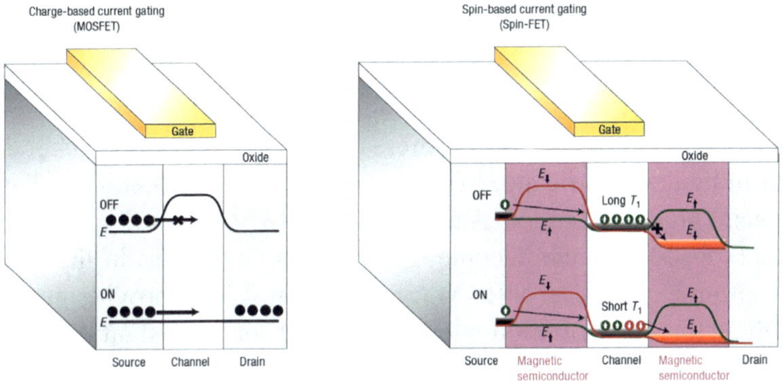

Figure 7.11. *The interplay between metal-oxide-semiconductor field-effect transistor (MOSFET) (left) and a spin field-effect transistor (Spin-FET) (right).*

Source: https://www.researchgate.net/profile/Michael-Flatte/publication/239030825/figure/fig4/AS:280095743856656@1443791437223/Schematic-structure-of-a-MOSFET-left-and-a-spin-FET-rightMOSFETs-work-by-raising-and.png.

The prospect of neuromorphic computing, which aims to emulate the operation and efficiency of biological neural networks, also highlights the strength of spintronic devices. As opposed to binary operation of conventional FETs, spintronic devices offer the possibility of multi-state processing via spin states. This capability allows for more advanced management of data, allowing complex computations approaching the ones performed by the human brain. Additionally, unraveling the purpose of phenomena such as magnons, as elucidated in current research, indicates further possibilities within spin-wave computing architectures that could revolutionize the way information receives processing in neuromorphic systems. Thus, the advantages of spintronic devices over FETs have the capability to redefine not only the terrain of data processing but also the very terrain upon which future models of computation are built, thus enriching the terrain of theoretical and applied computating disciplines.

7.4.3. Potential Applications in Electronics

Studies of spin transport and spin torque transfer processes have opened up new avenues in electronic device engineering. Such impacts make possible efficient spin manipulation and provide a qualitatively novel method of data processing and storage. Spin currents, once generated, can be converged to achieve particular effects, including reducing power consumption and boosting speed in electronic devices. The prospect of applying spin currents to create new memory and logic system architectures is full of promise. As researchers investigate these mechanisms, they use the core principles of spintronics to develop new technology, which bypasses the limitations of current electronic devices. This advancement paves the path towards more energy-efficient and quicker devices, thus confirming the significance of spin transport in next-generation electronics applications. Magnetic Random Access Memory (MRAM) and spin-field effect transistors (spin-FET) have become the cornerstone devices in the context of next-generation electronics. MRAM uses electron spin to provide non-volatile memory that stores information without the need for a constant power supply. The attribute provides radical improvement in power efficiency and device life as a response to increased memory needs in miniaturized electronics. Similarly, spin-FET technology uses electron spin to provide improved switching speed and low power consumption. With the globe inclined towards shrinking but capacitated devices, both MRAM and spin-FET layouts can change the process and storage of information. Application of these spintronic devices in existing architectures could possibly induce a massive increase in computation capability, thereby affecting various industries that rely on data processing. Furthermore, the spintronic technology is of relevance in the upcoming field of neuromorphic computing, where the neurobiological processes of the human brain are to be emulated. Leverage the unique features of the spintronic device, the scientists are investigating architectures that mimic neurobiological systems and operations, such as synaptic interfaces. Spin-based computation power efficiency is also in line with the neuromorphic design needs, which require low power dissipation to fuel advanced neural networks. As technology progresses, spintronic concepts' adoption in neuromorphic computing holds the potential to revolutionize artificial intelligence and machine learning. This meeting of spintronics and neuromorphic computing shows not only the potential of magnetic devices but also their relevance in mapping the future of smart systems, which holds a golden future of research and development in electronic applications (Shi et al., 2024; Irfan, 2021).

7.5. PROSPECTS FOR NEUROMORPHIC COMPUTING

The integration of spintronics with neuromorphic computing is a new path for future electronic devices, particularly with the conventional semiconductor technology reaching its physical limits. Spin transport and spin torque transfer mechanisms are now becoming increasingly identified for their capacity to model neurobiological operations of the human brain and consequently enhance computing capabilities. Spintronic devices such as magnetic random-access memory (MRAM) exploit spin currents' unique properties to store and process data in an efficient manner. Not only is it faster but also non-volatile, which is necessary to enable energy-efficient neuromorphic systems.

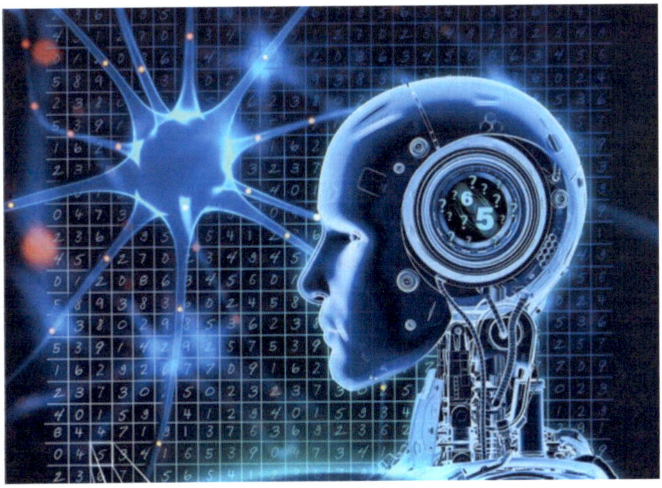

Figure 7.12. *The integration of advancement in neuromorphic computing.*

Source: https://www.hpcwire.com/wp-content/uploads/2023/06/bf20d015-a028-46d3-b386-db6dda01ea17-600x426.jpg.

As these systems evolve, they are becoming capable of processing data in real time and optimizing machine learning architectures to lead to more advanced applications of artificial intelligence. In order to realize the computational capability of spintronics, spin-field effect transistor (spin-FET) design is of critical importance for neuromorphic computing architectures. By enabling manipulation of spin polarization, they allow synaptic elements to be developed mimicking the biological neural networks and assist in realizing spiking neural networks capable of doing adaptive learning effectively. Recent advances in spiking neural network (SNN) algorithms reflect growing synergy between spintronics and neuromorphic computing, particularly in applications requiring rapid and efficient information processing (Figure 7.12).

Convergence also forms the basis of new architectures capable of handling the vast amount of data produced by modern electronic systems. The development of such technologies is not entirely hypothetical; the research today into these systems emphasizes the possible to revolutionize computing paradigms, particularly as these systems integrate further with sophisticated algorithms for supervised and unsupervised learning (Bucukovski et al., 2023). Research into the future of neuromorphic computing in the field of spintronics presents the potential for unrivaled advancements in smart technology. As noted in higher material analyses, oxide films and heterostructures are promising contenders for developing neuromorphic systems with enhanced functionality and efficiency (Althammer et al., 2019). Oxide films and heterostructures can potentially provide the needed flexibility and responsiveness to meet advanced computational models inspired by human intelligence. The inter-disciplinary innovation in recent work highlights the necessity of transcending the limitations of existing technology by using new materials and device structures. With researchers and engineers gathering their expertise on new solutions, the working paradigms of neuromorphic computing are ready to shift so that devices can learn and adapt by patterns following human cognition. The ongoing research of spintronics in this area not only holds significant implications for the computing future but also for a wide variety of applications across many sectors.

7.5.1. Definition and Significance of Neuromorphic Computing

As societies become increasingly dependent on advanced computing functionality, the advancement of neuromorphic computing has emerged as a revolutionary means at reinstating biological neural network structures. This paradigm revolution in computing architecture is crucial in particular because it holds the potential for enhanced efficiency and performance in handling complex tasks that traditional computing is incapable of handling effectively. Utilizing mechanisms akin to the operation of human brains, neuromorphic systems can perform tasks such as pattern recognition and sensory processing with remarkable power efficiency. Besides, integrating spintronics with neuromorphic computing will bring these systems to a whole new level since spintronic devices, like MRAM and spin-FETs, enable faster and more durable processing of data. The very existence of neuromorphic computing is founded on the application of devices with the potential to fully emulate the actions of biological neurons. Experiments have demonstrated that artificial neuronal devices can be created based on new materials with volatile switching behavior, thus emulating the dynamic nature of real neurons (Liu et al., 2023). Not only do such devices strengthen the computational capability of neuromorphic systems, but they also enable more realistic brain process emulations using various neuron models (Figure 7.13).

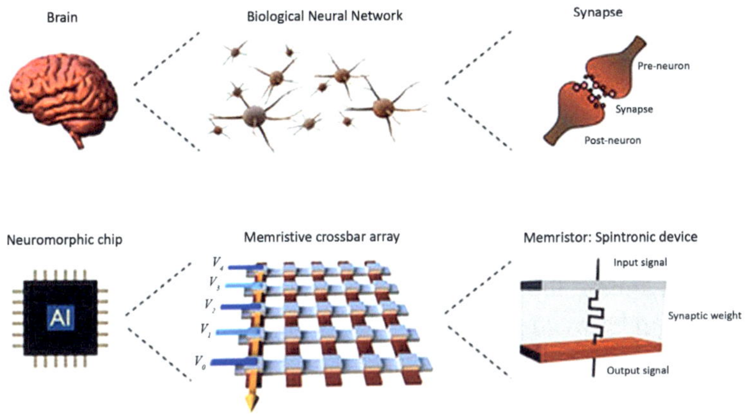

Figure 7.13. *Neuromorphic computing architecture inspired by a human brain.*

Source:https://www.researchgate.net/publication/366035852/figure/fig4/AS:1143128125672007l@1719566936081/Neuromorphic-computing-architecture-inspired-by-a-human-brain-a-The-brain-computing.jpg.

With the use of the spin transport and spin torque transfer concepts, researchers can be able to design architectures that take advantage of the benefits of spintronic and neuromorphic technology. From the cross-disciplinary approach, there will be new ways for promoting innovative computing systems that involve minimal energy input. In the future, the significance of neuromorphic computing rises with the increasing need of industries to embrace more complex algorithms and machine learning architecture. The integration of neuromorphic systems into spintronic devices holds the potential for a promising path to achieving higher performance in data-hungry applications with much less energy usage than conventional technology. The prospect of benefits extends beyond the faster processing speed to include increased robustness and fault tolerance because the neural network is inherently distributed. With advancing technology, neuromorphic computing, aided by spintronics innovations, could transform the manner in which we perform artificial intelligence and sensory processing applications, thus becoming a cornerstone for future technological innovations (Hayashi et al., 2023).

7.5.2. Role of Spintronics in Mimicking Neural Networks

Recent research advances in spintronics have opened up new frontiers for neural network simulation on the basis of the new properties of spin-based devices. One of the primary mechanisms of such emulation is spin transport and spin torque transfer, and they are instrumental to the functioning of various spintronic devices.

Through the manipulation of spin-polarized currents, these devices can mimic the dynamics of biological neurons quite effectively and in a predictable way. Modulation of electrical properties through magnetic fields and electric currents puts these devices in a prime position to be employed for hardware neural network design. Notably, experiments that examine the coupling of spin dynamics with the firing characteristics of neurons have led to models that efficiently mimic the operation of spiking neural networks (SNNs), thus bridging the gap between traditional computing models and more biologically inspired architectures. Moreover, the inclusion of materials like magnetic random-access memory (MRAM) and spin field-effect transistors (spin-FETs) has revolutionized the evolution of neuromorphic computation. The character of these devices allows them to store and process information similarly to the biological synapses and neurons.

Leveraging the different modes of operation inherent in MRAM technology, such as spin-transfer torque and spin-orbit torque, researchers have managed to create devices that can emulate the operations of synapses, such as potentiation and depression. This not only enhances the efficiency with which data is stored but also raises the speed at which computations for neural networks can be performed. The ability of such systems to be tuned, as they learn to adjust to novel inputs, speaks of their promise for emerging neuromorphic computing breakthroughs. Building on these advances, research into magnetic soliton-based devices has found that they hold promise in terms of their application toward simulating neuronal functionality. Magnetic skyrmion-based and domain-wall-based devices have proven highly promising, as synapses as well as neurons. By means of structural engineering of device properties, as envisioned in multilayer ferromagnetic system research, it is possible to integrate these spintronic elements into artificial neural networks effectively. In practical application, e.g., training and testing on the MNIST dataset, performance has demonstrated not only classification accuracies above 90% in artificial neural network configurations but also improved performance in spiking neural networks, achieving accuracies of approximately 97% (Lone et al., 2024, 2025). Such findings highlight the revolutionary promise of spintronic devices in facilitating energy-efficient, scalable neuromorphic computing architectures.

7.5.3. Future Directions and Research Opportunities

Investigating the intersection of spin transport and spin torque transfer holds major areas of future research in the spintronics field. As researchers gain a deeper understanding of the mechanisms that control these effects, the prospect for improved device performance becomes ever clearer. Studies of spin injection and manipulation efficiency can liberate the operating frequency and energy efficiency of spintronic devices. Most importantly, the spin-transfer torque phenomenon in magnetic tunnel junctions can be tailored with future directions to build robust memory technologies such as MRAM. As the technologies reach maturity, integrating them with spin field-effect transistors (spin-FETs) could make low-power, high-speed computing architectures possible. This convergence not just indicates the potential for more superior measures of performance but also propels the field into practical application, as the investigation of spin transport mechanisms becomes central to the future generation of magnetic devices. With further advances in magnetic device design and fabrication, advances in MRAM and spin-FET structures seem like necessary areas of research. Scalability and compatibility of such technologies with standard semiconductor processing may allow their inclusion in bulk computing systems, which could open the way to high-density and power-efficient memory devices. Further work must be done to optimize the materials used in these devices for improved thermal stability and reduced write-error rates. New tunnel barrier materials and their interactions with ferromagnetic layers could provide sets that significantly improve device performance. Moreover, the study of hybrid systems that integrate conventional charge-based devices with magnetic components can extend the functional capabilities of spintronic devices, and this opens a rich field for investigation that intersects multiple fields of materials science and engineering. Another productive field for future investigation in spintronics is the fast-evolving field of neuromorphic computing. For the sake of mimicking brain-like computing functions, neuromorphic devices can utilize spin-wave phenomena for efficient computational functions. Research on magnonics—where spin waves are controlled to facilitate data processing—is also hopeful in the area of high-speed processing from the GHz-to-THz scale, as discussed in existing research (Chumak et al., 2022). While researchers venture into new paradigms that encompass nonconventional computing architectures, the integration of spintronic devices can enable paradigms in energy-efficient data processing. Such technologies not only have the potential to increase the application of artificial intelligence but also push the evolution of advanced complex systems capable of performing tasks replicating mental processes. Given the challenges and experimental nature of this field, these opportunities are imperative to the ongoing success of spintronics and magnetic devices in future computing systems.

7.6. CURRENT RESEARCH AND INNOVATIONS

Advances in the field of spintronics continue to broaden the horizon for magnetic devices, particularly with advancements in spin transport and spin torque transfer processes. Researchers are actively exploring nonconventional spin-orbit torques (SOTs), which have shown great promise towards better magnetization control without the need for external magnetic fields. The method enables energy-efficient as well as scalable design, particularly critical for magnetic recording and storage applications, e.g., magnetic random-access memory (MRAM).

Figure 7.14. *Innovations in the field of spintronics.*

Source: https://www.innovationnewsnetwork.com/wp-content/uploads/2020/07/INN2-EURAMET-1088-atl-image-1.jpg.

Of special interest, some more recent works have delineated the employment of chiral antiferromagnets in combination with low-symmetry van der Waals materials to achieve remarkable improvements in magnetic switching performance, as supported by a high switching ratio of 90% in the field-free process (Zhou et al., 2025). These advances illustrate how the introduction of new materials can facilitate fundamentally enhanced efficiency and potential of next-generation spintronic devices. The crossroads of spintronics and device architecture has brought forth revolutionary ideas such as the spin field-effect transistor (spin-FET), which exploits spin polarization to regulate electronic signals. This groundbreaking solution has the potential to surpass the limitations of traditional electronic devices, opening the door to a new world of ultra-fast and low-power computing solutions. Current research highlights the potential that spintronics has to offer memory storage technology, in this case, MRAM, in terms of better data retention and read/write speeds. Furthermore, research on 2D metallic altermagnets has confirmed their potential in improving tunneling magnetoresistance, which is crucial for improved performance in magnetic tunnel junctions (Sattigeri et al., 2025). By continued exploration of the unique properties of such materials, researchers are establishing the foundation for future-state applications in data processing and data storage (Figure 7.14).

Future prospects for integrating spintronic ideas into neuromorphic computing systems hold the key to revolutionizing artificial intelligence and machine learning technologies. New developments leverage the operations of spintronic devices to replicate synaptic activities for fast processing with reduced energy consumption. The possibility to utilize magnetic materials in network systems can be equated to advancements in developing bio-inspired computing architectures with learning and associative memory capabilities. By immersing spintronics within the neuromorphic paradigm, researchers take advantage of both magnetic device switching speed and synapse analogity, ultimately resulting in next-generation smart systems. Ongoing research, with the complementarity of spintronics and neuromorphic ideas, depicts the effective synergy between material science and computing technology, pointing to an innovation-filled future.

7.6.1. Recent Advancements in Spintronic Materials

New breakthroughs in spintronic materials have indicated vast potential for enhancing spin transport and spin torque transfer methods. These breakthroughs have increased the effectiveness with which spin-polarized currents can be controlled, required for efficient data storage and processing in devices. Of particular significance has been the research on Heusler alloys, which exhibit tunable functionalities capable of significantly influencing spintronic applications. With the alteration of their compositions and structures, researchers have engineered these alloys to maximize their magnetic and electronic properties, thus enabling magnetoresistive applications to gain in performance. With this innovation, aside from having the potential for storing more data and faster, there is also the optimization of energy consumption over traditional electronic devices, eventually leading to the innovation in magnetic technology and future computing uses (Kumar et al., 2025). Another landmark advancements reports the incorporation of two-dimensional (2D) materials into the fabrication of future spintronic devices, namely in the areas of magnetic random-access memory (MRAM) and spin field-effect transistors (spin-FETs). Although majority of the 2D materials are naturally non-magnetic, new approaches are being explored to impart ferromagnetic properties to 2D materials through techniques such as heteroatom doping. Experiments have proven that such a process can create localized magnetic moments in such systems, enabling functionalities not made possible before in non-magnetic 2D materials. The implications of these findings transcend theoretical models; they prospect for the use of MRAM and spin-FET devices that are more efficient and denser in terms of storage, hence revolutionizing memory devices in contemporary computing environments (Wang et al., 2025) (Figure 7.15).

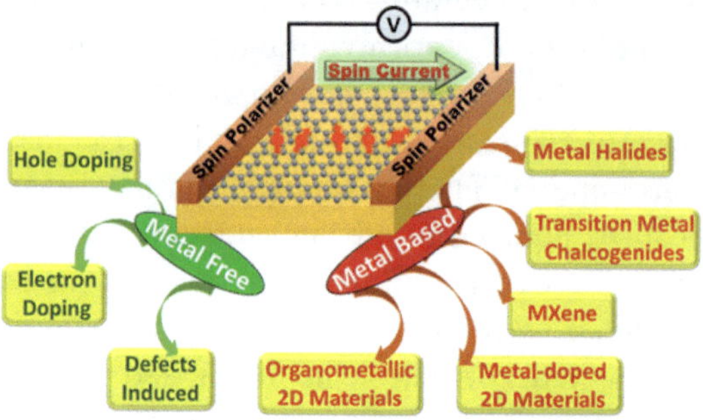

Figure 7.15. *The contemporary advancements in the field of spintronic materials.*

Source: https://pubs.acs.org/cms/10.1021/acs.chemmater.9b02243/asset/images/medium/cm9b02243_0015.gif.

Besides, advancement in spintronic materials also promises neuromorphic computing, a paradigm that emulates the architecture of the human brain to lower computational complexity and enhance adaptability. The integration of magnetism in plastically modifiable materials might enable spin-based synaptic devices, rather than charge-based, to be implemented, which would facilitate quicker and more energy-efficient computation. Furthermore, advancements in materials with tunable magnetic properties will likely equip new neuron-like devices with the capability to mimic synaptic behavior, which would be crucial in driving the development of artificial intelligence systems further. Since researchers continue investigating the intersection between materials science and computation, the future of neuromorphic computing will benefit immensely from recent breakthroughs in spintronic materials, creating new opportunities for intelligent, responsive machines (Kumar et al., 2025).

7.6.2. Breakthroughs in Device Fabrication Techniques

Advances in the device fabrication techniques have significantly impacted spintronic technologies development and scalability, particularly in spin transport and spin torque transfer domains. Advances in these techniques have enabled researchers to come up with new magnetic materials with engineered properties essential to enhancing spin coherence and manipulation. These developments allow the fabrication of intricate structures that can exploit spin dynamics for improved device performance. For instance, the combination of atomic layer deposition and molecular beam epitaxy allows precise control of thickness and uniformity of layers in magnetic tunnel junctions (MTJs). Such developments not just improve the efficiency of spin transport mechanisms but also provide new routes for the design of hybrid devices making use of both charge and spin, thereby paving the way for next-generation spintronic technology (Sud et al., 2025). In the development of memory technologies with high efficiency, innovation in fabrication methods has played a critical role in improving magnetic random-access memory (MRAM) as well as spin field-effect transistors (spin-FETs).

The emergence of low-dimensional materials (LDMs), transition metal dichalcogenides, and graphene is a revolutionary step towards optimizing device interfaces and removing defects that are inherent in traditional fabrication methods. These materials provide enhanced tunnel magnetoresistance (TMR) as a result of their unique electronic and magnetic properties. As a result, recent MRAM devices have better storage capacities and switching speeds at room temperature under working stress (Thapa et al., 2025). Moreover, the improvement in the MTJ structure and design has improved their scalability and reliability and made them suitable candidates for future computing paradigms. The promise of neuromorphic computing has also been significantly boosted through developments in device fabrication technology in the spintronics research area. With materials and geometries that mimic the synaptic function of natural systems, researchers can take advantage of spintronic devices in neural network structures. The addition of spintronic phenomena, such as spin-orbit torques and long-lived helix states of spin, enables low-power and high-speed operation essential for neuromorphic approaches. Technologies like van der Waals (vdWs) heterostructures also exhibit promising potential for enhancing spin injection and processing efficiencies. This intersection of device engineering and materials science provides a novel approach to neuromorphic computing that is beyond the confines of traditional architecture, hence setting the future direction of both artificial intelligence and information processing (Sud et al., 2025; Thapa et al., 2025).

7.6.3. Collaborative Efforts in Academia and Industry

In the context of new emerging technologies, the interface between research in academia and application in industry is more important than ever before, especially with spintronics. There is a need for cooperation to drive knowledge forward and create innovative solutions that will lead the next generation of magnetic devices. With partnerships, scientists will have access to the capabilities and knowledge within industry and will be able to bring theoretical ideas into practice. For instance, spin transport and spin torque transfer mechanisms, central aspects of spintronics, benefit significantly from the highly controlled experimental conditions that can be provided by industrial partners. Such collaboration not only accelerates research but also accelerates the probability of emerging technologies in information technology and consumer electronics. Furthermore, the innovation and production of Magnetoresistive Random Access Memory (MRAM) and spin-FET (spin field-effect transistor) devices are the greatest indications of the possibility to create breakthroughs from industry-academia collaborations. Universities and research institutions typically innovate theoretical concepts and novel materials, and companies convert this into voluminous manufacturing techniques. Collaboration enhances the understanding of material properties and behavior in various conditions, which can facilitate advanced designs that would be impossible in an isolated environment. An example is the combination of advanced research modeling with industrial applied prototyping producing novel solutions tailored to market needs, thus fostering a renewed and active way of spintronics technological development. In the future, neuromorphic computing potential is yet another powerful incentive for industry-academia collaboration. As needs for computing systems rise in efficiency, the ability to replicate neural processes using spintronic devices becomes a desirable research area. While academic researchers can explore theoretical underpinnings and new architectures, business companies can develop high-performance computing systems based on these concepts. This collaboration can yield technologies that utilize the unique aspects of spintronic behavior for higher computational speed, lowered energy efficiency, and improved data-processing ability. Crossing ideas from various disciplines brings about a dynamic setting where the development of spintronics can play a huge role in the technology industry, leading to groundbreaking innovations in the field of academia and industry (Research Institute of Electrical Communication Tohoku University, 2020, 2024).

SUMMARY

In considering the developments and consequences of spintronics, it is clear that combining spin transport and spin torque transfer mechanisms is not a theory for theory's sake but is instead an initial step towards the practicalities. These mechanisms are significant in that they have the capability to transform data storage and processing powers by virtue of using electron spin. For instance, devices like MRAM illustrate the promise of these ideas to provide solutions to energy consumption problems inherent in conventional electronic systems. Continued research to optimize spin-FET (field-effect transistor) architecture highlights the technology's readiness to compete with established CMOS-based designs, which leads to enhanced performance and reduced power consumption, thereby mitigating urgent issues regarding energy efficiency within current paradigms of computing. These advances underscore the promise of spintronic technology as a component of next-generation electronics. Spintronics research has provided broad avenues of opportunities for the development of next-generation computing paradigms, particularly with regard to neuromorphic computing. Neuromorphic architectures aim to replicate the functioning of biological neural networks through exploitation of the efficiency and adaptability of these designs. The interfacing of spintronic principles, especially in memory and processing devices, is very strong in this field. Advances in hybrid magnetic/silicon technology, for example, that have the potential for logic-in-memory implementations is welcome news for neuromorphic use cases. By the use of spintronic effects, the technology can potentially support higher speeds of processing and less energy consumption, putting it squarely within the operating tolerances of neuromorphic computing. As the technology matures, the marriage of spintronics with neuromorphic architectures could usher in a paradigm shift in how computational problems are solved. In the long term, the future of spintronics and magnetic devices holds a time of promise for computational ingenuity. As highlighted in recent perspectives, the underlying phenomena responsible for spintronics such as giant magnetoresistance and spin-transfer torque have allowed for the design of advanced units of memory beyond the limitations of standard architectures. Moreover, investigation on concepts such as the chirality-induced spin selectivity highlights the ongoing quest to understand and harness the complexity of electron spin to achieve utilitarian functions. With continuing technological progress and the expanding use of such advancements, the promise of future computing is enormous, with profound effects on industries based on sophisticated memory and processing (Barla et al., 2021; Evers et al., 2022). Accelerating work in this field not only promises increased efficiency but also introduces new paradigms, leading the way to earth-shattering breakthroughs in electronics and beyond (Barla et al., 2021; Evers et al., 2022).

Spintronics' advancements owe much to the physics of spin transport and spin torque transfer. The phenomena play central roles in controlling electron spins in devices effectively and, in doing so, enhancing their performance over traditional electronics based on charge transport. Spin torque transfer enables magnetic state switching at a faster speed, which underlies the MRAM operation. This memory technology exploits the unique properties of electron spins to achieve better access times and lower energy consumption, providing seeming advantages over conventional RAM. The ability of data storage in a non-volatile manner also mirrors the real-world applications of efficient spin transport, allowing for long-term data retention even through power loss, thus widening the scope and application of MRAM in current computing infrastructure. Spin-based devices also include designs such as the spin-FET, where the interdependence between charge and spin currents is used. Through spin integrated very effectively in semiconductor technology, spin-FETs hope to provide higher logic functionality with features such as low power consumption and increased speed. Spin-based devices fundamentally change information processing, leading to the possibility of integrating magnetic properties at the nanoscale. This ability has significant implications for next-generation computing architectures, which require more efficient and speedier devices. Spintronics' design optimization not only promises entry points for new memory technologies but also drives efforts to study hybrid devices that retain magnetic and electronic characteristics to exhibit novel behavior. Spintronics has tremendous potential in the future for neuromorphic computing that attempts to replicate how biological brains process things. The unique characteristics of spintronic devices enable the use of artificial neurons and synapses that have the ability to perform complex calculations in a manner akin to human thought. The integration of spintronics into neuromorphic architecture can lead to advances in machine learning and artificial intelligence with a foundation for systems that are both more efficient as well as learn from their environment. With more research, the intersection of spintronics and emerging paradigms for computing can redesign computing technologies to offer new methods that harness the physical characteristics of materials. These breakthroughs mark an essential milestone towards the creation of smarter and more perceptive systems that push the boundaries of what is technologically possible, hence transforming various applications across various fields.

REFERENCES

1. Al Saidi, W., Bhatti, S., Piramanayagam, S., & Sbiaa, R. (2024). Annihilation mechanisms for interacting skyrmions in magnetic nanowire. *Journal of Physics D: Applied Physics, 57*. https://www.semanticscholar.org/paper/da1b7553762e209c912ee6daf8aca587cf57a792 (accessed on 27th September 2025).
2. Althammer, M., Bibes, M., Boschker, H., Calleja, F., et al. (2019). Towards oxide electronics: A roadmap. https://core.ac.uk/download/199214191.pdf (accessed on 27th September 2025).
3. Bao, M., Galatsis, K., Khitun, A., Nikonov, D., et al. (2007). Feasibility study of logic circuits with spin wave bus. http://arxiv.org/abs/0704.2862 (accessed on 27th September 2025).
4. Barla, P., Joshi, V. K., & Bhat, S. (2021). Spintronic devices: A promising alternative to CMOS devices. *Journal of Computational Electronics, 20*, 805–837. https://doi.org/10.1007/s10825-020-01648-6 (accessed on 27th September 2025).
5. Barman, A., Gubbiotti, G., Ladak, S., Adeyeye, A. O., Krawczyk, M., Gräfe, J., Adelmann, C., et al. (2021). The 2021 magnonics roadmap. *Journal of Physics: Condensed Matter, 33*, 413001. https://doi.org/10.1088/1361-648x/abec1a (accessed on 27th September 2025).
6. Baumgartner, M. (2018). Time- and spatially-resolved magnetization dynamics induced by spin-orbit torques in thin film devices. https://www.semanticscholar.org/paper/eefb407a56958b7096dde205fdb0038dc5623055 (accessed on 27th September 2025).
7. Bordoloi, A., Garcia-Castro, A. C., Romestan, Z., Romero, A. H., & Singh, S. (2024). Promises and technological prospects of two-dimensional Rashba materials. *Journal of Applied Physics*. https://www.semanticscholar.org/paper/df2e1c660757119afc99009e625c6565c98e5a66 (accessed on 27th September 2025).
8. Brataas, A., Kamra, A., Korenivski, V., Polishchuk, Y., et al. (2019). Anisotropic and controllable Gilbert-Bloch dissipation in spin valves. http://arxiv.org/abs/1811.00020 (accessed on 27th September 2025).
9. Bucukovski, J., Carlson, E., Sima, L., Yien, T., et al. (2023). Advanced computing and related applications leveraging brain-inspired spiking neural networks. http://arxiv.org/abs/2309.04426 (accessed on 27th September 2025).

10. Chumak, A. V., Kaboš, P., Wu, M., Abert, C., Adelmann, C., Adeyeye, A. O., Åkerman, J., et al. (2022). Advances in magnetics roadmap on spin-wave computing. *IEEE Transactions on Magnetics, 58*, 1–72. https://doi.org/10.1109/tmag.2022.3149664 (accessed on 27th September 2025).

11. Evers, F., Aharony, A., Bar-Gill, N., Entin-Wohlman, O., Hedegård, P., Hod, O., Jelínek, P., et al. (2022). Theory of chirality induced spin selectivity: Progress and challenges. *Advanced Materials, 34*. https://doi.org/10.1002/adma.202106629 (accessed on 27th September 2025).

12. Fina, I., Jungwirth, T., & Martí, X. (2015). Prospect for antiferromagnetic spintronics. https://core.ac.uk/download/45448865.pdf (accessed on 27th September 2025).

13. Firdous, S., Malik, G. F. A., Parveen, N., Khanday, F. A., & Pandey, I. K. (2024). First principal investigation of Sr and Eu doped La_2NiMnO_6: Structural, electronic, optical, magnetic, and spintronics properties. *MatSci Express*. https://www.semanticscholar.org/paper/14cb78a1c1bd12a371345384930dc95c225ff99b (accessed on 27th September 2025).

14. Ghosh, B., Ghosh, T. K., & Kamra, A. (2010). Spin relaxation due to electron-electron magnetic interaction in high Lande g-factor semiconductors. http://arxiv.org/abs/1008.0739 (accessed on 27th September 2025).

15. Gu, X., Guo, L., Qin, Y., Yang, T., Meng, K., Hu, S., & Sun, X. (2023). Challenges and prospects of molecular spintronics. *Precision Chemistry, 2*, 1–13. https://www.semanticscholar.org/paper/9069f375ae267c64f09a588f43ddcbc6bc6eccea (accessed on 27th September 2025).

16. Guo, S., Zhang, Y., Ge, Y., Zhang, S., Zeng, H., & Zhang, H. (2019). 2D V-V binary materials: Status and challenges. *Advanced Materials, 31*. https://doi.org/10.1002/adma.201902352 (accessed on 27th September 2025).

17. Hamzehbahmani, H., Sabariego, R., & Ducharne, B. (2024). A knowledge-based analysis of interlaminar faults for condition monitoring of magnetic cores with predominant focus on axial offset between the fault points. *IEEE Transactions on Instrumentation and Measurement, 73*, 1–10. https://www.semanticscholar.org/paper/055efd198020b16e07ae14862245e2b7fa52acde (accessed on 27th September 2025).

18. Hayashi, H., Jo, D., Go, D., Gao, T., Haku, S., Mokrousov, Y., Lee, H. W., et al. (2023). Observation of long-range orbital transport and giant orbital torque. *Communications Physics, 6*. https://doi.org/10.1038/s42005-023-01139-7 (accessed on 27th September 2025).

19. He, W., Kong, L., Zhao, W., & Yu, P. (2022). Atomically thin 2D van der Waals magnetic materials: Fabrications, structure, magnetic properties and applications. *Coatings, 12*, 122. https://doi.org/10.3390/coatings12020122 (accessed on 27th September 2025).

20. Ioannou, M. (2023). The role of ferromagnets and antiferromagnets for spintronic memory applications and their impact in data storage. *Emerging Minds Journal for Student Research*. https://www.semanticscholar.org/paper/6b10d6919067830e65b9ea0de6856f4045290421 (accessed on 27th September 2025).

21. Irfan, B. (2021). Spintronics: Overview on spin-based electronics and its potential applications. *International Journal of Materials Science and Applications*. https://www.semanticscholar.org/paper/73ed27d8194d0d0240c8ec27d24106fcac565044 (accessed on 27th September 2025).

22. Ito, K., Kurniawan, I., Shimada, Y., Miura, Y., Endo, Y., & Seki, T. (2024). Giant tunability of magnetoelasticity in Fe_4N system as a platform to unveil correlation between magnetostriction and magnetic damping. *Communications Materials*. https://www.semanticscholar.org/paper/4fc7bf128347b8cf6c9ea99776f0f71b9e56b5f1 (accessed on 27th September 2025).

23. Jia, Z., Zhao, M., Chen, Q., Tian, Y., Liu, L., Zhang, F., Zhang, D., et al. (2025). Spintronic devices upon 2D magnetic materials and heterojunctions. *ACS Nano, 19*, 9452–9483. https://www.semanticscholar.org/paper/d432fffe57f42310221ba87dbf73e03b594c591c (accessed on 27th September 2025).

24. Khan, R., Rahman, A. U., Zhang, Q., Kratzer, P., & Ramay, S. M. (2021). First-principles computational exploration of ferromagnetism in monolayer *GaS* via substitutional doping. *Journal of Physics: Condensed Matter, 33*, 314003. https://doi.org/10.1088/1361-648x/ac04ce (accessed on 27th September 2025).

25. Kumar, R., & Gupta, S. (2025). Novel Heusler materials for spintronic applications: Growth, characterizations and applications. https://www.semanticscholar.org/paper/7fe3147d69dacad451774adf7032095a6252a311 (accessed on 27th September 2025).

26. Liu, H., Qin, Y., Chen, H. Y., Wu, J., Ma, J., Du, Z., Wang, N., et al. (2023). Artificial neuronal devices based on emerging materials: Neuronal dynamics and applications. *Advanced Materials, 35*. https://doi.org/10.1002/adma.202205047 (accessed on 27th September 2025).

27. Liu, S., Tomasello, R., Wu, Y., Fang, B., Chen, A., Zheng, D., Zhang, B., et al. (2025). Topological skyrmion-based spin-

torque-diode effect in magnetic tunnel junctions. *Advanced Electronic Materials*. https://www.semanticscholar.org/paper/d1ebeca30e90bcb71885de49f6174f3744fafe44 (accessed on 27th September 2025).

28. Lone, A. H., Tang, M., Rahimi, D. N., Zou, X., Zheng, D., Fariborzi, H., Zhang, X., et al. (2025). Spintronic memtransistor leaky integrate and fire neuron for spiking neural networks. *Advanced Electronic Materials*. https://www.semanticscholar.org/paper/48c901a70ab4e065def1f5824fe3166b2deb042a (accessed on 27th September 2025).
29. N/A. (2020). *RIEC Newsletter No. 6*. https://core.ac.uk/download/287358514.pdf (accessed on 27th September 2025).
30. N/A. (2024). *Annual Report of Research Institute of Electrical Communication, Tohoku University 2023*. https://core.ac.uk/download/622507588.pdf (accessed on 27th September 2025).
31. Okuno, T., Kim, D. H., Oh, S. H., Kim, S. K., Hirata, Y., Nishimura, T., Ham, W., et al. (2019). Spin-transfer torques for domain wall motion in antiferromagnetically coupled ferrimagnets. *Nature Electronics, 2*, 389–393. https://www.semanticscholar.org/paper/4ffda67119880df8f38190963941588d7d6ecdc3 (accessed on 27th September 2025).
32. Parvini, T. S., Paz, E., Böhnert, T., Schulman, A., Benetti, L., Oberbauer, F., Walowski, J., et al. (2023). Enhancing spin-transfer torque in magnetic tunnel junction devices: Exploring the influence of capping layer materials and thickness on device characteristics. *Journal of Applied Physics*. https://www.semanticscholar.org/paper/798d97ccf9c39aba183bc9c02c2af7cf79ff5c1f (accessed on 27th September 2025).
33. Qi, W., Zhang, H., Chen, L., Du, A., Zheng, D., Xiao, Y., Tian, D., et al. (2024). Antiferromagnetic spintronics in magnetic memory devices. *IEEE Transactions on Materials for Electron Devices, 1*, 23–35. https://www.semanticscholar.org/paper/b4fd9dfc8e994f7ef4371904877f0dc9a7d07629 (accessed on 27th September 2025).
34. Rabelo, R., Toma, L., Bentama, A., Stiriba, S. E., Ruiz-García, R., & Cano, J. (2024). Exploring spin-crossover cobalt (II) single-ion magnets as multifunctional and multiresponsive magnetic devices: Advancements and prospects in molecular spintronics and quantum computing technologies. *Magnetochemistry*. https://www.semanticscholar.org/paper/6c27b8f46455507e9e0a75497ac87d26c1ae632a (accessed on 27th September 2025).
35. Roche, S., van Wees, B. J., Garello, K., & Valenzuela, S. O. (2024).

Spintronics with two-dimensional materials and van der Waals heterostructures. *2D Materials, 11*. https://www.semanticscholar.org/paper/ddfaf2248c4b6f94fd2eac9ec24f40edeb846720 (accessed on 27th September 2025).

36. Samy, O., Zeng, S., Birowosuto, M. D., & El Moutaouakil, A. (2021). A review on MoS_2 properties, synthesis, sensing applications and challenges. *Crystals, 11*, 355. https://doi.org/10.3390/cryst11040355 (accessed on 27th September 2025).
37. Sattigeri, R. M., Gong, X., Fakhredine, A., Autieri, C., & Cuono, G. (2025). Dirac edge states of two-dimensional alter magnetic topological crystalline insulators. https://www.semanticscholar.org/paper/e9f6438fccbd8a97a3646193bc6eaeb6277c01ef (accessed on 27th September 2025).
38. Shao, Q., Li, P., Liu, L., Yang, H., Fukami, S., Razavi, A., Wu, H., et al. (2021). Roadmap of spin–orbit torques. *IEEE Transactions on Magnetics, 57*, 1–39. https://doi.org/10.1109/tmag.2021.3078583 (accessed on 27th September 2025).
39. Shi, G., Huang, N., Qiao, J., Zhang, X., Hu, F., Hu, H., Zhang, X., et al. (2024). Recent progress in two-dimensional magnetic materials. *Nanomaterials, 14*. https://www.semanticscholar.org/paper/cd4b82618c8200ace367d62a8b3b6e5e6af94653 (accessed on 27th September 2025).
40. Šmejkal, L., Sinova, J., & Jungwirth, T. (2022). Emerging research landscape of alter magnetism. *Physical Review X, 12*. https://doi.org/10.1103/physrevx.12.040501 (accessed on 27th September 2025).
41. Sud, A., Kumar, A., & Cubukcu, M. (2025). Novel magnetic materials for spintronic device technology. https://www.semanticscholar.org/paper/6e889ef5ba434f4adf32adb5996f218f430a0154 (accessed on 27th September 2025).
42. Thapa,A.,&Sharma,B.(2025).Areviewonnovellow-dimensionalmaterials basedmagnetictunneljunctions:Opportunities,challenges,andapplications. *Advanced Materials Technologies*. https://www.semanticscholar.org/paper/a0ab84f1d1f85030d34d0584480f19c95e395995 (accessed on 27th September 2025).
43. Tian, J., Li, J., Liu, H., Li, Y., Liu, Z., Li, L., Li, J., et al. (2024). Spin-layer coupling in alter magnets multilayer: A design principle for spintronics. https://www.semanticscholar.org/paper/52fc345328887acee993681e333846cde486dd65 (accessed on 27th September 2025).

44. Usami, T., Sanada, Y., Fujii, S., Yamada, S., Shiratsuchi, Y., Nakatani, R., & Hamaya, K. (2024). Artificial control of giant converse magnetoelectric effect in spintronic multiferroic heterostructure. *Advanced Science, 12*. https://www.semanticscholar.org/paper/47e34d660c279513efbb84abe5cf6897660aad2b (accessed on 27th September 2025).

45. Wang, S., Lin, J., & Zhao, Y. J. (2025). Challenges in magnetic modulation of 2D materials: Exemplified with MoS_2. *Journal of Physics: Condensed Matter, 37*. https://www.semanticscholar.org/paper/17d91b0240fbe03a4a34641e69cad28025062bba (accessed on 27th September 2025).

46. Wang, Y., Zhang, L., Luo, R., Lu, Y., Fang, B., & Zeng, Z. (2025). Artificial neuromorphic devices based on magnetic tunnel junctions. *Chinese Science Bulletin*. https://www.semanticscholar.org/paper/7f7bead693075f6e22242d30c7037eea86e376ba (accessed on 27th September 2025).

47. Yasuda, T., Horiuchi, T., & Suemasu, T. (2023). Fabrication of highly oriented Mn_4N/Pt epitaxial bilayer structure on *MgO(001)* for spintronics applications. *AIP Advances*. https://www.semanticscholar.org/paper/2c3c57dc06bda2b92030743cb4f2b9c945e1ff75 (accessed on 27th September 2025).

48. Yuan, H. Y., Cao, Y., Kamra, A., Duine, R. A., & Yan, P. (2022). Quantum magnonics: When magnon spintronics meets quantum information science. *Physics Reports, 965*, 1–74. https://doi.org/10.1016/j.physrep.2022.03.002 (accessed on 27th September 2025).

49. Yun, H., Lyu, D., Lv, Y., Zink, B. R., Khanal, P., Zhou, B., Wang, W., et al. (2024). Uncovering atomic migrations behind magnetic tunnel junction breakdown. *ACS Nano*. https://www.semanticscholar.org/paper/5b4ef587df3f59366700845ae4f168fae6a96f21 (accessed on 27th September 2025).

50. Zhang, X., Gong, P. W., Liu, F., & Zhu, S. (2023). Spin-gapless van der Waals heterostructure for spin gating through magnetic injection devices. *Physical Chemistry Chemical Physics (PCCP)*. https://www.semanticscholar.org/paper/cf5bee32f767ea029265c0ac29995786f64371f1 (accessed on 27th September 2025).

51. Zhang, Y., Feng, X., Zheng, Z., Zhang, Z., Lin, K., Sun, X., Wang, G., et al. (2023). Ferrimagnets for spintronic devices: From materials to applications. *Applied Physics Reviews*. https://www.semanticscholar.org/paper/97ae64c64c45eeaa532c8fcfca26c8439a74a441 (accessed on 27th September 2025).

52. Zhou, J., Li, S., Wang, C., Jin, H., Xu, S., Xiong, Z., Jacobsen, C., et al. (2025). Nanoscale quantum imaging of field-free deterministic switching of a chiral antiferromagnet. https://www.semanticscholar.org/paper/7c6c47de4d60157f17cba4539094726db8a6b055 (accessed on 27th September 2025).

CHAPTER 8

PHOTONIC AND OPTOELECTRONIC DEVICES

Contents

Introduction 311
8.1. Fundamentals of Photonics 316
8.2. Photonic Integrated Circuits (PICs) 321
8.3. Quantum Dot (QD) Lasers 326
8.4. Photodetectors 331
8.5. Silicon Photonics 336
8.6. Challenges in Photonic and Optoelectronic Devices 341
Summary 345
References 348

INTRODUCTION

Technological development is inseparably connected with advancements in photonic and optoelectronic devices. With the growing dependence on speedy data communication and energy-efficient computing, photonics being integrated into electronic circuits becomes extremely important. Photonic integrated circuits (PICs), where several operations are integrated into one chip, are revolutionizing the field of information processing and communication. This technology allows optical devices to be miniaturized, making for lighter, faster, and more efficient equipment. Furthermore, PICs allow high-speed data links to be created, which far surpass traditional electronic technologies in bandwidth as well as in power efficiency. These breakthroughs refer to the importance of embedding light manipulation in electronic systems as well as the opening of a new horizon for optoelectronics that propels telecommunication as well as other applications. Recent developments in quantum dot lasers and photodetectors are another case in point in optoelectronics innovation. Quantum dots, which are quantum-confined semiconductor particles with new optical properties, continue to gain prominence in laser technology.

The devices have the potential to offer better performance characteristics, such as greater efficiency and wavelength-tunable emission. Similarly, the enhancement of photodetectors through quantum dots provides the potential to develop more efficient and compact sensors, leading to applications in fields like biomedical imaging and environmental monitoring. Application of such technology in photonic devices is a testimonial to higher operational capacity and efficiency, thus making them unstoppable in today's technological application. The potential of silicon photonics is another area that indicates how important light is in electronic systems. Using existing silicon fabrication technologies, researchers are able to build high-speed optical data links with a significant increase in bandwidth while being backwards compatible with current electronic infrastructures. Silicon photonics not only facilitates the faster transfer of data, but also reduces power consumption, which is in harmony with global sustainability targets. But the task there is one of integrating active optical components with passive silicon waveguides, a task that needs more R&D. As the distinction between electronic and photonic technologies continues to blur, it becomes increasingly evident that the future of computing and communication is about unfettered interconnection of these disparate domains with a promise of new technological advancements and applications.

Definition of Photonic and Optoelectronic Devices

Where physics and technology converge, photonic and optoelectronic devices take center stage in modern technology. These devices use principles of optics and electromagnetism to control photons for use in a variety of applications ranging from communications to sensing. PICs, for example, enable the integration of various optical functionalities onto a single chip, significantly enhancing the efficiency and compactness of optical systems. With the application of materials like silicon, such circuits provide fast data transfer at high rates, which is essential in today's data-driven world. Silicon photonics have made it possible to create high-bandwidth systems with improved performance compared to traditional electronic devices. Such developments are reflective of the importance of optical technologies in supporting growing demands for higher data rates and lower power consumption in various applications (Figure 8.1).

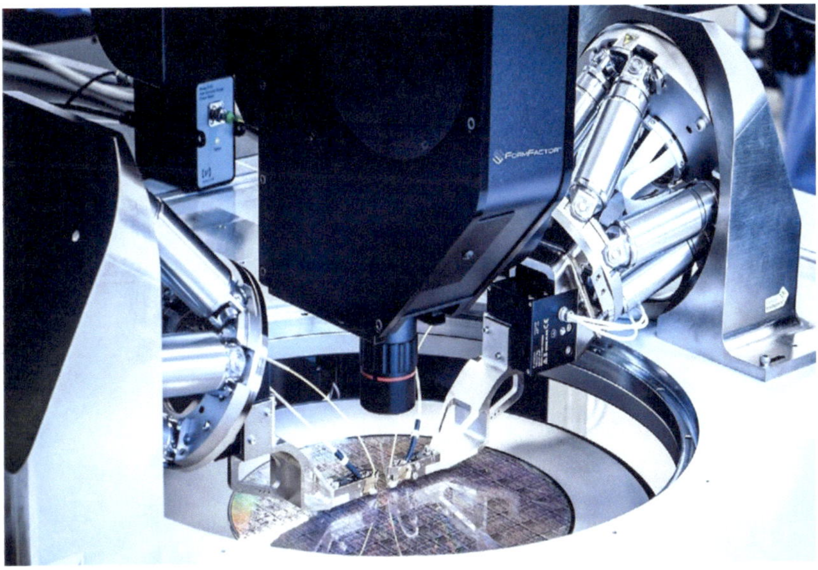

Figure 8.1. Utilization of photonic devices in modern technology.

Source: https://www.azom.com/images/Article_Images/ImageForArticle_19254_15898730431299588.png.

Piling on the significance even further, quantum dot lasers and photodetectors are also gaining significance as major players in the field of optoelectronics. Quantum dot lasers, characterized by their quantum-mechanical anomalous properties, offer the advantages of low threshold current and enhanced temperature performance. Their ability to adjust emission wavelengths with high accuracy makes them particularly suitable for telecommunications and data storage. Similarly, photodetectors based on quantum dot technology exhibit high efficiency and sensitivity, enabling weak signal detection with high accuracy. Such detectors are the backbone of a range of applications including remote sensing and imaging systems, where high-precision, fast detection is the demand. Their fabrication and integration into current systems mark a revolution in the use of optical signals, with a gate opened to enhanced performances in such applications as medical diagnosis and environmental observation (Figure 8.2).

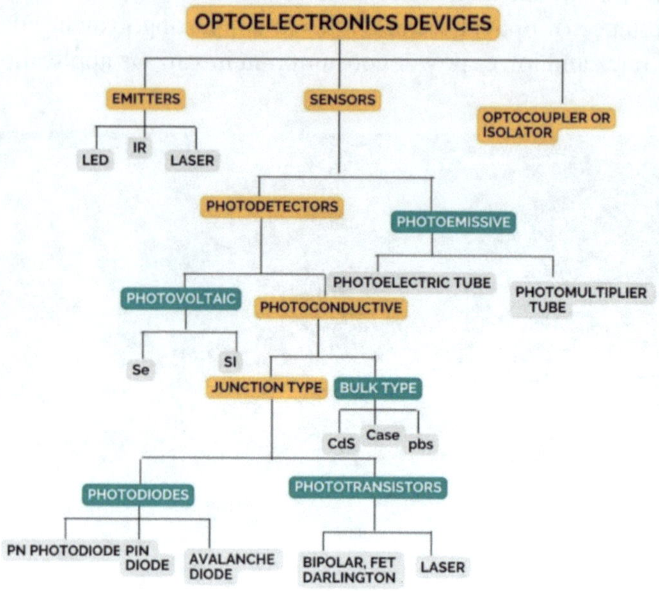

Figure 8.2. Types of optoelectronic devices.

Source:https://lumenci.com/wp-content/uploads/2023/07/Types-of-devices.webp.

With some applications, the use of novel materials such as InAsSbP/InAs LED technology has shown the potential of photonic devices in delivering tailored light emissions. Tests confirm that such LEDs can simulate a multitude of thermal backgrounds, and this is beneficial in thermal emissions being realistically simulated in surroundings such as military simulations and environmental studies (Malyutenko et al., 2006). The variability of optoelectronic devices even extends to the microwave field, with optical fibers being essential components in low-noise microwave delay lines. The capability of recording a significant delay at high stability offers potential for increased performance in measurement networks and oscillators, enabling leading-edge research in telecommunication and precision sensing (Brendel et al., 2008). This flexibility evidences the great contribution of photonic and optoelectronic devices toward advancing technology advancements in different industries.

Importance in Modern Technology

The union of photonics and technology is demonstrated in the developments of PICs, a technology enabler for modern communication networks. By integrating numerous photonic functions onto a single chip once physically cut apart, these circuits permit revolutionary miniaturization and power reduction at the same time that they improve performance and bandwidth. The PICs' provision of integration allows for faster data transmission rates required in this era of connectivity with billions of devices communicating with one another at nearly all times. Consequently, the development of PICs has not only transformed telecommunications but also ushered in sensing, imaging, and quantum computing applications, demonstrating their criticality in a broad array of technological fields, demonstrating their role as critical drivers of innovation and productivity in today's infrastructure. Another important feature of the landscape of present-day technology is in the field of quantum dot lasers and photodetectors. These nanoscale semiconductor devices take advantage of the unique quantum properties that are intrinsic to their nanostructure, providing enormous advancements over traditional devices. Quantum dot lasers, for example, exhibit greater performance features such as reduced threshold current and improved temperature stability, leading to increased efficiency and lifespan. Simultaneously, quantum dot-based photodetectors bring high sensitivity and sensitivity at various wavelengths, central to a variety of applications ranging from telecommunications to medical imaging. Their integration into optoelectronic systems is a significant advance, heralding the arrival of a new generation of accuracy and miniaturization in technology that meets the increasing demands of high-performance applications, potentially reshaping entire industries in the process, further solidifying their position in modern-day technological environments. For high-speed data connections, silicon photonics is a game-changing influence, bringing optical communication to the forefront of modern technology. Especially, silicon photonic devices have the prime mandate of developing power-efficient, high-bandwidth interconnects required for data center connectivity and telecommunication equipment. Recent developments in optical modulators, from MHz up to multi-GHz bandwidth, attest to the exceptional advancement in this field (Gardes et al., 2010). The quest for improved performance is relentless, the challenge of maximizing aspects such as energy efficiency and device footprint never fading. The ongoing innovation in silicon photonics is the remarkable advancement being made in optical technology, symbolizing the intricacies and intrinsic nature of such advances in satisfying the growing demands for speed and capacity in global communications, and underscoring their central role in current technological landscapes.

Overview of Applications

PICs' development is a significant advancement in light technology, optically combining a number of photonic elements onto a single chip. The novel approach adds functionality with a reduced footprint and power consumption and is particularly well-suited for telecommunications and information processing. The incorporation of functionalities such as waveguiding, modulation, and detection within PICs enables the cost-effective handling of light signals, finally leading to increased data rates and reduced latency. In addition, the compatibility of PICs enables them to be adopted easily into existing silicon technology, paving the way for their widespread application within high-speed data links and communication networks. Applications span across various sectors, including medicine, where PICs can be implemented in diagnostic devices, pointing towards a revolutionary impact on consumer electronics as well as sophisticated industrial processes. Quantum dot photodetectors and lasers offer examples of such versatility and performance enhancement possible in photonics applications. These devices exploit quantum confinement effects to achieve best efficiency and emit tunable wavelengths, which are critical in applications for optical imaging and communication. Their capability to set the wavelength with precision facilitates best operation for a broad range of conditions, including telecommunications, where specific frequencies are critical. In addition, the quantum dot dimension allows for hybrid integration with other materials, increasing the output of miniaturized optoelectronic devices. This hybridization of technologies provides enhanced capability in information capture and processing, further solidifying quantum dot devices as core components in the next-generation photonic applications due to their inherent scalability and tunability. Silicon photonics is also a fast-moving area, closing the gap between electrical and optical systems. More importantly, silicon photonic technology utilizes silicon's matured fabrication methods to establish high-speed data links that significantly outclass traditional copper connections. By converting the electrical signals into optical signals to be transmitted, silicon photonics provides the bandwidth needed for modern data-intensive applications like cloud computing and AI. Besides, photonic integration on silicon wafers enables low-cost manufacturing in large volumes with maintaining performance enhancements. Most recent developments, such as the implementation of silicon-based lasers and modulators, have further boosted silicon photonics functionality, promising its role in revolutionary applications ranging from telecommunications to sensor technology. Silicon photonics' performance and flexibility are poised to highlight its significance in the forthcoming optoelectronic device and information technology future.

8.1. FUNDAMENTALS OF PHOTONICS

It is necessary to learn about the basics of photonic integrated circuits (PICs) for continued work in contemporary optoelectronic devices. These circuits utilize optical components to guide light, providing more functionality compared to electronic circuits. With the use of materials such as silicon, PICs are able to move data at high speeds without inordinate loss of energy. Integrate various optical functions—like modulation, detection, and routing—on a single chip and it is a major leap in photonics technology. This level of miniaturization is achieved by precise engineering, especially as the demand for effective and compact systems grows. These breakthroughs enable extremely sensitive communication equipment capable of catering to digital networks' growing bandwidth needs, showing the manner in which the integration of photonic concepts is revolutionizing data communication infrastructure. Moreover, quantum dot laser and photodetector development is testimony to the profound impact of nanotechnology on photonics. Quantum dots, having their own electronic characteristics by virtue of quantum confinement, enable lasers with reduced thresholds and emission spectra that can be tuned. Such versatility qualifies them for applications from telecommunications to medical imaging (Figure 8.3).

Figure 8.3. Principles behind photonic.

Source:https://ocw.mit.edu/courses/6-974-fundamentals-of-photonics-quantum-electronics-spring-2006/e56304d7cbda6c2fcf2a1c45476aad6e_6-974s06.jpg.

Quantum dot materials can be utilized in photodetectors to realize dramatic sensitivity increases, which is a necessity in conditions where there is minimal light. In addition, integrating these building blocks into existing photonic devices allows more compact and efficient devices, which drives innovation in optoelectronic devices. With the advancement of technology, they will drive momentum toward the development of more complicated, smarter, and multidimensional devices. Silicon photonics is one of the leading technologies in high-speed data link applications that combines the virtues of both the electrical and the optical realms. This technology leverages silicon's high availability and compatibility with existing semiconductor processes to create cost-effective, scalable photonic devices. Data transmission at light speed and by using standard silicon fabrication processes makes production low-cost and efficient. New developments highlight novel architectures, like mode-multiplexed modulators, that enhance the functionality-per-area of these systems. These innovations provide the environment for tiny photonic devices that reliably double or even triple their functions, which address the increasing demands of bandwidth and speed in contemporary communication systems. Additionally, research into novel materials such as graphene can supplement these silicon-based platforms, bringing additional flexibility to next-generation optical devices (Calogeracos et al., 2014; Das et al., 2016).

8.1.1. Basic Principles of Light and Optics

The interaction of light with various materials is used as the foundation for the creation of photonic and optoelectronic devices. As an electromagnetic wave, light has laws of reflection, refraction, and diffraction, which form the foundations of understanding the functionality of optical devices. For instance, photonic integrated circuits (PICs) employ these optical principles to manipulate light at a nanoscale level, transmitting information in a more efficient way than electronic circuits.

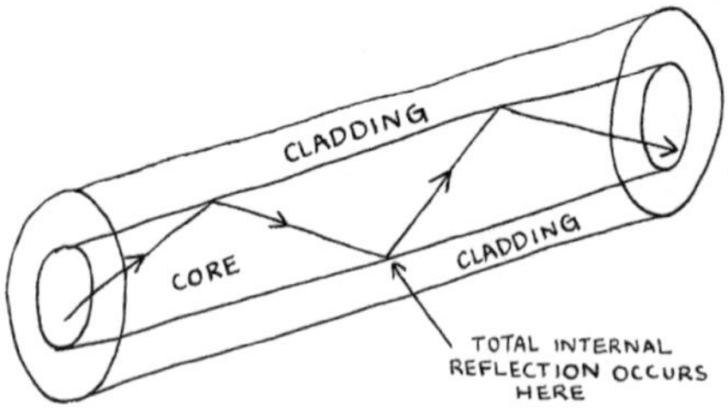

Figure 8.4. Basic principles of fiber optics series: refraction.

Source: https://www.truecable.com/cdn/shop/articles/Cladding_Diagram_1024x1024.png?v=1705555989.

By coupling several photonic components onto one chip, PICs improve functionality and minimize size, exemplifying the effective use of light in today's technologies. These developments highlight the significance of knowledge of basic optical principles in constructing high-performance devices that are capable of addressing the requirements of high-speed data communication and sophisticated processing tasks that are crucial in today's digital arena. Polarization is key in the field of manipulation of light, particularly in application areas like quantum dot lasers and photodetectors. Polarized light has the capability to have a significant effect on the performance and efficiency of devices that are engineered to utilize polarized light (Figure 8.4).

By employing material selective transmission or reflection of light based on the state of polarization, one can design systems with maximum signal integrity and minimum losses. It is quite fascinating to note how structural design accuracy can yield remarkable efficiency—98.5% reflectance for one polarization and 94.3% transmission for the other (Deng et al., 2016). These advances not only establish the significance of polarization in optics but also provide a potential for the inclusion of such technology in integrated photonic devices, further increasing the possibilities for future advanced optical communication systems. Additionally, the nature of two-dimensional materials, particularly graphene, indicates some interesting prospects for the future optoelectronics. Graphene's transparency along with high light-matter interaction makes it an extremely prospective candidate for photonics usage. This material is able to effectively trap photons, altering their wavelengths, which is central in the production of future plasmonic devices. Moreover, graphene's nonlinear optical response when exposed to an electric field offers new possibilities in the fabrication of new transistors and logic devices. Studies focused on inducing a band-gap in graphene through techniques like chemical doping showcase its potential and impact in high-frequency optical systems (Calogeracos et al., 2014). With continuous research work, incorporation of graphene in a variety of photonic devices can significantly make the device more efficient and functional, particularly in high-speed communication networks.

8.1.2. Key Components of Photonic Devices

The development of photonic devices is mainly supported by the incorporation of numerous components that make it possible to better manipulate and create light. One of the notable advances is the development of photonic integrated circuits (PICs), which integrate multiple photonic functions on a single integrated chip. Integration leads to improved performance and smaller physical area, ultimately achieving greater efficiency in communication systems. PIC design usually includes designing waveguides, modulators, and filters that interact together to manipulate and direct light signals. Here, silicon-like materials are significant due to their compatibility with existing semiconductor technologies, which allow them to seamlessly integrate into processing systems. The need for increased data transmission speed ensures that the performance of integrated circuits will be the priority, and novel design and material choices will follow in order to maximize light propagation and interaction. In addition to silicon-based technology, quantum dot lasers and photodetectors represent a key component in modern photonic devices.

Quantum dots are nanocrystals of semiconductors that are small enough to exhibit quantum mechanical properties. They exhibit interesting advantages such as size-tunable emission wavelength and enhanced efficiency in light emission and detection. As lasers, the nanostructures can emit highly coherent light and can find applications in telecommunications and biomedical sensing. Besides, their application as photodetectors in devices permits the detection of low light intensities and, with greater efficiency, where other detectors would fail to perform. The union of quantum dots and traditional photonic platforms has the potential to not only enhance the elementary knowledge of light-matter interactions but also address engineering problems in high-speed data communications and sensing technologies, rendering them relevant in the sector of photonic devices. Besides, silicon photonics has transformed the area of high-speed data links with a solution to address the ever-increasing bandwidth demands of modern communication networks. By capitalizing on silicon's intrinsic optical characteristics, scientists have been able to create devices that enable the transmission of data at record rates with the additional benefit of capitalizing on the ubiquitous infrastructure of existing silicon-based electronics. This compatibility enables effective interaction between optical and electrical signals on the same chip, minimizing conversion loss and overall efficiency. Developing designs, such as waveguide-integrated modulators and miniaturized photonic components, extend silicon photonics' capabilities further. In addition, the integration of two-dimensional materials, including graphene, with traditional silicon devices has also been promising in enhancing light-matter interaction and widening operational bandwidth, which demonstrates the promise of silicon photonics as a fundamental component of future photonic and optoelectronic devices, which will facilitate novel communications technology (Calogeracos et al., 2014; Grigorenko et al., 2017).

8.1.3. Role of Photons in Information Transfer

The formation of photonic integrated circuits (PICs) is on the cutting edge of existing information transmission, providing quicker and more reliable communications technology. Through the integration of multiple photonic devices onto one chip, these circuits make it possible to route and manipulate photons, which are vital information carriers in optical fibers and free-space systems. Preeminently, devices like graphene-silicon p-i-n photodiodes are the cornerstone of this technology, having enhanced speed and efficiency in photodetection. An ultrafast interfacial carrier transfer is employed in such a device to record impressive signal-to-noise ratios even under zero bias conditions. As emphasized by Ding et al. (2018), performance and scalability of these two-dimensional materials pave the way for new means of developing high-speed compact photonic systems required for data transmission in current networks. In addition to PICs, quantum dot lasers and photodetectors also have a significant role in information transmission through the accuracy and efficiency with which they are able to produce and detect light. Quantum dots make tunable light emission at various wavelengths possible, and this is particularly beneficial for minimizing loss in transmission of data over long distances. By incorporating the semiconductor objects on the nanoscale into optoelectronic devices, researchers have been able to advance significantly in the development of light sources with precisely controlled spectral properties. Sophisticated doping techniques that stimulate emissions have led to enhanced optical gain properties, according to Ho et al. (2018).

Therefore, the hybridization of silicon platforms and quantum dot technology will enable hybrid optoelectronic devices, which leverage both material strengths to eventually optimize data transfer rates in telecommunication infrastructure. Silicon photonics has also emerged as a primary area for high-speed data links, employing photons to deliver record-high data rates for networking applications. The ability to source light within silicon structures is not only compatible with existing CMOS technology but also creates enormous possibility for interfacing with standard electronics. Effective light emission from silicon is being addressed by new approaches such as employing Er-doped GaN for creating room-temperature light sources within telecommunication optimal wavelength range. These developments underscore the role of photons in shaping the future of information transfer, paving the way for applications in the future in fields ranging from communications to health devices. Through the fusion of these technologies, the vision for latency-free data communication with optimal efficiency is at hand, highlighting the transformative role of photons in optoelectronics.

8.2. PHOTONIC INTEGRATED CIRCUITS (PICS)

The development of PICs is important, since the devices have facilitated huge advances in numerous applications, including telecommunications and data processing. Optics placement on a single chip not only maximizes space efficiency but also maximizes performance by minimizing optical interconnect losses. New technologies such as quantum dot lasers and photodetectors are some examples of cutting-edge technology that exploits the unique characteristics of such nanostructures to optimize light emission and detection. Integrating quantum dot technology into PICs is also likely to introduce more compactness and multifunctionality, enabling better control of light properties. As a result, these advanced devices are poised to increase the performance of systems from high-speed data links to advanced sensing platforms, expanding the scope of operation of photonic devices. Silicon photonics is gaining popularity since it has compatibility with proven semiconductor fabrication processes such that PICs can be utilized extensively in data-dense environments. The integration of silicon waveguides with active devices such as modulators and lasers enables light manipulation on the nanoscale, resulting in faster data transmission rates. Most significantly, the extensive silicon device research has demonstrated high-speed data links that can achieve terabits per second, much greater than traditional electronic approaches (Figure 8.5).

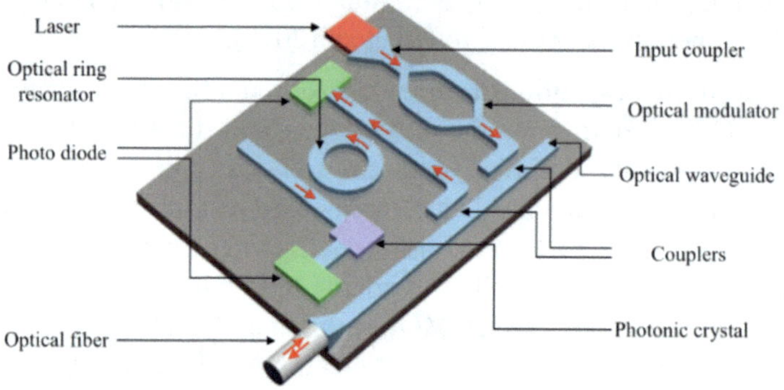

Figure 8.5. Photonic integrated circuit (PIC).

Source:https://lh7-us.googleusercontent.com/yPKFWcmd9ddAMqKNOND8vWM-kUUpWiPxR7TmO7PV8GO1N0jSFMz-WJhj1fgn3Yhx1n1dOD-G9A4pl7hbVM-fyIRha9-l8Uf3_q7TmOrO74u3hfFzlNMMlKZt4q-uQUPGmE0X0OznFKnYr-rCf98T5cuw6E.

The widespread deployment of such high-speed systems in data centers is a testament to the growing demand for efficient solutions in data processing and communication. Moreover, silicon's unique properties render it an ideal substrate for the development of hybrid devices, thus spurring innovation at the photonics-electronics interface. In order to further expand PICs' functionalities, novel methods of coupling two-dimensional (2-D) material into photonic circuitry have now been created. These glasses, with their excellent optical qualities, introduce new functionality in PICs, for instance, modulation and saturable absorption. By employing chalcogenide glass, scientists have introduced a new integration concept, which simplifies the complexity of fabrication, improving the yield and enabling complex device geometry that is optimized for optimal light-matter interaction. This approach not only preserves the unique properties of 2-D materials but also offers a path toward fabricating a range of high-performance photonic devices, including ultra-broadband polarizers and energy-efficient thermo-optic switches. As studies into this integration method continue, the use of 2-D materials is anticipated to dominate future photonic integrated circuits, paving the way for novel concepts and enhanced device functions (Grigorenko et al., 2017; Morral et al., 2017).

8.2.1. Definition and Structure of Photonic Integrated Circuits (PICs)

With continuously changing current technology, the demand for faster, more efficient methods of data processing has increased to unforeseen heights. This need has spurred the development of PICs, which combine multiple photonic devices on a shared semiconductor substrate. PICs transfer information using light rather than electrical signals and therefore have much greater speed and bandwidth. The fundamental building block of a PIC would have various components, such as lasers, modulators, detectors, and waveguides, all of which are arranged diligently to function in unison. With the search for more efficient communication equipment, PIC design has become more powerful and smaller in size to serve such diverse uses as telecommunications and high-end computing systems so as to allow the restructurization of data streams within contemporary digital environments. In the realm of PICs, quantum dot lasers and photodetectors play a pivotal role in advancing performance and efficiency to an unprecedented level.

Quantum dot lasers provide better characteristics compared to conventional semiconductor lasers, including reduced threshold current and enhanced temperature stability, factors that play a major role in the reliability of optical communication systems. Quantum dot lasers are normally integrated into PICs, giving the circuits the capability to produce high-frequency optical signals needed for effective data transmission. Similarly, photodetectors in PICs convert the light signals into electrical signals to enable the flow of information seamlessly through a photonic circuit. Besides adding to the performance of PICs by integrating such advanced photonic components, it also establishes an environment that can cater to growing demands for speed and efficiency across numerous applications such as artificial intelligence and high-performance computing (de Lima et al., 2017). Silicon photonics is an innovative platform for PIC realization driven by the requirements of high-speed data links and scalability (Figure 8.6).

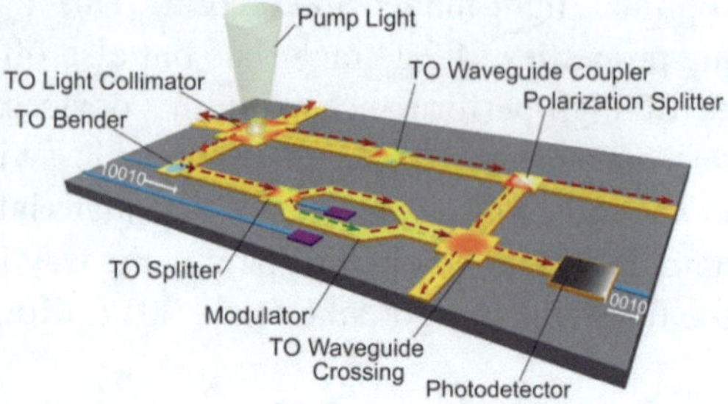

Figure 8.6. Structure of photonic integrated circuits (PICs).

Source: https://media.springernature.com/full/springer-static/image/art%3A10.1038%2Flsa.2012.38/MediaObjects/41377_2012_Article_BFlsa201238_Fig1_HTML.jpg.

It leverages the mature semiconductor manufacturing infrastructure to deliver low-cost large-scale production of photonic devices. The combination of photonic and electronic devices in silicon photonics enables the fabrication of complex integrated systems capable of supporting effective data traffic management, particularly in communication networks and data centers. The ongoing development of open-access silicon photonics platforms has provided opportunities for innovations, and researchers and industries have been made accessible to state-of-the-art technology for fabricating PICs. As such, the technology landscape for photonic integrated circuits continues to evolve, presenting fresh opportunities for telecommunications and computing technology, ultimately reshaping the information processing dynamics of the digital age.

8.2.2. Advantages of Using Photonic Integrated Circuits (PICs) in Technology

The integration of photonic elements into a compact structure is a key milestone in the technology sector primarily through the use of PICs. These circuits capitalize on the unique properties of light to enable data transmission, much more than in conventional electronic circuits. By making use of material like silicon and embedding quantum dot lasers, PICs enable high-speed data links with reduced power consumption. This efficiency is essential for today's communications networks, where higher-speed, more reliable data transmission continually grows in demand. In addition, PICs simplify device structure by reducing the number of discrete components needed, leading to lower manufacturing cost as well as simpler integration into existing technology. Moreover, the ability to integrate two-dimensional (2-D) materials into PICs presents opportunities for new device functionality. As in the current research, the atomically thin materials possess excellent optical properties that are well suited for application within various photonic devices such as modulators and detectors.

The integration processes through chalcogenide glass, as in Grigorenko et al. (2017), indicate a revolutionary approach that provides direct patterning and depositing of the materials. This method successfully increases fabrication yields and can be employed to fabricate intricate device geometries with optimized light-matter interactions. Therefore, PICs with 2-D materials are on the verge of transforming telecommunications, sensing, and imaging applications and making devices perform significantly better than their traditional counterparts. Finally, the scalability and versatility of PICs allow them to be utilized across a wide range of various technologies. They not only constitute an essential part of telecommunications but are also of promise in other uses like consumer electronics, healthcare devices, and environmental monitoring.

By their capacity to integrate multiple functions on a single chip, PICs reduce size and weight with greater performance. This revolution is seen in technological advancements like high-performance glass-on-graphene devices that have power-efficient switches and high-end photodetectors, as explained in Grigorenko et al. (2017). Additionally, the broad scope of applications of PICs underscores their importance in facilitating innovation, with profound implications for the Internet of Things (IoT) and smart technologies. Continued growth of PIC technology is at the brink of initiating phenomenal advances in various sectors and is therefore an important area of research for researchers and engineers.

8.2.3. Applications of Photonic Integrated Circuits (PICs) in Telecommunications

Telecommunications advances are increasingly reliant on new technologies taking advantage of the power of light. A very promising area is the application of PICs, which are now essential in high-bandwidth environments. Combining a number of photonic devices on a single chip allows higher performance with drastically lower physical size and power consumption compared to traditional electronic systems. As telecommunications networks continue to evolve with greater data needs, PICs are crucial to allowing data to be transmitted at high speeds, especially using technology like silicon photonics. Silicon technology provides a platform for greater efficiency and reliability that enables data links to run at record speeds with minimal amounts of loss in energy transmission. Quantum dot lasers and photodetectors are also crucial in the telecom industry. These chips provide impressive performance in efficiency and speed, properties that are crucial for new-generation communications systems.

Integration of quantum dot technologies into photonic circuits will bring about tremendous improvements in emission qualities of lasers and response speed of photodetectors. This capability can lead to functionality breakthroughs such as wavelength division multiplexing (WDM), where several messages are transmitted on a single fiber optic cable simultaneously. The result is a revolutionary increase in bandwidth without the need for additional physical infrastructure.

Thus, uses of such novel materials and structures hold the key to future telecommunication networks requiring more capacity and less delay to manage the rising data load. Additionally, emerging technologies such as integrated microwave photonic (IMWP) filters are a new generation of telecommunications technology. These systems play a vital role in handling microwave signals and providing optimum performance in radio-frequency applications. With an integrated photonics design, IMWP filters offer capabilities that are not available with conventional RF electronics, such as reconfigurability and tunability over a wide frequency range. Their compact size, lightweight character, and low power dissipation make them particularly appealing for incorporation into existing telecommunications infrastructures. As these kinds of systems become more sophisticated, they will certainly affect the telecommunications business, offering the possibility of improved system performance and reduced operational costs in many applications in actual usage (Liu et al., 2020; Moody et al., 2021).

8.3. QUANTUM DOT (QD) LASERS

The latest advances in photonic integrated circuit (PIC) technology highlight the significant role played by QD lasers in modern optoelectronic devices. Through the unique properties of quantum dots, researchers have succeeded in making laser devices with significantly lowered power thresholds, enabling efficient operation in a large variety of applications. For instance, novel fabrication methods have created electrically pumped photonic crystal lasers with historically low thresholds while exhibiting exemplary performance. These advancements signal that quantum dot lasers are increasingly viable for incorporation in coming PICs, particularly for application where speedy data transfer is required. This is required for coping with the rising bandwidth demands in communications systems, hence opening up the immense impact quantum dot technology has on the development of photonic devices as a whole. In addition, quantum dot lasers' compatibility with silicon photonics is a huge step toward high-speed data links (Figure 8.7).

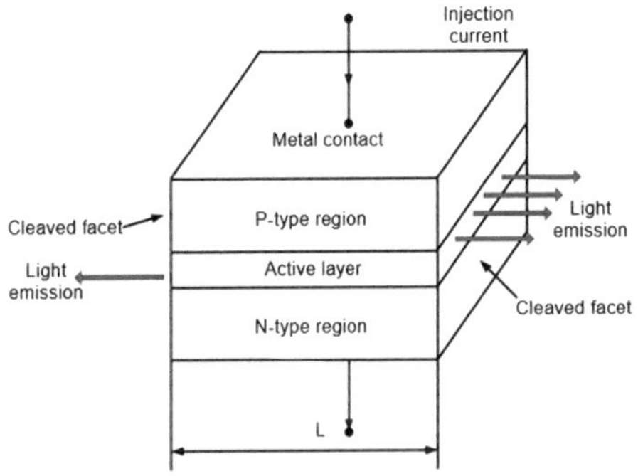

Figure 8.7. Quantum dot (QD) laser technology.

Source: https://cdn.gophotonics.com/community/5_1_1__638338303121645359.jpg.

Success in developing electrically pumped, room-temperature continuous-wave distributed feedback lasers demonstrates the compatibility of QD gain material with silicon substrates. Such integration is necessary as it will enable the unification of conventional silicon microelectronics with high-end optical functionality for enhanced overall performance of photonic devices. The experimentally determined performance parameters such as low threshold current and high side mode suppression ratios clearly establish the ability of such lasers to meet the stringent specifications of coarse wavelength division multiplexing (CWDM) systems. It is evident that QD lasers not only pave the way for new designs in optical networks but also help enable the facilitation of silicon-based photonic circuit scalability. Finally, the advancements made in quantum dot photodetectors are supporting the advancement realized in laser technologies, enhancing the bidirectional nature of optoelectronic systems. Current developments in two-photon absorption techniques in photodetectors show higher efficiency and sensitivity, which are essential in ensuring high data transmission via optic networks. The ability to derive ultrafast electrical signals from photonic devices ensures that their performance is optimized for various applications, most importantly in telecommunication. Coupling of these detectors with QD lasers creates a unified platform that eases the process of converting electrical signals to optical data and vice versa, ultimately paving the way for the creation of high-speed communication systems. Thus, the mutual interaction of quantum dot lasers and photodetectors raises their stakes in future photonic and optoelectronic device development.

8.3.1. Explanation of Quantum Dot Technology

Quantum dot technology has revolutionized photonics with the promise of new applications in integrated devices. Central to the technology is nanoscale semiconductor particles with unique optical properties based on quantum confinement effects. These properties allow quantum dots to emit light at specific wavelengths when excited, very much controlled by size and composition. This tunability makes quantum dots particularly desirable for the production of high-efficiency light sources, such as quantum dot lasers, and advanced photodetectors. When integrated with photonic circuits, they enable miniaturization of devices and enhance the functionality of devices, thereby enabling the development of photonic integrated circuits (PICs) that can integrate several functionalities into one chip. One of the principal applications of quantum dot technology is in the fabrication of quantum dot lasers and photodetectors, which are key components of current optical communication systems.

Quantum dot lasers are distinguished by their low threshold current and high power output capacities, and thus they are highly suitable for high speed data transmission. For instance, coupling of InAs/GaAs quantum dots gives rise to outstanding performance characteristics, such as continuous-wave operation and mode suppression ratios meeting the stringent specifications of coarse wavelength division multiplexing (CWDM) systems. Such efficiency and reliability are necessary in supporting high-speed data links, especially as global data needs keep on growing. Moreover, the quantum dot photodetectors' sensitivity and response time further enhance the overall data transmission rate, emphasizing the technology's potential for future communication systems. Silicon photonics' contribution as a complement to the technology shows a synergy-based approach to overcoming the limitations of the traditional electronic devices.

The un-disruptive compatibility of silicon technology with quantum dot materials offers avenues for the fabrication of electrically-pumped room-temperature-operating lasers, hence lowering complexity in fabrication. There have been notable advancements showing prospects of realization of low-threshold high-performance lasers grown directly on silicon substrates. These developments offer viable application avenues, such as the integration of quantum dot lasers into present silicon microelectronics systems, leading to efficient energy and miniaturized photonic devices. With the constant quest for increasingly rapid data transfer, the marriage of silicon photonics and quantum dot technology provides a platform for a new generation of optoelectronic devices, poised to impact a broad range of applications from telecommunications to data centers.

8.3.2. Comparison with Traditional Laser Technologies

The innovation of photonic integrated circuits (PICs) is a big step for the development of optoelectronic devices, particularly in terms of their potential to outperform traditional laser technologies. In traditional systems, the single laser components are limiting as far as miniaturization and integration are concerned, resulting, in most instances, in block-like configurations that hinder overall performance. In contrast, PICs allow for the integration of various photonic components on a single chip, significantly improving efficiency and reducing signal loss. Nanoscale manipulation of light enhances such performance indicators as higher bandwidths for data transmission. Notably, the incorporation of new materials such as graphene, as exemplified in research on two-dimensional (2-D) materials, offers a platform on which new ideas can emerge, with an apparent opportunity for the development of devices that operate over wider wavelengths and have improved immunity to environmental conditions (Grigorenko et al., 2017). The trend addresses the major advantages of the application of advanced photonic technologies over their traditional counterparts. Quantum dot lasers represent another major departure from conventional laser technology, signaling the potential of semiconductor nanostructures among the optoelectronic field. Unlike typical semiconductor lasers, quantum dot devices exploit unique carrier confinement characteristics, which exhibit lower threshold currents and improved temperature stability. These advantages are the result of the quantized energy levels introduced by quantum dots, which contribute to higher efficiency and performance. Moreover, quantum dot photodetectors are complementing traditional technologies and providing complementary capabilities in light scattering detection and amplification. The value of such advancements remains to be substantially amplified in uses such as telecommunications and data centers where high-speed links are of primary importance. Comparison with traditional laser systems enables us to understand how quantum dots not only surpass current limitations, like decreased efficiency at high temperatures, but provide a gateway to the next generation of optoelectronics, which pushes the boundaries of performance in integrated systems. Silicon photonics is one such groundbreaking method when comparing photonic technologies, particularly when measured against traditional laser systems. Silicon integration with photonic devices facilitates the high-speed data links that are in excess of what can be offered by conventionally implemented electronic interconnects. Unlike traditional lasers, which often involve complex assembly and alignment to be used in optical networks, silicon photonic devices make manufacturing easier through the established semiconductor processes. This not only reduces production cost but also provides better scalability and reliability. In addition, silicon lasers may be successfully combined with existing microelectronic devices and circuits, thereby synergizing operations and enhancing operating efficacy. As shown by the success triggered by interdisciplinarity research centers, silicon photonics technologies hold vast potential for broadening the scope of applications in telecommunications, sensing, and high-performance computing (The Boston University Photonics Center, 2013). Through an investigation of these comparative strengths, it is apparent that future optoelectronic devices will increasingly rely on creative solutions incorporating the potential of newer technologies.

8.3.3. Applications in Telecommunications and Medical Devices

Optoelectronic devices are significantly enhancing telecommunications, particularly through the application of photonic integrated circuits (PICs). Such circuits allow hundreds of photonic functions to be integrated onto a single chip, leading to reduced size, weight, and power requirements compared to traditional optical systems. PICs are meant to facilitate high-speed data transfer over telecommunication networks, offering greater bandwidth and improved signal integrity. Also, their ability to operate at various wavelengths provides the ability to have more versatility in wavelength division multiplexing, thus allowing greater capacity and efficiency of fiber-optic communications. Noteworthy is that advances originating from research programs, such as those conducted at the Boston University Photonics Center, serve to better represent the development of this technology, showing how PIC technology continues to advance and re-make the face of telecommunications today (The Boston University Photonics Center, 2013). In medical devices, the application of quantum dot lasers and photodetectors has brought revolutionary new technologies in imaging and diagnostic technologies. Quantum dot lasers emit light at precise wavelengths that are crucial for specific medical applications such as optical coherence tomography and fluorescence imaging, which provide high-definition and detailed images of biological tissue. Quantum dot photodetectors have displayed remarkable sensitivity and efficiency and have been used effectively to detect extremely weak signals in a plethora of medical diagnostics. The application of the devices makes possible non-invasive surgery and enhances the diagnostic capability of medical devices. Scientific research has led to groundbreaking techniques utilizing the unique characteristics of quantum dots, and ultimately, it has enhanced the patients' outcomes as well as the medical imaging systems (Alexander et al., 2009). Besides, silicon photonics has become a revolutionary technology for telecommunications and medical devices by making it possible to create photonic elements in silicon wafers. The integration not only benefits the telecommunication sector with increased speed and efficiency in data links but also makes it possible that medical devices can achieve full communication ability in healthcare systems. By the exploitation of silicon, researchers are creating inexpensive and compact devices that produce minimal thermal noise, a key element for sensitive medical application. The ongoing research on silicon photonics is indicative of the miniaturization and energy efficiency trends in technology as a whole, towards the day when data at high speeds and stable performance of medical devices become increasingly intertwined (The Boston University Photonics Center, 2013).

8.4. PHOTODETECTORS

Recent research in photonics has significantly impacted the fabrication and efficiency of photodetectors, which are essential elements in contemporary optoelectronic devices. These advances are especially notable in the integration of two-dimensional materials, e.g., graphene and transition-metal dichalcogenides (TMDCs), with photonic integrated circuits (PICs). Such integration increases the operational functions of photodetectors, enabling utilization in high-speed data links and telecommunications. Use of these materials has enabled the creation of ultrafast photodetectors capable of handling broad bandwidths and having high responsivities, thus meeting the increasing demand for efficient optical interconnects in digital communication networks.

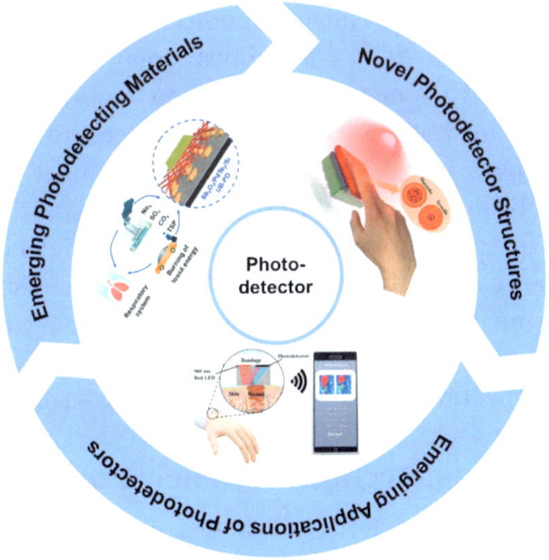

Figure 8.8. Design and application of photodetectors.

Source: https://www.mdpi.com/micromachines/micromachines-15-01249/article_deploy/html/images/micromachines-15-01249-g001.png.

These advances make way for a future full of hope where better photodetectors can bring about dramatic improvement to optical system performance parameters, thus to the entire development of photonic and optoelectronic technology. Behind this technological development is the design of new photodetector architectures that leverage the unique properties of two-dimensional materials to push the limits beyond those of standard semiconductor photodetectors. Incorporating high-quality graphene in a hexagonal boron nitride matrix, for instance, has led to on-chip photodetectors with unprecedented speed, as evidenced by a cutoff frequency of up to 42 GHz. In addition, these devices have exhibited photocurrents that are controlled by hot electron-mediated processes, expanding their application in high-speed optics. At the same time, novel findings into vertical van der Waals heterostructures such as the ones combining $MoTe_2$ and graphene have demonstrated increased external responsivity and bandwidth performance, positioning them suitable for precision tasks in telecommunications, e.g., ones operating within the standardized telecom wavelengths. This marriage of materials science and design is facilitating a new generation of photodetectors (Figure 8.8).

Additionally, difficulties brought about by the low, intrinsic mobilities of TMDCs have compelled researchers to drive device designs to their extremes to optimize for maximum performance. Optimizations like minimizing carrier transit path lengths and bias voltage optimization have brought enormous breakthroughs, with bandwidth capabilities going up to 50 GHz. These advances are facilitated by new designs having built-in waveguides, which allow for engineered interaction with the light sources. Studies have indicated that material thickness and layering of heterostructures can be controlled significantly to establish the responsivity and operating speeds of these devices, demonstrating the finesse involved in realizing speed, efficiency, and sensitivity in harmony. With the continued evolution of the photodetector technology, these discoveries aid in meeting the overall aim of developing high-performance optoelectronic devices, ultimately contributing to enhancing photonic systems in various applications, ranging from data centers to consumer electronics.

8.4.1. Types of Photodetectors and Their Functions

The development of photodetectors is all about extending various optoelectronic applications, particularly under the umbra of photonic integrated circuits (PICs). These circuits enable the integration of multiple photonic functions on a single substrate, both enhancing performance and cost-effectiveness. Of the most common types of photodetectors, photodiodes and phototransistors are utilized for other purposes. Photodiodes, especially silicon ones, are the core of optical communication systems due to high sensitivity and speed.

Phototransistors give more amplification and are thus the best to use in low-level light signal detection. These strengths are merged to facilitate high-speed data links required in modern telecommunication systems, where speed and effectiveness of data transmission matter most. Thus, photodetectors are not only mechanically involved but also play crucial roles in coherently developing technology like PICs, ultimately defining the limits to data transmission rates. Beyond traditional technologies, the introduction of quantum dots into photodetector structures introduces innovative thinking. Quantum dot lasers and photodetectors exploit the novel electronic properties of such nanostructures for more efficient absorption and emission of light. Such tunability renders the photodetectors wavelength-tunable, enhancing the versatility of photodetectors in a variety of applications from medical imaging to environmental monitoring. Moreover, the scalability and compactness of quantum dots allow small photodetector geometries, enabling the prospect of applications in miniaturized devices (Figure 8.9).

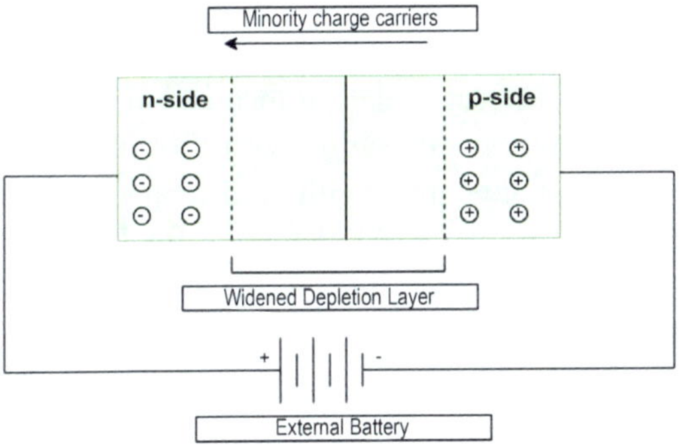

Figure 8.9. Function and mechanism of photodetectors.

Source: https://media.geeksforgeeks.org/wp-content/uploads/20240202154929/Untitled-Diagram---2024-02-02T154812908.webp.

This is in line with the new world requirements for electronic components to be faster, smaller, and more efficient. Hence, two-dimensional materials like graphene are coming into focus more for the potential of the advantages that it holds for photodetections, and this indicates directions of research for the field of optoelectronic devices, particularly in highly niche applications and conditions. One of the interesting developments in the area of photodetectors is the work on silicon photonics that offers excellent advantages for high-speed data transmission. Because silicon is a common material for the semiconductor industry, scientists can construct cost-effective and efficient photonic components using silicon. Silicon photodetectors offer high sensitivity in the near-infrared range that makes them particularly suitable for communications using optical fibers. The coupling of these detectors with conventional silicon electronic circuits adds value, creating electronic and light signal processing synergies. The integration of advanced materials, such as graphene with silicon structures, enhances the performance of these devices in operational terms, creating new applications across cloud computing platforms and data centers. Last but not least, the ongoing development in photodetector technology, and indeed in silicon photonics, marks a landmark in the pursuit of efficient and high-bandwidth optical networks, which will ultimately pave the way for more global connectivity.

8.4.2. Importance in Optical Communication Systems

The advancement of communication technology has, by and large, transformed the process of transmission and reception of information, making photonic devices the height of the revolution. Photonic integrated circuits (PICs) are an important leap, as they increase data transmission capacity by integrating multiple photonic functions on one chip. With this integration, it is feasible to reduce size and power demands with higher performance and stability over traditional electronic systems. As demands for higher bandwidth grow, incorporating PICs into optical communication systems without interruption is downright essential, especially in uses ranging from data centers to long-haul telecoms. Beyond ease of use for communication protocols, the advantages of PICs also open the door to new circuit and materials innovations down the road, advancing the field to increasingly higher levels of efficiency and speed in data transmission. In optical communications, quantum dot lasers and photodetectors offer innovative solutions with the emerging requirement for high-speed data links. The devices exploit the unique properties of quantum dots to achieve precise control over light emission, thereby offering highly efficient and tunable lasers. Furthermore, the use of quantum dot technology in conventional optical communications has the potential to significantly enhance performance metrics such as signal purity and scalability. This is particularly crucial in applications where there is a high requirement for reliability since it enables the manipulation of vast data streams without the degradation that is typical with conventional semiconductor lasers. The ongoing R&D efforts in quantum dot technology are indicative of the necessity of innovative approaches to overcome the challenges in contemporary communication, thus highlighting their importance in the larger perspective of photonic and optoelectronic technologies. Emerging silicon photonics technologies are also reshaping the infrastructure of optical communication systems, providing scalable solutions for data links at high speeds. By taking advantage of the unique optical and electronic properties of silicon, researchers are developing photonic devices that can be seamlessly integrated into present-day electronic infrastructures. Such integration is crucial for today's telecommunication systems, enabling expanded rates of data transport and broader bandwidth uses without requiring complete retooling of present technologies. Besides, the ability to miniaturize and lower the cost of silicon photonic devices is now a significant advantage, considering that consumers and businesses alike require more efficient means of communication. The convergence of silicon photonics and traditional electronic systems illustrates how imperative such technologies are for enhancing the efficiency of optical communication to enable the infrastructure to be sustainable for the growing demand for higher speed and reliability in business and personal use.

8.4.3. Emerging Technologies in Photodetection

Photodetection technology has recorded tremendous developments in the recent past that have contributed to the evolution of photonic integrated circuits (PICs) towards the integration of optical and electrical components on a single chip. This combines the efficiency and performance of light-based communication networks. The development of two-dimensional (2D) materials, such as graphene and transition metal dichalcogenides, into PICs has opened up new avenues in the creation of highly sensitive photodetectors. These materials contain amazing properties including broadband optical responses and strong light-matter interactions that are necessary to improve data speed and accuracy in communication networks. With researchers making efforts towards enhancing these materials in PIC architectures, the probability of revolutionizing multichannel data communication and monolithically integrating large photonic systems becomes highly realistic (Chang et al., 2021). Moreover, the integration of quantum dot technology has prompted the development of new lasers and photodetectors that take advantage of quantum confinement effects. The quantum dot lasers exhibit a particular advantage through tunable wavelength emissions and high efficiency, thereby enhancing photonic device performance in applications. By integrating quantum dots into traditional semiconductor material, researchers are able to produce hybrid devices that maximize the benefits of both realms—increased performance and yet remain compatible with established fabrication techniques. This achievement paves the way for the application of new photodetectors that have the ability to work in numerous wavelengths, addressing the needs of a wide variety of applications, from telecommunications to sensing technology. The ability to develop these devices for specific applications puts quantum dot-based technologies at the forefront of photonics tomorrow (Tan et al., 2020). In ultra-high-speed data links, silicon photonics is a disruptive technology that holds out the promise of a scalable solution to meeting growing bandwidth demands in data communications. By employing silicon, an element which is already well established in mass-market semiconductor technology, designers can take advantage of existing manufacturing capabilities to fabricate efficient photodetectors and light devices. The integration of silicon photonics with next-generation 2D materials enhances further the photodetector functionality, offering efficient response rates and increased sensitivity. The dual-faceted approach allows strong optical signal processing systems to be developed, which form a critical element of next-generation internet infrastructure and data centers. With the development of these technologies becoming more sophisticated, the possibility of applications of silicon photonics in new fields such as cloud computing and AI holds huge potential for enhancing the efficiency and speed of modern communication networks (Chang et al., 2021).

8.5. SILICON PHOTONICS

The integration of photonic technologies with silicon photonics has opened up unprecedented possibilities in the area of optical computation and communication. The photonic integrated circuits (PICs) form the backbone of this technology, which helps to miniaturize complex optical systems with enhanced performance parameters such as speed and efficiency. By taking advantage of the unique properties of silicon, such as its positive refractive index and compatibility with standard semiconductor fabrication processes, researchers are able to develop devices that can manipulate light at micro and nanoscale resolutions. This has led to enormous advances in high-speed data links, wherein silicon photonics can manage bandwidths larger than standard electronic systems and thereby address the growing demands for faster and more reliable communication networks. The invention of new materials and device structures further adds to the revolutionary nature of silicon photonics. For instance, the integration of quantum dot lasers and photodetectors within silicon systems has provided new fields of optoelectronic applications (Figure 8.10).

Quantum dots, due to their electronic characteristics being size-tunable, are highly flexible in the emission of light from a broad spectrum of wavelengths and can be engineered to optimize performance in varying device configurations. Such developments facilitate the realization of sensitive detectors and efficient light emitters critical for applications ranging from telecommunications to sensor technologies. Thus, silicon photonics stands at the forefront of optoelectronic integration with possible breakthroughs capable of having dramatic effects across technology domains. The integration of silicon photonics with new materials such as chalcogenide phase-change materials (PCMs) is also a testament to the fields flexibility and potential for breakthrough applications. The advancement of non-volatile photonic memory, proven by the findings such as scandium-doped antimony telluride (SST) thin films, reflects the pace at which innovation is occurring. Not only do they enhance the memory speed, but also enable functionalities that mimic biological behaviors, such as synaptic functionality in artificial neural systems. Such prospects make silicon photonics a central player towards the advancement of neuromorphic computing, photonic computing, and other emerging optical applications. As research in this domain continues, it is obvious that silicon photonics will continue to engineer the future of photonic and optoelectronic devices for many years to come.

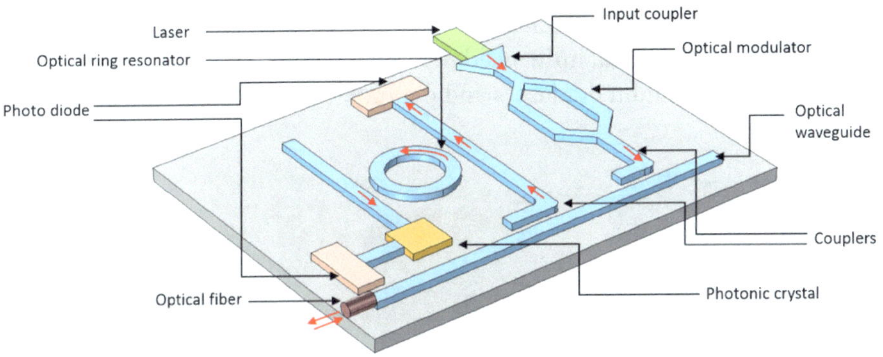

Figure 8.10. Silicon photonics: Designing and prototyping.

Source: https://cdn.comsol.com/wordpress/2017/09/Photonic-integrated-circuit_schematic.png.

8.5.1. Overview of Silicon Photonics Technology

The integration of photonic devices on silicon platforms has revolutionized the field of optoelectronics, significantly enhancing the functionality and performance of various devices. Photonic integrated circuits (PICs) are a new technology in this domain that enables photonic functions to be miniaturized and integrated into a single chip. Silicon's favorable optical properties are utilized to achieve efficient manipulation of light from modulators and lasers to detectors, ending in optical systems that are miniaturized and high-performance. This multipurpose functionality integration not only makes manufacturing processes easier but also expands the scalability of optical networks used in data communication as well as responds to the increased demand for fast data links in today's information age. Advances in material science also assist silicon photonics technology, particularly through the integration of two-dimensional (2D) materials. These materials, such as graphene, possess excellent optical properties, and these may significantly contribute to the performance of silicon photonic devices. In fact, the integration of 2D materials with silicon platforms holds out the prospect for improved light emission and modulation capabilities and for new device geometries. Taking advantage of these characteristics, scientists have realized state-of-the-art devices, such as ultra-broadband on-chip polarizers and high-efficiency thermo-optic switches. These advancements not only ensure enhanced light-matter interaction but potentially lead to new functionalities in terms of telecom and data processing businesses, bearing testament to the revolutionary potential of fusing 2D materials with silicon photonics technology (Grigorenko et al., 2017). Furthermore, the future of silicon photonics will be influenced by the exploration of novel photonic devices, such as quantum dot lasers and photodetectors (Figure 8.11).

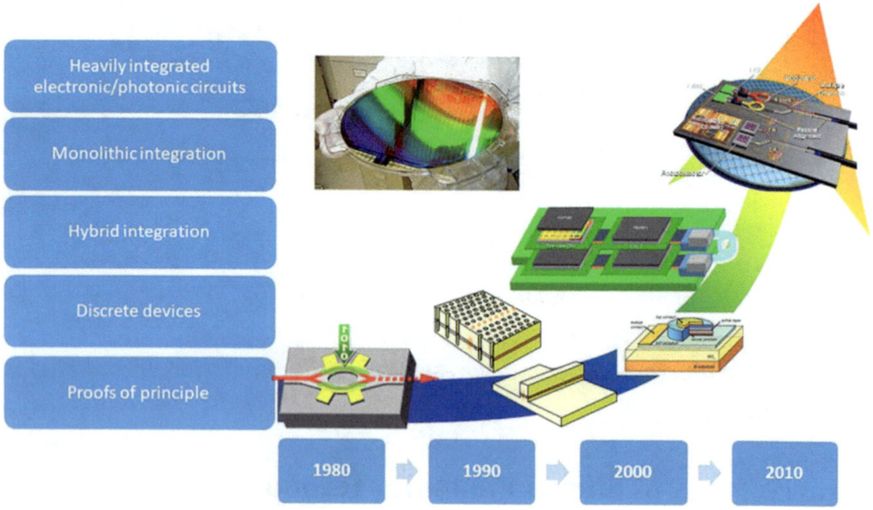

Figure 8.11. The integration of photonic devices onto silicon platforms.

Source: https://www.frontiersin.org/files/Articles/786028/fphy-09-786028-HTML/image_m/fphy-09-786028-g001.jpg.

The devices exploit the unique quantum properties of semiconductor nanostructures to offer unparalleled amounts of performance in terms of speed as well as efficiency. For instance, the integration of silicon waveguides with quantum dot technologies leads to ultrafast photodetectors that exhibit high responsivity and potential for low-power optical interconnects. Such technologies, besides breaking records in data rate of transmission in communications networks, also enhance the performance in optical sensing and imaging applications. As demand for more bandwidth and less latency grows, silicon photonics technology stands in an ideal position to backup the future optoelectronic device environment (Assefa et al., 2015).

8.5.2. Benefits for High-Speed Data Links

The development of photonic integrated circuits (PICs) has significantly transformed the age of high-speed data communication into a faster and more efficient era of data transmission. With the utilization of the potential of numerous optical frequency lines, these integrated circuits enable modulation of individual channels, thus optimizing encoding modes. Specifically, any given channel can be optimized to operate at high speeds with data transmission velocities of up to 6.5. Gb/s. The financial feasibility of PICs, particularly when produced using regrowth-free processes, renders them ideal for deployment in data centers. The stability offered by such circuits, especially in systems demanding high performance like inter-satellite communications, guarantees their relevance in raising data throughput without much additional cost, thus guaranteeing their application in the modern-day telecommunications architecture (Zhu et al., 2016). Another revolutionary field in high-speed data links is the application of quantum dot lasers and photodetectors, which enhance optical signal processing and communication efficiency. The incorporation of quantum dots offers more precise control over device properties such as wavelength tuning and emission efficiency, critical for applications where accuracy is required in frequency assignment. Furthermore, the enhanced performance in lower threshold current and temperature stability by these lasers also renders them more viable for implementation in high-speed network applications. Therefore, quantum dot technology is able to support high data rates that are commensurate with the evolving needs of the contemporary communication networks while, simultaneously, being less energy intensive than the traditional semiconductor laser geometries. The integration of these advanced technologies promises a revolutionary enhancement of the strength and reliability of high-speed data transmission systems (Achouche et al., 2016). Moreover, application of silicon photonics for high-speed data links represents a milestone in lightwave communication technologies. The application of silicon as a substrate for photonic devices makes them suitable to be integrated with existing semiconductor processes easily, thereby reducing manufacturing cost and enhancing scalability. This compatibility is essential in delivering affordable solutions for applications with high volume like data centers, which must be able to utilize bandwidth efficiently. Silicon photonics enables the use of small modulators and detectors that are capable of managing the high demand for data rates because they are capable of operating within the optical regime. Thus, the technology not only accelerates transmission but also extends the geographical reach of optical networks and can support more users without compromising on performance. Constant development of silicon photonics keeps it at the leading edge of innovation in high-speed data link systems (Achouche et al., 2016).

8.5.3. Future Trends and Developments in Silicon Photonics

Future research in the area of silicon photonics is the development of photonic integrated circuits (PICs) that put multiple photonic components on a single chip. Not only does it provide enhanced performance, but the size of production and cost are also greatly reduced compared to traditional optical systems. The flexibility of the material as a substrate material allows it to be compatible with conventional semiconductor processing techniques, and therefore there is scope for mass production. Directions in the future in this area are towards a strong emphasis on completely integrated systems that combine light sources, detectors, and optical waveguides, leading to possibilities of application in telecommunications and data centers. Furthermore, the introduction of new materials, such as graphene, can transform the very essence of PICs so that they are more beneficial with the incorporation of new optical behavior and superior functionalities compared to the conventional silicon-based devices (Calogeracos et al., 2014). As silicon photonics and photonic technology have advanced, quantum dot lasers and photodetectors are also being contemplated to boost photonic device efficiency and performance. Quantum dots with their discrete energy levels allow control of light emission and absorption very accurately and thus are the ultimate candidates for future sources of light. Such devices are extremely small in size and can be employed to emit light in numerous wavelengths at room temperature, meeting the requirements of high-speed data transmission. Future research will likely focus on further integration of silicon substrates with quantum dot-based devices, potentially resulting in revolutionary increases in data rate and energy efficiency. Moreover, hybrid device development combining quantum dots on silicon platforms could lead to enhancement in the facet of attaining low power with high performance in order to meet the growing need for energy-efficient optoelectronic devices to drive the data-oriented world (Brouckaert et al., 2010). Silicon photonics will transform the performance parameters of optical communication networks in high-speed data links. Devices such as InGaAs PIN photodetectors on silicon-on-insulator waveguides exhibit the potential of obtaining efficient evanescent coupling and high responsivity at zero external bias, which is critical in minimizing power consumption. The ability of the photodetectors to cover extensive communication bands with low dark current is critical, marking a step closer to more reliable and faster communication systems. As the demand for high-bandwidth data transfer increases, silicon photonics will be at the heart of the ability of optical interconnects to enable next-generation data centers and cloud computing systems. Continuously advancing fabrication techniques and materials science will surely result in further performance optimizations, making silicon photonics one of the pillars of future optoelectronic device technologies (Brouckaert et al., 2010).

8.6. CHALLENGES IN PHOTONIC AND OPTOELECTRONIC DEVICES

The integration of photonic devices with existing electronic systems presents a variety of challenges that hamper the creation of photonic integrated circuits (PICs). With increasing demand for high-speed data transmission, compact and efficient components are increasingly in demand. One of the greatest challenges is light-matter interaction, which must be engineered at the nanoscale with great care to allow optimal performance.

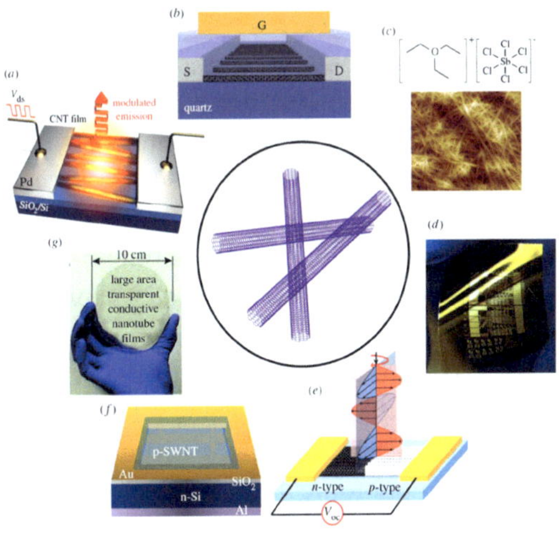

Figure 8.12. Representative photonic and optoelectronic devices using CNTs.

Source:https://www.researchgate.net/publication/332080652/figure/fig2/AS:1165394899410944@1654863204450/Representative-photonic-and-optoelectronic-devices-using-CNTs-a-A-current-driven.tif.

For instance, the development of materials that can effectively manage light propagation with low losses is a condition for PICs to operate. Additionally, the thermal management of devices is another issue, considering that excessive heat can ruin performance and reliability. To address issues such as these, new materials like graphene have been proposed to enhance light-matter interaction, thereby improving the performance and efficiency of devices (Assefa et al., 2015). These developments reflect the complexity of engineering issues in the area of modern optoelectronics. Another complexity in the area of photonic devices is the functioning of quantum dot lasers and photodetectors. While these devices offer significant advantages such as increased efficiency and tunability, they pose new challenges related to fabrication and scalability. It is crucial to realize uniformity in the size and distribution of the quantum dots to have reproducible spectral properties, which in large-scale production is sometimes difficult to obtain (Figure 8.12).

Secondly, integration of the quantum dots into existing semiconductor platforms requires advanced techniques to ensure compatibility and minimize losses. Current studies show the potential of hybrid approaches that combine quantum dots with traditional materials to combat these issues. Of specific interest are the developments in two-dimensional materials for photonics use, which hold the potential of overcoming the issues faced by quantum dots, including enhancing absorption characteristics and operating stability (Amin et al., 2017). The intricacy of silicon photonics as a high-speed data link technology also highlights some problems that require solution. Seamless integration is possible due to the compatibility of silicon technologies with existing electronic circuitry, but signal integrity and bandwidth issues overrule compatibility. As higher bandwidth is the need, efficient modulators play a key role and researchers look for alternatives such as plasmonic electro-optic modulators. Based on the design of the optical mode in accordance with material properties like graphene, researchers have been successful with energy-efficient and compact modulator designs (Amin et al., 2017). It has now become a requirement to advance towards ultra-dense photonic integrated circuits. Therefore, ongoing research is necessary in overcoming these challenges in pursuit of the improvements necessary to next-generation photonic and optoelectronic devices.

8.6.1. Technical Limitations and Hurdles

Ongoing development in photonic and optoelectronic devices continually throws open avenues for various technical challenges and obstacles to their complete realization. Integration of photonic integrated circuits (PICs) is one of the primary challenges, which, although holding out the promise of increasing data transmission, is slowed by manufacturing consistency and performance reliability. Waveguide material uniformity and exacting fabrication procedures are an issue that researchers are trying to overcome. The intricacy of aligning optical components on a microscopic scale can lead to signal loss and additional noise, ultimately diminishing the efficiency of such devices. Additionally, the scaling up of PICs to be used in extensive applications is often hindered by material constraints and compatibility concerns with existing electronic systems. Therefore, the search for novel fabrication methods and improved materials continues to be the key to enhancing the scope of the performance of PICs. The fabrication of quantum dot lasers and photodetectors also exhibits a set of inherent challenges that must be addressed in order to optimize their functionality. Quantum dots, with their size-dependent optical properties, promise extraordinary things for the future of photonic devices. However, the complexity of producing high-quality quantum dots and establishing uniformity between batches makes one question their scalability and integration. Moreover, the design of efficient coupling schemes to other systems is a technical challenge because misalignment will cause very high optical losses. Moreover, whereas they have the potential to prove useful in biosensing and display technology applications, current solid-state conditions may exacerbate performance degradation further, particularly as a result of fluctuating thermal and electrical conditions. Thus, solving these quantum dot-problematic issues becomes critical in making the transition to functional use in photonic systems. In the field of silicon photonics, the development of fast data links is outweighed by several important limitations that hold back progress. Foremost among these is silicon optical component integration with traditional electronic processing circuits due to differences in material properties and processing. Silicon photonic devices currently struggle to provide the bandwidth and energy efficiency required for high-speed uses. Furthermore, the thermal sensitivity of silicon-based devices makes it hard to operate at higher data rates, inducing distortion in the signal. Efforts are underway to enhance the modulation speed and reduce power, but new solutions in materials and design have to be found. Here, successful bridge closing of these gaps in silicon photonics will be imperative to the realization of stable and efficient optical communication networks as well as the scaling of photonic technologies to mainstream applications in contemporary computing architecture.

8.6.2. Economic Factors Affecting Development

The progress of photonic and optoelectronic devices is also driven significantly by economic drivers that make them developed and implemented. Research and development investment is critical, as countries see the promise of Photonic Integrated Circuits (PICs) in transforming communications and data processing.

Nonetheless, the amount invested tends to be in proportion with a country's economic stability and policy environment. For example, when governments invest in high-tech industries, at the same time, they encourage innovation, thus making such cutting-edge technologies as silicon photonics for high-speed data links more accessible. This financial assistance not only facilitates rapid progress in technology but also makes nations competitive within the global economy, effectively translating economic health into technological advancements (Ozaktas et al., 1997). Apart from direct investment, the global need for frontier photonic devices also creates a far-reaching web of economic interaction that encourages even more development. The advent of quantum dot lasers and photodetectors is an example of how market demand can stimulate innovation. Markets ranging from telecommunications to consumer electronics provide the market forces needed that encourage companies to improve their products and seek new applications. Economic trends can be anticipated by companies so that they can anticipate and react to market changes, facilitating proactive development strategies. Such economic sensitivity implies that photonic and optoelectronic technology innovations are not only responsive but are actually founded on consumer expectations and demands, thereby driving ensuing economic growth (Arias et al., 2019). Lastly, how globalization affects local economies dictates the future of photonic technologies. As countries ally with other countries internationally, resource and knowledge sharing continues to drive innovation across the board. Such a culture of collaboration may be followed by the establishment of specialized industries around photonics, with collective aims and shared investments. For instance, a place that establishes local research capability in silicon photonics may entice multinationals, which subsequently invest and infuse money and resources to further build local research capability. Thus, economic factors extend beyond the funding; they create an interdependency environment in which growth is sustained by local as well as global forces, hence boosting the total strength of resilience in the photonic and optoelectronic sectors (Arias et al., 2019).

8.6.3. Environmental Considerations and Sustainability

The rapid development of photonic and optoelectronic devices has paved the avenue for a new generation of technology, but also requires renewed scrutiny of the sustainability and environmental impacts. The manufacture and installation of parts like photonic integrated circuits (PICs) and quantum dot lasers are energy-consuming and tend to use materials that can be environmentally destructive if not handled carefully. While the need for high-speed data links is on the rise, particularly in data centers, so is the related energy consumption that becomes a source of concern regarding carbon footprints and resource depletion. Therefore, the integration of green materials and processes in the life cycle of these devices becomes imperative to nullify unwanted effects and maintain an eco-friendly means. Thus, producers and scientists must prioritize eco-friendly alternatives in their process and designs to assure technological advancement while conserving the planet. Additionally, silicon photonic device production fosters the movement towards greener processes. Use of silicon as a substrate, which is recyclable and abundantly available, has a significantly lower cost to the environment compared to most other materials commonly employed in photonic devices. With increased research, development towards energy-efficient manufacturing and low power consumption of these devices must be emphasized. Monitoring material usage throughout manufacturing can help minimize waste and further enhance the recyclability of silicon photonic devices, thus enhancing the use of eco-friendly technology practices. In this regard, the use of epitaxy's innovations, such as those outlined in new studies, can provide ways for improved optical properties and efficiency and reduced reliance on materials that would be harmful to the environment, as Wangila et al. (2024) observe. Moreover, the consideration of the entire lifecycle of photonic and optoelectronic devices highlights the need for sustainable operations outside so-called manufacturing efficacies. As designers embrace quantum dot lasers and photodetectors, it is pertinent to consider not just the operational lifespan and energy consumption but also the end-of-life recyclability of the components. A shift to circular economies for the electronics sector can minimize e-waste resulting from outdated or defective devices by a large margin. By utilizing cutting-edge epitaxial growth techniques to produce high-quality material on environmentally friendly substrates, the potential for producing fully recyclable devices is possible. This is in line with sustainability goals in that the progress within photonics will be good for the environment as well as further innovation in the industry, states N/A (2017).

SUMMARY

The creation of photonic and optoelectronic devices has been a technological milestone, particularly with the introduction of PICs. These circuits exhibit integration of numerous photonic components onto one chip, thereby raising speed and efficiency with reduced space. With demands for high-speed data transfer increasing, technologies such as quantum dot lasers and enhanced photodetectors are essential in filling gaps of traditional semiconductors. With their ability to effectively emit and detect light, quantum dot technologies provide for improved performance and scalability in applications ranging from telecommunications to medical diagnostics. Current R&D efforts in the field indicate PICs' potential for revolutionizing industries by offering integrated solutions to electronic circuitry, thus paving the way for future breakthroughs in photonic integration. In exploring the use of these innovations, industries and the scientific community have now begun to look at silicon photonics as a viable platform for improved high-speed data links. Silicons mundane role in microelectronics coupled with its properties optically makes it a unique platform where integration can occur with less cost and mass production capability. The integration of these technologies enables ultra-fast data transfer beyond conventional electronic systems. Furthermore, devices such as plasmonic electro-optic modulators provide hopeful prospects, where engineers have made use of the optical modes to greatly enhance performance figures. This has made devices not only energy-efficient but also sensitive to the demanding needs of modern computing environments, paving the way for more sophisticated applications across different fields. In summary, integration of the different technologies is showing a future with highly complex and highly efficient photonic and optoelectronic devices. Successful demonstration of high-frequency mode-locking in semiconductor lasers, as demonstrated through experiments with monolithic structures having photonic-bandgap mirrors, demonstrates one path to faster communications and improved imaging devices. Advances such as cooling schemes to achieve higher performance values further highlight the dynamic nature of the technology. As shown in studies, the integration of cutting-edge materials like graphene with plasmonic devices is able to realize nearly perfect modulators with very short device lengths, representing a significant milestone towards ultra-dense photonic circuits. Collectively, these developments portend the rosy future of photonic technologies, which can potentially transform not only communications systems but also more extensive applications in medicine and sensing technology as they continue to progress.

Advances in photonic and optoelectronic devices highlight the transformative potential of materials like graphene and silicon photonics to change technology. The unique characteristics of graphene, such as transparency and high interaction with photons, position it as a prime candidate for next-generation optoelectronic devices. Researchers have begun exploring how to introduce a band-gap in graphene, which is a prerequisite for its application in logical devices. Methods such as doping with chemicals and engineering structures, i.e., the introduction of carbon nanotubes (CNTs), form the core of contemporary research. The advantages of CNTs, especially their ballistic transport potential, form the core of transistor engineering and open the doors to the full promise of the materials in revolutionizing the optoelectronic device design architecture as well as their functionality across many applications, above all in photonic integrated circuits (PICs) (Calogeracos et al., 2014). Another important area of research is in silicon photonics, an important paradigm for high-data-rate processing and transmission. Taking leverage of light signals instead of electric current, silicon systems have the ability to get things done faster and better. By integrating optical components—modulators and waveguides—onto a single piece of silicon, huge advances can be made in optical networking capability and miniaturization. Moreover, the use of quantum dot lasers and photodetectors ensures better performance parameters such as increased sensitivity and lower power consumption. As silicon photonics progresses, researchers are investigating the integration of devices to develop more sophisticated systems, showing the growing demand for enhanced bandwidth support and better data management in current communication systems (Chang et al., 2016). Lastly, zero-index photonic media bring with them radical possibilities for the future of optoelectronic devices. Such new media, capable of bending light to previously unimaginable levels, promise functionalities still to be unexploited. Some significant advancements encompass the fabrication of zero-refractive-index lenses using silicon nanopillar arrays that provide anomalous focusing effects without the adverse impact of longitudinal spherical aberration. These innovations not only enhance optical performance but also provide the possibility of utilizing these materials in integrated circuits for practical applications. The integration of these zero-index photonics into the existing silicon-based technologies offers a promising avenue towards enhancing optical communication, sensing, and modulation, and is a milestone advancement in harnessing the underlying physics of photonic technologies (Chang et al., 2016).

REFERENCES

1. Achouche, M., Bauwelinck, J., Blache, F., Demeester, P., et al. (2016). An asymmetric high serial rate TDM-PON with single carrier 25 Gb/s upstream and 50 Gb/s downstream. https://core.ac.uk/download/55742543.pdf (accessed on 27th September 2025).
2. Alexander, R. R., Assamoi, P. J., Childs, D. T. D., Groom, K., et al. (2009). Quantum well and dot self-aligned stripe lasers utilizing an InGaP optoelectronic confinement layer.
3. Amin, R., Khan, S., Ma, Z., Sorger, V. J., et al. (2017). Temperature dependence of a sub-wavelength compact graphene plasmon-slot modulator. http://arxiv.org/abs/1709.01465 (accessed on 27th September 2025).
4. An, J., Zhao, X., Zhang, Y., Liu, M., Yuan, J., Sun, X., Zhang, Z., et al. (2021). Perspectives of 2D materials for optoelectronic integration. Advanced Functional Materials, 32. https://doi.org/10.1002/adfm.202110119 (accessed on 27th September 2025).
5. Assefa, S., Efetov, D. K., Englund, D., Gao, M., et al. (2015). High-responsivity graphene-boron nitride photodetector and autocorrelator in a silicon photonic integrated circuit. http://arxiv.org/abs/1507.00426 (accessed on 27th September 2025).
6. Aubert, T., Brainis, E., Hens, Z., Van Thourhout, D., et al. (2015). Fabrication and characterization of on-chip silicon nitride microdisk integrated with colloidal quantum dots. https://core.ac.uk/download/55807720.pdf (accessed on 27th September 2025).
7. Avrutin, E. A., Marsh, J. H., McDougall, S. D., Street, M., et al. (2001). Ultrafast harmonic mode-locking of monolithic compound-cavity laser diodes incorporating photonic-bandgap reflectors. https://core.ac.uk/download/58807.pdf (accessed on 27th September 2025).
8. Baets, R., Bogaerts, W., & Spuesens, T. (2018). Open-access silicon photonics: Current status and emerging initiatives. https://core.ac.uk/download/188637333.pdf (accessed on 27th September 2025).
9. Brendel, C., Cibiel, G., Cundiff, S. T., Rubiola, E., Friis, H., Giordano, V., Sauvage, G., et al. (2008). Applications of the optical fiber to the generation and to the measurement of low-phase-noise microwave signals. http://arxiv.org/abs/0807.3494 (accessed on 27th September 2025).
10. Brouckaert, J., He, S., Liu, L., Sheng, Z., et al. (2010). InGaAs PIN photodetectors integrated on silicon-on-insulator waveguides. https://core.ac.uk/download/55842545.pdf (accessed on 27th September 2025).

11. Canal Arias, J. M., Canal Barnils, C., Cernák, M., Cvelbar, U., et al. (2019). White paper on the future of plasma science and technology in plastics and textiles. https://core.ac.uk/download/187265494.pdf (accessed on 27th September 2025).

12. Chang, C., Chen, W., Chen, Y., Chen, Y., Chen, Y., Ding, F., Fan, C., et al. (2021). Recent progress on two-dimensional materials. Acta Physico-Chimica Sinica, 0, 2108017. https://doi.org/10.3866/pku.whxb202108017 (accessed on 27th September 2025).

13. Chang, M. L., Deng, S. Z., Dong, J. W., He, J., et al. (2016). Realization of zero-refractive-index lens with ultralow spherical aberration. http://arxiv.org/abs/1610.05442 (accessed on 27th September 2025).

14. Chen, S., Li, W., Liao, M., Liu, Y., et al. (2018). Monolithic quantum-dot distributed feedback laser array on silicon. https://core.ac.uk/download/159611142.pdf (accessed on 27th September 2025).

15. Chen, X., Xue, Y., Sun, Y., Shen, J., Song, S., Zhu, M., Song, Z., et al. (2022). Neuromorphic photonic memory devices using ultrafast, non-volatile phase-change materials. Advanced Materials, 35. https://doi.org/10.1002/adma.202203909 (accessed on 27th September 2025).

16. Das, S., Fardad, S., Hui, R., Kim, S., et al. (2016). Nanophotonic modal dichroism: Mode-multiplexed modulators. http://arxiv.org/abs/1705.07954 (accessed on 27th September 2025).

17. de Lima, T. F., Nahmias, M. A., Peng, H. T., Prucnal, P. R., et al. (2017). Principles of neuromorphic photonics. http://arxiv.org/abs/1801.00016 (accessed on 27th September 2025).

18. Deng, Q., He, P., Kenney, M. G., Li, et al. (2016). Ultracompact high-efficiency polarizing beam splitter based on silicon nano brick arrays. https://core.ac.uk/download/42371131.pdf (accessed on 27th September 2025).

19. Ding, Y., Grassi, R., Gu, T., Hone, et al. (2018). Spatially controlled electrostatic doping in graphene p-I-n junction for hybrid silicon photodiode. https://core.ac.uk/download/189890176.pdf (accessed on 27th September 2025).

20. Ellis, B., Haller, E. E., Harris, J., et al. (2012). Electrically driven photonic crystal nanocavity devices. http://arxiv.org/abs/1201.0964 (accessed on 27th September 2025).

21. Emboras, A., Flöry, N., Leuthold, J., et al. (2019). Waveguide-integrated van der Waals heterostructure photodetector at telecom band with high speed and high responsivity. http://arxiv.org/abs/1904.10287 (accessed on 27th September 2025).

22. Fontcuberta i Morral, A., Smith, C. L. C., Bermúdez-Ureña, E., García-Vidal, F. J., Tutuncuoglu, G., Hunsperger, R. G., Cuerda, J., et al. (2017). Plasmonic waveguide-integrated nanowire laser. https://core.ac.uk/download/84004543.pdf (accessed on 27th September 2025).

23. Gardes, F. Y., Mashanovich, G., Reed, G. T., et al. (2010). Silicon optical modulators. https://core.ac.uk/download/397036.pdf (accessed on 27th September 2025).

24. Gopalan, K. K. (2018). Optical surfaces for mid-infrared sensing. https://core.ac.uk/download/185528365.pdf (accessed on 27th September 2025).

25. Grigorenko, A., Urich, A., Yadav, A., Lee, B., Eggleton, B. J., Ye, C., Huang, C. C., et al. (2017). Chalcogenide glass-on-graphene photonics. https://core.ac.uk/download/323947279.pdf (accessed on 27th September 2025).

26. Ho, V. X., Jiang, H. X., Lin, J. Y., et al. (2018). Room-temperature lasing action in GaN quantum wells in the infrared 1.5-micron region. http://arxiv.org/abs/1802.10456 (accessed on 27th September 2025).

27. Hossein-Zadeh, M., & Vahala, K. J. (2010). An optomechanical oscillator on a silicon chip. https://core.ac.uk/download/4884077.pdf (accessed on 27th September 2025).

28. Kelleher, E., & Woodward, R. (2015). 2D saturable absorbers for fiber lasers. https://core.ac.uk/download/77010163.pdf (accessed on 27th September 2025).

29. Kuznetsov, A. I., Brongersma, M. L., Yao, J., Chen, M. K., Levy, U., Tsai, D. P., Zheludev, N. I., et al. (2024). Roadmap for optical meta surfaces. ACS Photonics, 11, 816–865. https://doi.org/10.1021/acsphotonics.3c00457 (accessed on 27th September 2025).

30. Lin, C. C., Wu, Y. R., Kuo, H. C., Wong, M. S., DenBaars, S. P., Nakamura, S., Pandey, A., et al. (2023). The micro-LED roadmap: Status quo and prospects. Journal of Physics Photonics, 5, 042502. https://doi.org/10.1088/2515-7647/acf972 (accessed on 27th September 2025).

31. Liu, Y., Choudhary, A., Marpaung, D., & Eggleton, B. J. (2020). Integrated microwave photonic filters. Advances in Optics and Photonics, 12, 485–485. https://doi.org/10.1364/aop.378686 (accessed on 27th September 2025).

32. Malyutenko, V. K., & Zinovchuk, A. V. (2006). Mid-infrared LEDs versus thermal emitters in IR dynamic scene simulation devices. https://core.ac.uk/download/12086995.pdf (accessed on 27th September 2025).

33. Moody, G., Sorger, V. J., Blumenthal, D. J., Juodawlkis, P., Loh, W.,

Sorace-Agaskar, C., Jones, A. E., et al. (2021). 2022 roadmap on integrated quantum photonics. Journal of Physics: Photonics, 4, 012501–012501. https://doi.org/10.1088/2515-7647/ac1ef4 (accessed on 27th September 2025).

34. N/A. (2013). The Boston University Photonics Center Annual Report 2013–2014. https://open.bu.edu/bitstream/2144/29752/1/Photonics-Center-Final-Annual-Report-2013-2014-v2.pdf (accessed on 27th September 2025).

35. N/A. (2017). SciTech News, 71(1). https://core.ac.uk/download/83049835.pdf (accessed on 27th September 2025).

36. Ozaktas, H. M. (1997). Toward an optimal foundation architecture for optoelectronic computing. 1. Regularly interconnected device planes. https://core.ac.uk/download/52921873.pdf (accessed on 27th September 2025).

37. Rifat, A. A., Rifat, A. A., Rifat, A. A., Paul, B. K., Bariáin, C., Ahuja, D., Bykov, D., et al. (2018). Development of photonic crystal fiber-based gas/chemical sensors. http://arxiv.org/abs/1801.06170 (accessed on 27th September 2025).

38. Tan, T., Jiang, X., Wang, C., Yao, B., & Zhang, H. (2020). 2D material optoelectronics for information functional device applications: Status and challenges. Advanced Science, 7. https://doi.org/10.1002/advs.202000058 (accessed on 27th September 2025).

39. Wangila, E. (2024). Development of III–V and IV epitaxy on sapphire technology. https://core.ac.uk/download/620867123.pdf (accessed on 27th September 2025).

40. Zhu, H., Marsh, J. H., Xu, J., Han, L., Hou, L., Deng, Q., & Liang, S. (2016). Fully integrated multi-optoelectronic synthesizer for THz pumping source in wireless communications with rich backup redundancy and wide tuning range. https://core.ac.uk/download/42372375.pdf (accessed on 27th September 2025).

CHAPTER 9

QUANTUM AND SINGLE-ELECTRON DEVICES

Contents

Introduction 353
9.1. Fundamentals of Quantum Mechanics 356
9.2. Qubits in Quantum Computing 361
9.3. Single-Electron Transistors (SETs) 366
9.4. Quantum Dots and Their Applications 371
9.5. Cryogenic Electronics Design Considerations 376
9.6. Current Research and Developments 381
Summary 386
References 389

INTRODUCTION

Quantum and single-electron devices development is a significant revolution in the area of modern electronics, offering novel opportunities for computing and information processing. Such technologies are based on the principles of quantum mechanics, employing quantum bits or qubits that can exist in a superposition of states. Various realizations of qubits are proposed, including superconducting qubits, semiconductor qubits, and topological qubits, each with unique characteristics and challenges. Research into these qubit forms holds importance in fault-tolerant quantum computation and has the promise to lead to exponentially accelerated computation compared to classical systems. As methods for qubit control and coherence are refined by researchers, the potential for application in areas such as quantum cryptography, simulation, and hard problem-solving becomes increasingly viable, allowing for a new age of computational technology. The operation of single-electron transistors (SETs) is unique in quantum devices since they utilize the control of single electrons to facilitate functionality. SETs operate based on Coulomb blockade mechanisms wherein charge transport is managed by individual electrons' energy states and their couplings with a supporting gate structure. This novel operating process allows SETs to exhibit unparalleled sensitivity and low power consumption, which makes them promising candidates for next-generation electronic devices. Moreover, with the requirement for increasingly small and efficient devices growing, the use of SETs in cryogenic electronics also helps in further performance enhancement by diminishing thermal noise. By studying interactions between SETs, qubits, and cryogenic environments, researchers aim to develop scalable solutions that bridge the gap between quantum computing and practical applications in the electronic industry. Cryogenic electronics design considerations can't be overstated, particularly for integrating quantum and single-electron devices. Proper thermal management is now a concern since quantum systems should be operated at temperatures near absolute zero to preserve the fragile quantum states and avoid decoherence. This requires advanced cooling technologies and materials capable of withstanding the harsh conditions and operate at the optimal level. Furthermore, the circuitry should also overcome signal integrity and power dissipation challenges at low temperatures, hence the requirement for accurate designing techniques and materials. With the onset of cryogenic technology and materials science, there is the potential for greater fidelity in qubit operations and improved performance of single-electron devices, which would be a turning point in the evolution of quantum electronics and their eventual commercialization and integration into mainstream technologies.

Definition of Quantum Devices

Quantum devices have promoted the advent of a new era where classical technology limitations are breached, presenting unexplored functionalities and efficiencies. At the heart of this ecosystem are qubits, the quantum information carriers, implemented with various physical realizations in the form of superconducting circuits, semiconductor-based systems, and topological architectures. Each of these qubit flavors has its own particular strengths;

for instance, superconducting qubits have the benefit of rapid operations and compatibility with microwave technology, whereas topological qubits have the benefit of enhanced error tolerance due to their natural robustness against local perturbations. With progress in the area, the search for higher fidelity and lower decoherence remains the driver field, with resulting research activities toward improved qubit construction and fault-tolerant quantum computing procedures. In addition to qubits, SETs are now basic building blocks of quantum devices, facilitating one-electron manipulation in semiconductor nanostructures. SETs exploit the principle of Coulomb blockade, where electron transport is regulated at the quantum regime. The scalability and integration of these devices into big circuits are of major interest due to their potential to render quantum computing architectures tiny and efficient. Secondly, SETs can also be utilized to measure quantum systems with high resolution and sensitivity. The development of SET technology is a major milestone towards the achievement of practical quantum devices that can perform real-world calculations, thus widening the applicability and scope of quantum research and applications. The cryogenic electronic design concerns are of utmost importance while designing quantum devices that will operate at very low temperatures where thermal fluctuations are minimal and quantum coherence is preserved. Materials possess unique electronic and thermal characteristics at such low temperatures, which influence the device's performance. Sustaining effective heat management, noise reduction, and instrumentation at cryogenic temperatures are central challenges in this technology. In addition, emerging trends in cryogenic design continue to unravel, enhancing the performance of quantum architecture. With progress in both theoretical and practical engineering components, the interplay of quantum and single-electron devices will grow deeper, opening doors for the next generation of computation and information processing technologies, demonstrating the innovative nature of present-day physics research (Ozawa et al., 2019; Kühne et al., 2020).

Importance of Single-Electron Devices

Creation of technological applications in quantum computing has brought to prominence the pivotal role of single-electron devices, particularly through the creation of SETs. The devices function through control of individual electron flow, hence offering control over electrical currents in a manner never seen before. This accuracy allows for the utilization of qubits, a required element of quantum computing, which are superconducting-based, semiconductor-based, or topologically based. The ability to localize charge carriers in such devices not only enhances the efficiency of computing but also allows for the introduction of novel quantum algorithms that exploit qubit performance.

Researchers and engineers are, therefore, more and more exploring SETs as a future solution to computing architecture, where the nuanced interplay between electron dynamics provides the foundation for improvements in processing speed and capacity. In the realm of quantum sensors, the significance of single-electron devices becomes even more pronounced. These devices, noted for their high sensitivity, are essential in precision metrology, especially when exploiting quantum entanglement. Traditional classical sensors often limit the potential for accuracy, but with single-electron devices like SETs, one can harness quantum properties to achieve sensitivity levels previously deemed unattainable. The applications of the devices in quantum sensing applications make it possible to conduct extensive explorations of physical phenomena, challenging previous known limits of comprehension in areas such as time measurement and magnetic field detection. Current studies believe that the incorporation of single-electron devices in quantum sensor implementations is set to revolutionize scientific experimental metrics and therefore make the devices vital in the experimental physicist's arsenal (Degen et al., 2017). The cryogenic electronics design challenges are another critical aspect of single-electron devices, particularly when it comes to their operating environment. Cryogenic temperatures are needed for maintaining quantum state coherence, which is a top priority in the mitigation of thermal interference and noise. In this system, hardware and circuit design have to be adapted for compatibility with the unique demands of low-temperature operation. The intricate interplay between materials and engineering solutions within SETs serves to overcome the challenges brought about by thermal fluctuations and enhance the overall quantum system stability. With this, the development of advanced single-electron devices not only pushes the frontiers of quantum computing but also stimulates material science innovation, as researchers are constantly seeking to improve the functional properties of these groundbreaking technologies (Hu et al., 2014).

9.1. FUNDAMENTALS OF QUANTUM MECHANICS

The finding of quantum mechanical principles that are universal has immense implications for modern electronic devices, notably for qubits. Quantum bits may be realized in various physical systems, e.g., semiconductor devices, superconducting circuits, and topological media. Various forms of qubits offer varying strengths and vulnerabilities, with implications for coherence times, operational purity, and prospects for scaling for quantum computations. Superconducting qubits are distinguished by their high operation speeds and widespread experimental application, while semiconductor qubits are looked upon as a prospect of being incorporated into existing electronic technologies.

Topological qubits, leveraging the robustness of topological invariants, have the potential to ultimately overcome some constraints of traditional types of qubits, such as decoherence. It is significant to be familiar with these distinctions in order to more vigorously continue developing quantum devices in the context of quantum and single-electron. Other than qubits, single-electron transistors (SETs) cannot also be ignored when discussing quantum mechanics impact on electronic design. SETs use quantum tunneling to control the transit of single electrons and are extremely sensitive and are best employed in cases that require good measurements (Figure 9.1).

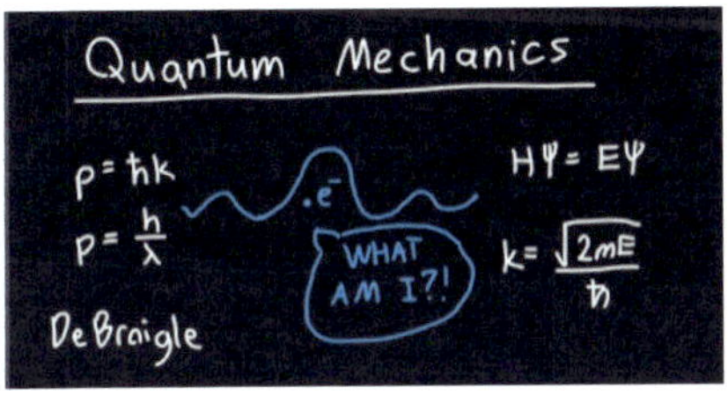

Figure 9.1. *Method of quantum mechanics.*

Source: https://i.ytimg.com/vi/rir3U4Eg1F0/hq720.jpg?sqp=-oaymwEhCK4FEI-IDSFryq4qpAxMIARUAAAAAGAElAADIQj0AgKJD&rs=AOn4CLDJNl7hLuApCOozj6IWNjQrImDnqg.

The unique operational mode of SETs stems from the fundamental assumptions of quantum mechanics, in which Coulomb blockade effects regulate electron transport at the nanoscale. This sensitivity renders SETs particularly appropriate for applications in sensing and metrology, where charge changes at the nanoscale can be detected. As researchers go on to refine SET designs and integrate them into larger quantum technologies, an understanding of the quantum mechanics underlying will be crucial to refining their operation and integration in real applications. Finally, cryogenic electronics design illustrates how ideas within quantum mechanics shape practical engineering solutions. Cryogenic conditions are necessary to maintain qubit coherence and allow SETs to operate because lower temperatures decrease thermal noise. At cryogenic temperatures, the material properties and electron dynamics are radically changed, calling for special design considerations in interconnections and circuit components. This involves the choice of materials possessing the needed electrical and thermal properties, as well as the incorporation of suitable shielding and microfabrication technology. Lastly, with further advancement in quantum technologies, the combination of quantum mechanics and the realities of practical design in cryogenic electronics will be essential in maximizing the potential of quantum circuits and the ability of single-electron devices.

9.1.1. Key Principles of Quantum Theory

The exploration of the microscale regime is accompanied by new phenomena that challenge classical intuition about particle and wave behavior. Central to quantum theory is matter's dual nature in which particles possess both wave and particle properties. The duality takes on particular importance when it comes to qubits, the building blocks of quantum information. Various physical realizations of qubits such as superconducting qubits, semiconductor qubits, and topological qubits rely on these quantum principles to manage and leverage quantum states for computation. Every type of qubit possesses certain advantages; superconducting qubits, for instance, are more immune to external noise, while topological qubits promise increased immunity from decoherence, a major problem for quantum computing machines. The research on such critical issues is crucial in building and refining quantum devices that possess the ability to outperform classical devices in computation and processing capacity. Single-electron transistors (SETs) are a prime example of how quantum principles are being applied to create innovative electronic devices. SETs operate on the single-electron tunneling mechanism that allows them to be very low in power consumption and scalable. The quantum nature of the electron is essential in this application because the Coulomb blockade limit hinders electron conduction until sufficient voltage is provided. This allows SETs to function efficiently at the nanoscale, where classical physics is no longer applicable (Figure 9.2).

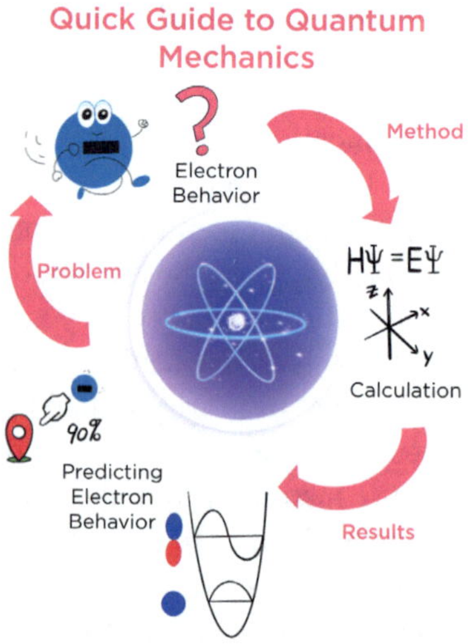

Figure 9.2. *The working principle of quantum mechanics.*

Source: https://d20khd7ddkh5ls.cloudfront.net/untitled-artwork_3.png.

The control of individual electrons not only plays a key role in reducing energy dissipation in transistors but also offers promise as a path towards quantum information processing and storage. The working regimes that define SETs provide the need for the incorporation of quantum mechanics in the continuing evolution of electronic devices, particularly because industries are directing themselves towards more advanced semiconductor technology. Cryogenic electronic design factors are a natural result of the peculiar characteristics provided by quantum systems. Under conditions of low temperature, thermal noise is greatly reduced, and quantum devices like qubits and SETs can be designed to operate more effectively. However, lowering such systems to the necessary cryogenic temperatures is an engineering task in itself, primarily maintaining steady operation and avoiding quantum decoherence. Material integration with desirable electronic characteristics at low temperatures is of central importance in imagining successful electronic components for use in quantum. In addition to material selection, also circuit designs must be compatible with the coherent control of quantum states with the need for sophisticated precise measurement and feedback systems. Addressing these design obstacles will ultimately shape the future landscape of quantum computing and electronic devices by offering reliable, scalable, and efficient systems that harness the peculiar properties of quantum mechanics.

9.1.2. Quantum Superposition and Entanglement

The complementarity of quantum superposition and entanglement underlies numerous quantum technologies, including single-electron devices. By manipulating qubits—superconducting, semiconductor, and topological types—researchers can apply quantum phenomena to new purposes. In this case, each qubit possesses numerous states simultaneously due to superposition, enhancing computing power. Superposition enables quantum computers to calculate information far more efficiently compared to their classical counterparts. Entanglement also enables the creation of strong correlations among qubits, enabling quantum communication and cryptography. Qubits can become entangled so that their states become a function of one another, and because of this, they are required for sophisticated networks that use quantum mechanics to ensure security and data integrity. Recent advancements in quantum devices are marked by the potential for quantum superposition and entanglement for use in areas like quantum cryptography and quantum networking. Quantum key distribution protocols based on such principles are therefore based on them, offering secure channels no classical system could possibly equal. These quantum cryptographic methods have also been rapidly evolving from theories to operational implementations, demonstrating that the synergy between superposition and entanglement is capable of producing robust security protocols. Device independence and overcoming satellite communications challenges as a case area of use indicate just how flexible quantum technologies are. As researchers move toward creating greater rate protocols based upon continuous variable systems, the practical applications to both secure encryption and transmission of data are transformative and deep (Pirandola et al., 2020).

As the quantum internet vision moves towards actuality, incorporating quantum entanglement and superposition into single-electron devices will be essential. These networks promise to revolutionize communications by connecting quantum processors across distances, enabling capabilities that today's classical technology cannot reach. Researchers are reporting development milestones toward this ambitious goal, noting experimental and theoretical advances essential to a functioning quantum internet. The ability to use entangled qubits for device-to-device communication is just one illustration of how quantum mechanics can enhance logistical structures in technology. By both its theoretical underpinnings and practical advances, the field is set to transform not only secure communication but entirely new forms of worldwide interaction founded on quantum phenomena (Wehner et al., 2018).

9.1.3. Implications for Computing and Electronics

With the advancement of quantum computing spurred on, the necessity for effective qubit designs—specifically superconducting, semiconductor, and topological qubits—has become a pivotal shift in electronics. Researchers are discovering that superconducting qubits, with minimal energy dissipation, offer viable paths to scaling quantum computers. This is paired in semiconductor qubits by advances in materials science and manufacturing technology that enhance performance and integration. Topological qubits, on the other hand, offer new opportunities for fault tolerance since they are robust in nature against local perturbations. The implications of these developments exceed computing capacity; they have the power to transform the landscape of electronic devices by facilitating computations previously impossible, thus bringing a paradigm change in designing and building electronic systems within the industry. No less significant in this regard is the development of single-electron transistors (SETs), which are promoted for energy efficiency in consumption and unpreceented scaling capabilities. SETs can operate at the quantum level, controlling the movement of individual electrons across a nanoscale junction. Due to this feature, they can function effectively in ultra-low-power devices, which are increasingly becoming a necessity in modern computing technology. With companies seeking green electronics, SETs' capacity for lowering power consumption provides them a good reason for application in future electronic systems. Furthermore, the ability of SETs to operate at cryogenic temperatures opens up new opportunities for their application in quantum computing applications, creating a synergy between advances in transistor technology and quantum devices. This convergence of technologies ushers an era of revolutionary electronics, underlining the necessity for multidisciplinary research. The consideration of cryogenic electronics design not only enhances the operational capability of qubits and SETs but also comes with unusual challenges that need innovative solutions. Ordinary electronic components at cryogenic temperatures can behave unpredictably, thus the necessity for careful design to ensure reliability and performance. This has generated interest in designing special materials and structures with capability to operate under extreme conditions. Of special interest are metasurfaces, which are becoming more and more part of the scene; they exhibit unique optical properties that can be tailored for use in quantum optics and computing with compact solutions meeting the industry's growing need for miniaturized components. These shifting capabilities also better explain how the convergence of new technologies can drive improvements in scientific research and manufacturing processes, propelling the computing and electronics field towards a new golden age of technology (Bauer et al., 2023; Kuznetsov et al., 2024).

9.2. QUBITS IN QUANTUM COMPUTING

Most of the big quantum computing breakthroughs depend on the control and manipulation of qubits, quantum information building blocks. Of numerous possible realizations, superconducting qubits are very serious candidates due to their very good control and coherence properties.

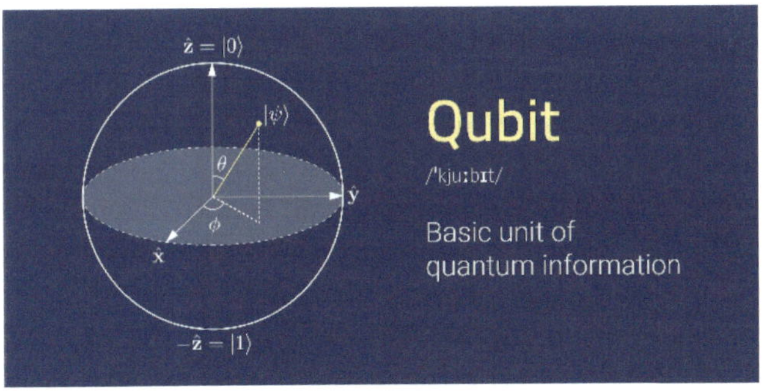

Figure 9.3. *Basic features of qubit.*

Source: https://www.mouser.in/blog/Portals/11/Figure%201%20Qubit%20Representation_1.jpg.

Superconducting qubits use circuits that benefit from superconductive material in order to support quantum states controllable with microwave photons. This approach has brought about circuit quantum electrodynamics (QED), which enhances the coupling of qubits to electromagnetic fields and makes room for sophisticated quantum operations. In addition, control of the qubits in superconducting circuits not only enables fundamental experiments but also scalable architectures needed for future quantum processors. The present research in this area focuses on the interaction between qubits and their surroundings, the central factor for knowledge of noise prevention and improvement in qubit performance in real-world applications (Bharti et al., 2022) (Figure 9.3).

Compared to superconducting qubits, semiconductor qubits present a different avenue for quantum computing, primarily through the use of single-electron transistors (SETs). SETs utilize the discrete character of charge in a single electron to store quantum states that can be controlled through gate voltage changes. It can be highly integrated, facilitating scaling in quantum systems while simultaneously leveraging mature semiconductor processing technologies. Semiconductor qubits also benefit from long coherence times, which are due to their low interaction with the thermal environment, especially when designed to be optimized for cryogenic operation. Cryogenic electronic technologies play a pivotal role in the functioning of such equipment as they enable one to control qubits at very low temperatures, which further reduces decoherence and enhances the operating efficiency. Such complementary advancements in semiconductor technology and quantum information science pave the way for robust and scalable architectures in quantum computing (Blais et al., 2021). Topological qubits represent a paradigm shift in efforts towards fault-tolerant quantum computation and differ from earlier proposals in being inherently noise and quantum decoherence resistant. Topological qubits arise from topological materials, which are defined by exhibiting topological phases, and quantum information is pinned in non-local degrees of freedom. The nature of the properties being utilized as the qubit basis lowers their susceptibility to local disturbances, hence presenting a path towards fault-tolerant quantum computation. Efforts towards the achievement of topological qubits have revealed new paradigms for qubit control and interaction, grounded in principles that originate from condensed matter physics. Taking the advantage of topological state benefits, scientists are creating devices that not only provide the integration of quantum at a scalable level but also ensure quantum operations' reliability. With the progress in the field, the coupling between theoretical expectations and experimental realizations in topological qubits continues to add to the general quantum information processing regime (Bharti et al., 2022; Blais et al., 2021).

9.2.1. Superconducting Qubits

Research on superconducting qubits represents a significant development in the field of quantum information processing, particularly due to their unique strengths and potential. These superconducting circuit-based qubits are highly controlled and scalable, two important necessities for realizing operational quantum computers. By combining superconducting qubits with microwave photons, the field of circuit quantum electrodynamics (QED) has emerged, which allows researchers to manipulate light-matter interactions on the quantum level. This has not only built theoretical understanding but also improved experimental techniques, facilitating coherent control of the qubits themselves. With the continued advancement of superconducting qubit technology, it is increasingly clear that integrating these systems with more complex quantum computational structures will enable advances in advanced quantum algorithms and simulations, putting them center stage in the race to practical quantum computing (Blais et al., 2021). Apart from their computational capabilities, superconducting qubits are also important in investigating phenomena of quantum communications and quantum optics. They have been made prime candidates for use in applications of quantum cryptography due to their strength and compatibility with microwave technologies (Figure 9.4).

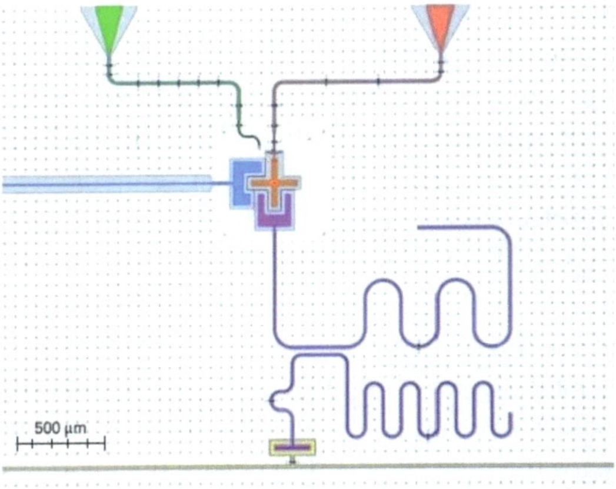

Figure 9.4. *The pathway of superconducting qubits.*

Source: https://www.azoquantum.com/image-handler/ts/20220215121012/ri/467/src/images/Article_Images/ImageForArticle_115_4460750708348385175.jpg.

Since quantum cryptography relies on secure information transfer, the coherent qubit-state control capability of superconducting circuits automatically translates to more secure protocols. Research work is currently concentrated on taking advantage of superconducting qubits together with the construction of quantum repeaters and networks, which will increase the scope of quantum communications and prevent the constraints that currently affect point-to-point communications. Finally, superconducting qubits not only assist in quantum computing but potentially transform secure data transmission methodologies within the increasingly dynamic information technology environment (Pirandola et al., 2020). Nevertheless, deploying superconducting qubits is no walk in the park, especially in cryogenic electronics design. Superconducting qubits run at temperatures near absolute zero and require sophisticated cooling systems to maintain their quantum coherence and prevent thermal noise from interrupting their delicate states. Control circuit design and measurement techniques must therefore be modified to operate optimally in such extreme conditions, demanding innovation in cryogenic engineering and materials science. Mechanisms of decoherence, as modeled in circuit QED, are critical to learn in order to optimize qubit performance within these systems. Resolving these problems does not just enhance superconducting qubit reliability but also encourages interdisciplinarity between quantum physicists and engineers, in the end furthering the body of quantum technology (Blais et al., 2021).

9.2.2. Semiconductor Qubits

The advent of quantum computing has seen many applications of qubits, and semiconductor qubits are among those that have generated tremendous interest for being incorporated into existing electronic technologies. Based on the technique of controlling electronic states in semiconductors, these qubits encode quantum information using electron spins or charge states localized in quantum dots (QDs). One of the key advantages of semiconductor qubits is their scalability potential; as semiconductor technology advances, bigger collections of quantum dots can be made. The arrays offer a solid foundation for building complex qubit structures and executing quantum algorithms. Having the right control over such qubits is no simple feat since precise electrostatic and dynamic voltages must be achieved to ensure the operational effectiveness of the qubit systems, which poses a significant challenge to classical control methods (Zwolak et al., 2023). This becomes even more complicated with more quantum dots, and hence advanced controls become necessary for optimal performance. Growing need for more automation and efficiency has led the way to using machine learning (ML) algorithms in conjunction with standard control algorithms. Scientists are able to automate gate voltage programming and dynamical parameter control with higher qubit fidelity and coherence times by employing algorithms that are capable of learning from experimental data. Besides, the employment of ML enables one to realize adaptive control schemes that are capable of adapting themselves based on environmental noise, a development crucial in the maintenance of the integrity of qubits in realistic quantum computing environments. The blending of existing computing techniques with sound semiconductor physics provides the foundation for new approaches to addressing the complexity in QD systems and renders semiconductor qubits more viable in quantum computing architectures (Zwolak et al., 2023). The significance of semiconductor qubits further resides in coupling to electric fields, specifically in spin-orbit coupled qubits where electric excitation can be used to control charged particle spin states. This gives rise to an important property for the rapid and efficient control of qubits, something that is necessary for the realization of helpful quantum gates. But this approach is not without limitations, such as susceptibility to electrical noise that can compromise coherence and overall qubit quality. Recent advances have been promising to overcome such issues by discovering specific operating conditions, also known as operation sweet spots, that reduce the impacts of noise. For instance, studies of hole spin qubits in natural silicon devices have produced stunning enhancements in coherence time, holding promises for scalable quantum processing. The findings highlight the importance of combining semiconductor technology and quantum mechanics to ensure that quantum systems are robust to environmental noise while remaining sufficient for quantum computing applications in the future.

9.2.3. Topological Qubits

The search for quantum computing has forced scientists to look into various implementations of qubits, and topological qubits are extremely promising due to the fact that they are naturally robust to noise. Unlike the conventional qubits, which are susceptible to outside influences, topological qubits take advantage of the properties of topology to encode information in a less error-prone manner. This immunity is largely guaranteed by non-Abelian anyons, exotic particles that exist within two-dimensional systems and obey new braiding statistics. The manipulation of these anyons ensures fault-tolerant quantum computation, which is crucial to the creation of stable and scalable quantum devices. The accomplishment is particularly important, as the field strives to create systems that can be relied upon to perform intricate calculations necessary in applications like quantum simulation and materials science, demonstrating the potential for unprecedented growth in technology and computation. Another essential aspect of recent research on topological qubits is their integration with other quantum systems and with their surroundings. Topological qubits must be interfaced with current technologies such as superconducting circuits or single-electron transistors (SETs), which are also considered for use as qubits. Superconducting qubits, according to circuit quantum electrodynamics rules, have shown excellent coherence times but are still plagued by decoherence issues like coupling with the environment, including various forms of noise (Blais et al., 2021). Topological qubits are poised to alleviate some of these issues because of their specific braiding characteristics and leave space for hybrid systems, which could utilize the best of each category of qubit. Creating successful quantum devices means meticulous cryogenic electronics design, especially when one takes into account the temperature ranges for the functioning of superconducting and topological qubits.

As science continues to progress, the vision for obtaining a quantum advantage through the use of topological qubits becomes increasingly plausible. The quest to utilize these qubits is not purely speculative; empirical advancements attest to their viability in solving complex issues in all fields of science. For instance, the ability to simulate quantum systems—a domain that Feynman identified as crucial—could see massive breakthroughs via topological qubits, particularly in fields such as materials science and chemistry, where simulating many-body interactions is key (Bauer et al., 2020). As quantum computing's future keeps evolving, identifying the intricate interaction between various qubit types and their applications will be critical to unleashing their full potential. The intersection of topological qubits with the development of coherent quantum devices is prelude to the future technological triumph, a time of quantum information processing revolution.

9.3. SINGLE-ELECTRON TRANSISTORS (SETS)

Electronic device advancement has been marked by the progressive reduction in size, with SETs being the climax of the trend. SETs are founded on the concept of the control of electron transfer individually, Coulomb blockade and quantum tunneling being the basic mechanisms. This unique ability makes SETs consume very little power, and thus they are particularly ideal for applications in quantum computing and low-power electronics.

On a further note, nanotechnology-based integration of structures such as quantum dots and metal nanoparticles enhances the efficiency of SETs by maximizing the energy barriers for electron tunneling. With the ongoing increase in the demand for extremely efficient transistors, SETs are potential candidates with the potential of expanding the frontiers of current electronic technology, paving the way towards future breakthroughs in the areas of nanoscale circuits as well as in information processing systems. Although they possess such huge advantages, experimental realization of SETs is faced with some engineering challenges that have to be addressed cautiously. One of the requirements is the operating temperature as SETs are often operated at cryogenic temperatures in order to mitigate thermal fluctuations which are sources of decoherence. One such necessity calls for very advanced materials science and engineering practices, especially while designing efficient cooling systems to obtain proper working conditions (Figure 9.5).

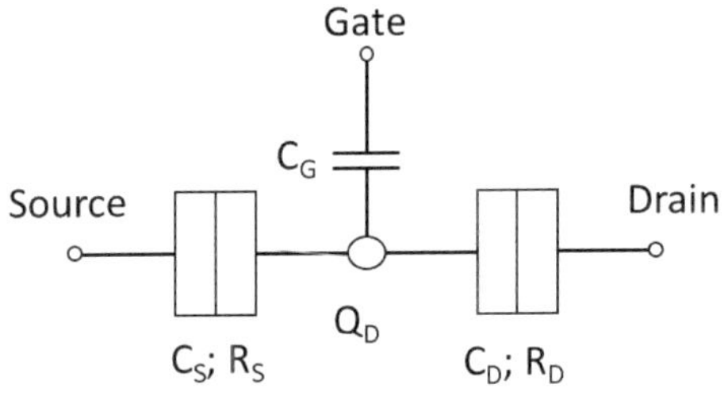

Figure 9.5. *Schematic of a basic single-electron transistors (SET) and its internal electrical components.*

Source: https://upload.wikimedia.org/wikipedia/commons/5/57/SET_schematic2.jpg.

Furthermore, the processes for manufacturing SETs must offer strict control over the quantum mechanical properties of the devices, which could include nanoscale patterning and lithography techniques. Resolutions to these challenges will be important in integrating SETs into large-scale quantum circuits in an effort to make them work in tandem with other types of qubits, including superconducting and topological qubits, to enhance the promise of large-scale quantum computing applications. The future of SETs also intersects with the overall environment of quantum technologies where their unique characteristics can be of paramount value to emerging fields of innovation within the Internet of NanoThings. As the technology of nanomaterials continues to advance, enhanced SET designs can enable more sensitivity and specificity enhancements, particularly in biosensing and environmental monitoring areas where the detection of slight biochemical interactions is the driving concern. The integration of such devices with cognitive communication environments, according to the ongoing research work, will further aid in the development of devices that coexist well in the networks of the future, like 6G. As the foundational columns of quantum computing and future biosensing technologies converge, SETs are poised to be at the forefront of revolutionizing electronics, ushering in a paradigm shift towards more integrated, efficient, and intelligent systems (Akyildiz et al., 2020; Naresh et al., 2021).

9.3.1. Operating Principles of Single-Electron Transistors (SETs)

Individual electrons in a quantum manner are the control that is at the core of the functional nature of SETs, which are essential devices in the quantum electronics fraternity. SETs utilize the process of quantum tunneling to enable effective charge conduction. SETs regulate electron movement by confining electrons in a quantum dot, and thus modulate the electrical current through the application of voltages. The small size of SETs allows for a regime under which Coulomb blockade effects become important, i.e., the conductance of electrons across the transistor is strongly based on the electrostatic interaction among the electrons themselves. This indicates the unique advantages that SETs can provide, especially for those applications that demand high sensitivity and precision, as small potential variations can result in large current flow variations. In addition, cryogenic electronics design specifications play a key role in optimizing the performance of SETs. Operating the devices at low temperatures considerably enhances stability and efficiency by reducing thermal noise, which otherwise would obscure the electron tunneling processes on which SETs rely. As the temperature approaches absolute zero, thermal fluctuations that otherwise disturb electronic performance reduce, and the quantum mechanical properties of the device dominate. Scientists are therefore inclined to turn more towards cryogenic temperatures to achieve optimal performance and fidelity in SETs, particularly in quantum computer systems where errors must be kept at the barest minimum (Figure 9.6).

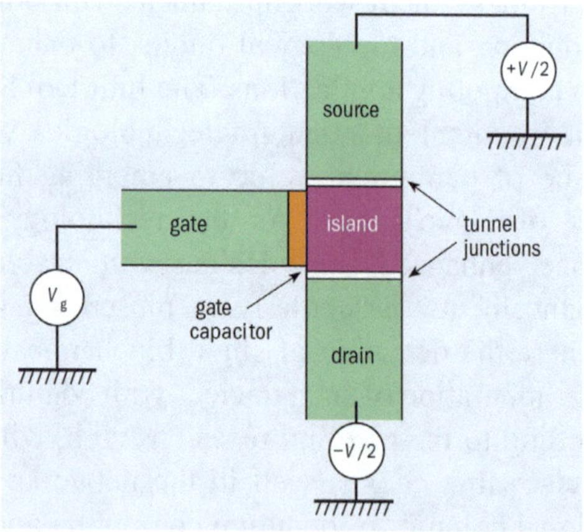

Figure 9.6. *The manipulation of single-electron transistors (SETs).*

Source: https://physicsworld.com/wp-content/uploads/1998/09/fig4.png.

The ability of maintaining qubits in a coherent state especially excels with these unique working conditions, boosting the potential of quantum technologies. Moreover, utilizing SETs with nanomaterials also represents a future path for advancing the performance of devices. Nanostructured materials are described by surface area and electrical conductivity, which can be beneficial when building highly sensitive biosensors and other electronic devices. For instance, on combining quantum dots or carbon nanotubes with SETs, the hybrid devices that are formed have the ability to selectively respond as well as possess amplified charge transfer characteristics. This blend not only extends the range of applications of SETs in sensing technologies, as outlined in Naresh et al. (2021), but also with the flexible, miniaturized electronics technologies conceived in the transformation towards the Internet of NanoThings and other future communication systems addressed in Akyildiz et al. (2020). These advancements highlight the revolutionary nature of SETs in quantum computing and sensing technologies, shaping the future of electronic devices.

9.3.2. Advantages of Single-Electron Devices

The development of single-electron devices has marked the beginning of a revolutionary era in nanotechnology, particularly in quantum computing and electronics. Single-electron transistors (SETs) introduce new functionalities via single-electron transfer control, hence allowing for significant levels of energy reduction. This is a feature with a clear advantage over conventional transistors, which make use of larger charge carriers and therefore have increased power dissipation. The quantum nature of electrons in SETs lays a foundation for ultra-low-power consumption, and for this reason, they are particularly ideal for portable and low-energy devices. Besides, single-electron control lends itself to more precise quantum bit (qubit) control, improving their potential as simple building blocks in quantum computing architectures. Above their energy efficiency, single-electron devices are capable of extreme miniaturization, opening up possibilities for emerging nanofabrication technologies. As sizes of electronic devices diminish, traditional scaling limits in minimizing semiconductor devices become dominant, such as increased leakage currents and performance fluctuation. SETs, in contrast, circumvent most of these issues, as their operating mode is intrinsic quantized charge tunneling rather than continuous conduction. Not only does this unusual mechanism allow SETs to work effectively at nanometer sizes, but it also enables the possibility of building complex integrated circuits that were not previously feasible. Besides, integration of single-electron devices into well-established semiconductor technology is also promising for the fabrication of hybrid systems that leverage the power of both classical and quantum properties, which further enhance performance and application in most areas of electronics. Cryogenic electronics design emphasizes the viability and advantages of single-electron devices in quantum applications. With the development of future quantum computing technology, ultra-low temperature operation becomes critical in order to maintain coherence and minimize decoherence of qubits. SETs have the capability to run well under such conditions because of their low-energy operating and scalability, through which the accurate control inherent in qubit configurations could be made possible. Moreover, materials such as $SrRuO_3$ with new quantum features could perform complementary roles in the integration of traditional and quantum electronics and enhance single-electron device effectiveness. Such achievements in the realization of the coupling between materials and device functionalities can lead to grand advances in the development of stable quantum systems as well as prove the necessity of researching new device geometries for the success of quantum electronics (Cuoco et al., 2022; Khan et al., 2017).

9.3.3. Applications of Single-Electron Transistors (SETs) in Modern Electronics

The novel character of SETs has placed them at the forefront of future electronics discussion, particularly when compared to quantum information processing and quantum computing. By the control of the charge transport of individual electrons, SETs offer a unique path to ultra-low-power devices. This is accomplished through their structure, which involves quantum tunneling across a nanoscale junction with the ability to manage charge states tightly. Such precise control becomes priceless when it comes to the fabrication of qubits, especially when it comes to superconducting and semiconductor qubits. The capability of SETs to act with little dissipation of energy is in accordance with the primary demands of having successful quantum computation, wherein losing energy should be kept to a minimum to ensure quantum coherence. Thus, SETs not only enhance performance parameters in normal applications but also provide entry into experimental applications of qubit systems on the basis of quantum mechanical concepts. Apart from their role in qubit technology, SETs have exhibited promising applications in cryogenic electronics, particularly in environments that find temperatures close to absolute zero. Effective operation at such low temperatures is critical for superconducting qubits, whose thermal management is efficient enough to avoid decoherence. Applying SETs in such a case enables greater sensitivity when monitoring electrical signal fluctuations, critical for the function of cryogenic sensors and read-out systems. For instance, their ability to amplify extremely tiny signals is critical in superconducting quantum interference devices (SQUIDs), which are a vital element in the majority of experimental setups in quantum research.

Also, advances in nanofabrication tools have fueled the development of hybrid systems using SETs in superconducting circuits, thereby expanding their use in quantum circuit designs. Their interaction with cryogenic temperature thus underlines their commonality in the new world of low-temperature electronics. Theogeneity of SETs also involves their potential incorporation into artificial intelligence and machine learning algorithms targeted for tasks of complex computation. With the demand for faster processing and lower power consumption increasing all the while, SETs can be used as the foundation for neuromorphic computing architecture that aims to replicate the behavior of biological neural networks. The inherent parallelism in SET-based circuits is well placed to be applied to high-efficiency processing just like that of computer networks traditionally implemented based on standard transistors. In addition, through additional research breakthroughs, the integration of organic semiconductors or two-dimensional materials with SETs may lead to record-breaking device performance and scalability. The ground that SETs lay puts not only quantum device innovation in fast-forward but also emphasizes their multifaceted nature in being capable of supporting numerous high-performance electronic applications in modern electronics, as evidenced in current material science trends (Zhang et al., 2022; Zou et al., 2021).

9.4. QUANTUM DOTS AND THEIR APPLICATIONS

Quantum dots have introduced unprecedented advancements in the technology of quantum and single-electron devices. For instance, in the development of qubits, the elementary components of quantum computation. Quantum dots, due to their discrete energy levels due to quantum confinement, are the most suitable for use in electron-spin-based qubits. Such a potential to manipulate individual electrons in quantum dots gives rise to the possibility of achieving scalable qubits that are also compatible with existing semiconductor technology. Furthermore, the integration of quantum dots into superconducting quduits provides very promising prospects for hybrid quantum systems with enhanced coherence times as well as overall operating efficiency. With the demand for increasingly advanced and powerful computing on the rise, quantum dots remain at the forefront of research seeking to advance the function of qubit systems. Another valuable application for quantum dots is with single-electron transistors (SETs), where quantum mechanical principles are used for charge transport. By allowing the control of single electron tunneling, SETs offer a good platform for the observation of basic quantum phenomena and for the realization of low-power electronics, which is important in the design of contemporary devices (Figure 9.7).

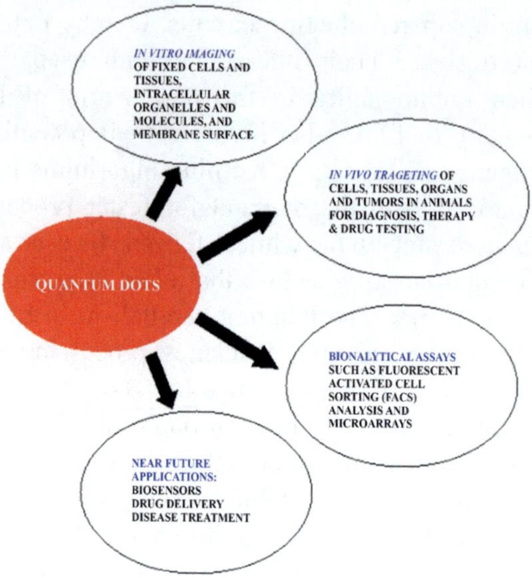

Figure 9.7. *Applications of quantum dots.*

Source: https://www.researchgate.net/profile/Sandeep-Kumar-Vashist-2/publication/284960734/figure/fig1/AS:457345143709696@1486050987506/Applications-of-Quantum-dots.png.

The technology for fabricating quantum dots is becoming more advanced, and it now allows for good control of the size and spacing of the dots, which directly affect the electronic characteristics of the devices. Recent advances point toward the feasibility of integrating SETs with silicon quantum dots, which could potentially disentangle the processing involved with present semiconductor technologies and provide less expensive and more efficient electronic devices. This not only enhances performance but also reveals new possibilities of nano-electronics and sensor technologies. As far as cryogenic electronics is concerned, the quantum dots' special properties pose distinctive design challenges because they operate optimally at ultra-low temperatures. Such temperatures reduce thermal fluctuations to their minimum levels, thereby extending the coherence time of quantum bits and making quantum logic operations reliable. This kind of environment necessitates innovative packaging as well as materials that maintain operational integrity at cryogenic temperatures without inducing decoherence. Also, research in this area indicates that use of quantum dots in SETs can result in achievement of quantum circuits that are stable and dissipate very low powers, a note for scalability of quantum computing. With advancing technology, ongoing research on the behavior of quantum dots with cryogenic states will be critical to resolve power consumption, speed, and reliability concerns, thus making them a key component in the future quantum and single-electron devices (Ansaloni et al., 2020; Harrison et al., 2016).

9.4.1. Structure and Function of Quantum Dots (QDs)

The versatility of QDs owes to their peculiar optical and electronic properties, which are dominated by their size and shape. Quantum dots, being semiconductor particles at the nanoscale, confine excitons, or electron-hole pairs, in three dimensions and thus enable discrete energy levels based on the size of the dots. This quantum confinement, or size effect, depends on the emission wavelength that can be tuned to a wide range of colors and renders QDs very useful in numerous applications like medical imaging, solar cells, and quantum information processing.

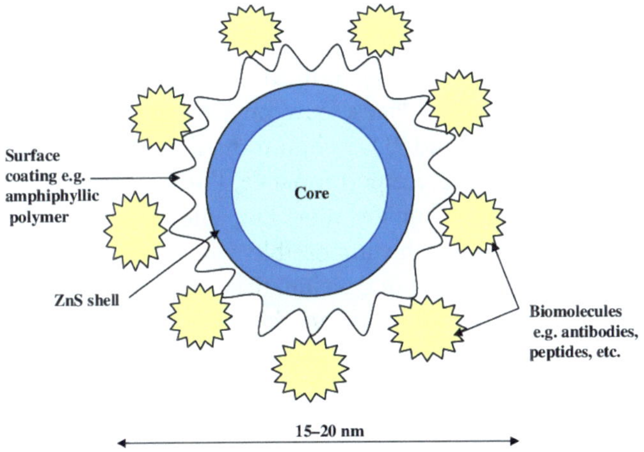

Figure 9.8. *Structure of a quantum dot.*

Source: https://www.researchgate.net/profile/Alexander-Seifalian/publication/51824404/figure/fig5/AS:668669940092933@1536434745201/Structure-of-a-quantum-dot.png.

Moreover, tunability of the nanomaterials also makes them integrable in single-electron devices, with the possibility to contribute to more efficient qubit designs required for the development of quantum computing technologies. Internal advantages of QDs in charge transport and manipulation under well-controlled conditions further enhance the performance of these devices, illustrating a critical interconnection between quantum electronics and nanotechnology. Advances in quantum dot structural formulation, along with innovative synthesis methods, have opened up new possibilities for their application in multifunctional nanocomposites. For instance, the doping of traditional semiconductors with quantum dots can lead to increased photocatalytic activity that bypasses the inherent limitations associated with large bandgap materials like TiO_2 (Figure 9.8).

The incorporation of graphene quantum dots (GQDs) into TiO_2 matrices enhances light absorption and decreases recombination losses for charge carriers, thus expanding the operation of these systems. These hybrid nanocomposites hold promise in low-temperature applications, including energy conversion and biomedical devices, where multifunctionality thereof can be used for sensors and photonic elements. However, despite the optimism about such structures, issues remain to optimize low-temperature nanocrystals for extensive low-temperature applications, emphasizing the importance of continued research and development towards quantum dot system synthesis and characterization (Bokare et al., 2021). In the area of quantum and single-electron devices, two-dimensional material and heterostructure research is a critical quantum dot functionality area. Through the utilization of individual 2D crystals to create novel heterostructure devices, new physics involving commensurate-incommensurate transitions and 2D exciton formation are released, which play a crucial role in enhancing device performance. For instance, newly developed tunneling transistors and resonant tunneling diodes exploit such unique properties in order to realize the impossible with traditional semiconductor devices. By incorporating quantum dots into these 2D materials, researchers are able to develop devices that are more efficient and possess new functionality, and therefore are capable of revolutionizing quantum computing and cryogenic electronic applications. This marriage of two-dimensional materials and quantum dots not only unveils the technological scope but also the door to future quantum devices that harness the synergies in between (Novoselov et al., 2016).

9.4.2. Role in Quantum Computing

The search for quantum computing has opened new avenues for information manipulation at a scale unprecedented, predominantly as varied qubit architectures. Among these, superconducting, semiconductor, and topological qubits have been the subject of research interest due to their distinguishing operation characteristics and advantages. Superconducting qubits utilize the zero-resistance property of materials near absolute zero, allowing rapid state manipulation. Semiconductor qubits, often realized through single-electron transistors (SETs), provide a domain for the marriage of quantum information processing with established semiconductor technology, favoring scalability.

Topological qubits, drawing on particle physics, promise decoherence resistance, opening up the way for fault-tolerant quantum computation. Each form of qubit brings a rich variety to an ecosystem, dictating the architecture to quantum algorithms that can execute demanding tasks in a manner unavailable to classical computing. Central to qubit architecture are SETs, which operate on the principle of tunneling by single electrons. SETs are very important building blocks of qubits because they can provide high-degree control and strong confinement of electron states. SETs' high sensitivity to their environment necessitates advanced cryogenic electronic design considerations as the operation of such devices can be highly influenced by thermal noise and other environmental sources. The integration of SETs into quantum devices not only aids qubit realization but also in measurement processes required for quantum state readout. As the field advances, the union of SETs with other quantum devices will undoubtedly play a role in enabling the deployment of scalable quantum information systems, capitalizing on their unique properties in realizing pragmatic simplicity in quantum computing applications. The complete potential of quantum computing is realized through advancements in fabrication technology and electronic function. Atomically precise fabrication techniques, such as hydrogen-based scanning probe lithography, enable the deterministic positioning of a single dopant atom, enabling the production of highly controlled electronic devices required for quantum applications. These advancements form the basis for the creation of devices that employ both SET technology and other quantum effects necessary to efficiently control quantum. Secondly, ongoing nanoscale optimization of electronic properties will enable more efficient coupling of qubits and better noise robustness, both of which are crucial for secure quantum operations. With advancements in these techniques, the field will be able to develop more secure quantum computing platforms based on the precise control of single-electron devices that eventually open up new possibilities in quantum information processing (Lyshevski, 2018; Wyrick et al., 2019).

9.4.3. Use in Photonic Devices

The quantum and single-electron device innovation has played a significant role in the evolution of photonic technologies. Of particular importance has been the application of single solid-state quantum emitters in combination with photonic nanostructures as a dominant tool in enabling the progress of photonic quantum applications. Random spatial placement of quantum emitters poses a limitation, but recent technological advancement has introduced various positioning strategies prior to nanofabrication. Techniques such as atomic force microscopy and confocal microscopy have been able to with hopeful success position these emitters precisely. The outcome has been a new generation of high-performance devices, including high-quality single-photon sources and entangled-photon pairs, which are the constituents of most quantum communication and computation protocols (Liu et al., 2021). Thus, the use of these positioning techniques not only enhances functionality but also provides scalability in quantum photonics. The strong coupling between quantum emitters and photonic modes, particularly using exciton-plasmon interaction, is another critical factor in the design of next-generation photonic devices. Recent experiments using methods like electron energy loss spectroscopy have illuminated the dynamics of such interactions. Achieving strong coupling with single plasmonic elements, such as bowtie antennas, establishes the feasibility to manipulate quantum states with high efficiency. Notably, coupling strengths observed in such systems are impressive, bringing them to the boundary of the strong coupling regime, which is beneficial to construct new quantum optical platforms. The ability to explore even the dark plasmonic modes expands the application of both quantum systems and photonic devices with new doors for research of quantum dynamics and chemical reactivity (Bitton et al., 2020). Not only do such interactions illustrate the complexity of quantum technologies but also their potential use in future photonic devices. With the development of photonic devices, matters of cryogenic electronics become essential in enabling them to function at optimal capacities. For instance, superconducting qubits and SETs must be used within controlled conditions to work at an optimum level, minimizing thermal noise and maximizing coherence times. This demands elaborate cryogenic designs that are capable of reliably sustaining millikelvin temperatures. Coupling photonic devices with SETs enables the production of complex quantum circuits, where signal transmission and processing are pivotal. The use of engineered photonic structures can also decrease noise further and increase the overall resilience of quantum operations. By expanding the interaction between the quantum processing components with photonic interfaces within cryogenic platforms, the discipline can make tremendous progress toward practical quantum computing applications. These advancements serve as a testament to the necessity of multidisciplinary solutions in the innovation of future photonic technologies (Bitton et al., 2020; Liu et al., 2021).

9.5. CRYOGENIC ELECTRONICS DESIGN CONSIDERATIONS

The electronics design in cryogenic environments is directly related to the success of quantum and single-electron devices during operation. As such technologies expand, namely superconducting qubits and SETs, the intricacy in maintaining function at near absolute zero grows. A key part of this design work is the careful selection of materials and components for the circuits that can withstand extreme cooling. As a case in point, traditional semiconductor materials will be inadequate because resistance changes radically at cryogenic temperatures, requiring the searching elsewhere for avenues that will meet both thermal and electrical stability. The ability to suppress thermal noise while enhancing signal integrity is also key to the successful operation of quantum devices, which would finally ensure qubit entanglement and coherence stability, critical to their applications in quantum computing and future sensing technologies. Aside from materials challenges, good thermal management principles are necessary for cryogenic electronics. Effective thermal insulation techniques should be employed to make sure that the external heat sources have no effect on the operation of essential components. Furthermore, sophisticated thermal link designs should be employed in order to facilitate the dissipation of heat from local devices while the overall temperature is kept constant (Figure 9.9).

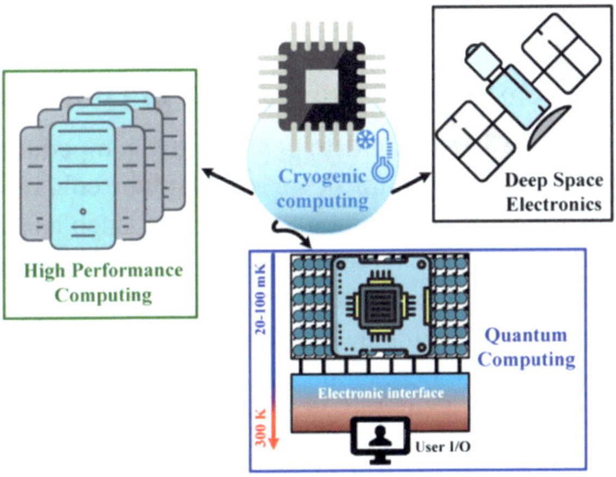

Figure 9.9. *Cryogenic applications: from high-performance computing and deep space electronics, to quantum computing.*

Source: https://www.mdpi.com/electronics/electronics-11-00061/article_deploy/html/images/electronics-11-00061-g001-550.jpg.

This stability becomes particularly critical for systems like trapped ion quantum computers, where such small thermal fluctuations can induce decoherence and loss of quantum information. Assuring electronic control systems that there is stable operation of qubits under a wide range of thermal conditions has direct bearing on the scalability of such quantum systems, and provides a platform to advance further in quantum technologies. The integration of new thermal management into the design horizons will establish a bridge in fulfilling existing constraints and future scalability. Last but not least, as the field of quantum computing continues to advance, so must the methods used on the design of such cryogenic electronics. New advancing technologies of integrated photonics offer new avenues towards fulfilling the extremely complex challenges of cryogenic environments. Quantum photonic integrated circuit (QPIC) developments, for instance, encourage smaller form factors and better performance metrics, ultimately paving the way to applications in quantum computing, communications, and metrology. Cryogenic electronics and integration technologies not only facilitate component integration but also spearhead multi-functional system implementation on a grand scale. This advancement requires ongoing interdisciplinary investment and collaboration to make the process from theoretical models to functional systems effective and efficient in addressing the complexities of quantum systems as presented in the present research framework (Moody et al., 2021) and the promising future of trapped-ion systems (Bruzewicz et al., 2019).

9.5.1. Importance of Low Temperatures in Quantum Devices

Quantum systems have been found to require low temperatures in order to enhance the efficiency of quantum devices. Quantum coherence, one of the fundamental properties that allow qubits to function, such as superconducting and semiconductor systems, is maximized at cryogenic temperatures. Cooling actually suppresses thermal noise that could otherwise disrupt the delicate quantum states required for computation. In single-electron transistors (SETs), to cite one, low-temperature operation curtails undesirable electron scattering, allowing for greater control of the individual electrons. Consequently, such conditions give rise to higher fidelity for quantum calculations, an essential factor in the advancement of quantum technology and in bridging the gap between theory and practice for quantum and single-electron devices. These advancements not only reveal the significance of thermal management in quantum physics but also illustrate the intricate interplay between temperature and device operation. Aside from facilitating qubit coherence, low temperatures also permit indispensable design implications in cryogenic electronics. At these lower levels of thermal condition, electronic materials possess lower resistivity and higher superconducting properties, which are indispensable for efficient device operation. The use of low-temperature superconductors allows the design of energy-efficient circuits, which are essential for scaling quantum technologies. Researchers indicate that through the use of the phenomenon of superconducting qubits, it is possible to design more robust quantum processors that are capable of conducting more operations and yet have low error rates. In addition, utilization of cryogenic conditions improves the accuracy of sensing technologies, as is indicated by advancements in quantum sensors based on qubits as part of their working architecture, which represents expanding interest in low-temperature quantum technologies for numerous applications (Degen et al., 2017). This is complemented by a mitigation of non-radiative energy losses within quantum devices through the use of low temperatures. As noted by recent studies, reduced thermal conditions might limit thermally triggered excitations that would otherwise lead to energy loss. For example, single-electron transistors at low temperatures demonstrate minimized non-radiative losses, which enhance device efficiency. Surprisingly, advancements in material design, such as the synthesis of chlorinated non-fullerene acceptors, focus on maximizing electronic properties to enable high-efficiency performance (Cui et al., 2019). Through the exact tailoring of the characteristics of quantum devices with low-temperature application, researchers continue to create new pathways toward enhanced performance. This synergy between device design and low temperatures is a key area of research, ultimately flowing into the broader development of quantum technology and its translation into practical applications.

9.5.2. Challenges in Cryogenic Design

The intricacies of cryogenic design are due to the particular demands of quantum and single-electron devices, which, in states of extreme low temperatures, perform best. At such temperatures, natural thermal noise is greatly reduced, making possible extended coherence times in qubits, particularly in superconducting and semiconductor forms. The problem lies, however, in establishing and maintaining such cryogenic states. Choice of materials becomes crucial since prevalent metals and semiconductors start to exhibit different properties under cryogenic temperatures, influencing quantum circuit operation and stability. Thermal insulation reliability and efficient thermal anchoring systems become essential to prevent heat transfer, which will ruin the meticulously engineered environments for quantum processing and SET operation. Also, having materials and manufacturing processes compatible at various temperatures adds even more complexity to the design process. In designing and developing cryogenic systems for quantum devices, there are various other challenges that have potential stumbling blocks. Integrating intricate wiring and circuit topologies of circuits, which need to function consistently under cryogenic temperatures without generating excessive electromagnetic interference, is one such significant consideration. The selection of superconducting materials for wiring is typically controlled by their critical temperature and field values, but in the use with standard semiconductor devices, thermal and electrical parameters must be carefully considered. In addition, the use of sophisticated readout schemes is inescapable in single-electron devices, which rely significantly on delicate measurement techniques that can be adversely influenced by even small temperature fluctuations. Together, these make the engineering environment difficult for developers and researchers to optimize cryogenic design. The impact of breakthroughs in material science is another critical in quantum and single-electron devices. The exploration of new materials such as those with topological property or atomically thin forms such as monolayer transition metal dichalcogenides (TMDs) provides exciting opportunities but also creates challenges. These materials have special electronic and optical properties that can benefit device operation; however, their response under cryogenic conditions must be well-studied. As more is known about exciton interactions and how they affect valleytronics and spin properties, suitable ways of integrating these materials into cryogenic designs need to be established. Lastly, grappling with the heterogeneous complexities of cryogenic design will be a deciding factor in realizing the full promise of quantum and single-electron devices in practical applications, as highlighted by the need to address thermal management and material compatibility for the entire system (Figure 9.10).

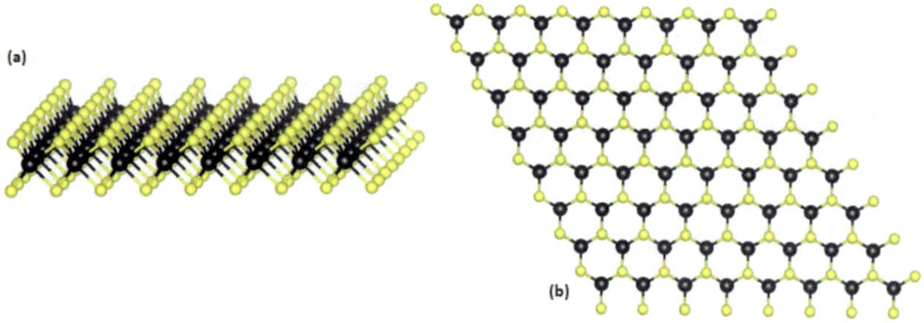

Figure 9.10. *(a) Structure of a hexagonal transition metal dichalcogenide (TMD) monolayer. M atoms are in black and X atoms are in yellow. (b) A hexagonal TMD monolayer.*

Source: https://upload.wikimedia.org/wikipedia/commons/a/a7/Monolayer_TMDC_structure.jpg.

9.5.3. Solutions for Effective Cryogenic Electronics

The development of cryogenic electronics has opened up new avenues to enhance the performance of quantum and single-electron devices. Specifically, superconducting qubit technology has taken advantage of cryogenic temperature to minimize thermal noise and increase coherence times both of which are essential for quantum computing to be successful. These circuit quantum electrodynamics-designed superconducting qubits demonstrate the potential for microwave photon-strong and tunable coupling, as outlined in Blais et al. (2021). Such coupling enables accurate control and measurement of quantum states, both of which are essential for efficient quantum information processing. Furthermore, the use of cryogenic cooling technologies ensures that such devices are within the stipulated thermal regimes to be able to attain superfluid states, thus maximizing performance and stability, ultimately pushing quantum computing beyond limits. Apart from superconducting qubits, the need for single-electron transistors (SETs) has also garnered significant interest in the field of cryogenic applications. Operating at cryogenic temperatures allows SETs to exhibit perfect control over single-electron charge transport, thus improving their performance and efficiency measures. Control over electron spin and charge in these low-temperature conditions transforms conventional electronic devices into effective spintronics devices that have the potential to improve memory storage and processing efficiency, as argued by Hirohata et al. (2020). The integration of SETs in quantum and single-electron device circuits requires innovative design considerations; overcoming low-temperature stability and the prevention of external interference is vital. These problems, in return, can allow for new applications, making nano-scale electronics more energy efficient. Design considerations for cryogenic electronics require utmost research of material characteristics and device architecture tailored for low-temperature application. These issues like thermal management, electromagnetic shielding, and interconnects also create severe difficulties that must be addressed in a proper manner to ensure ultimate performance. Superconducting materials, for instance, must not only be selected based on their superconductivity but also with respect to their longevity against environmental forces at cryogenic temperatures. Besides, with the advent of quantum information science, topological qubits, and new materials and device architectures, new functionalities and properties have been incorporated with higher fault tolerance and coherence times. The coupling of cryogenic electronics with semiconductor technologies offers the prospect of hybrid systems combining both superconducting and semiconductor qubits, to create a universal platform for the exploration of quantum computing. By giving emphasis to these design considerations, researchers can develop usable solutions to maximize the potential of cryogenic electronics in quantum and single-electron devices and further accelerate innovation in this rapidly emerging field.

9.6. CURRENT RESEARCH AND DEVELOPMENTS

Current advances in quantum devices have put the development of superconducting qubits at the forefront, with a lot of debate centered on their coherence characteristics and scalability potential. These qubits take advantage of superconductivity principles for fast quantum state control, allowing them to perform better than conventional electronic devices in computing systems. Scientists are investigating various semiconductor materials for generating qubits that not only sustain coherence but also fit directly into current semiconductor technology. This integration may bring quantum computing more in line with classical systems, making computing potentially more powerful. Also on the horizon is research into topological qubits, which are potential for being immune to certain types of errors due to their non-local encoding of quantum information. They are creating the foundation for more robust quantum systems. These breakthroughs are a testimony to the transformative potential of quantum devices in advancing the frontier of computing technology. Single-electron transistors (SETs) have also made major breakthroughs based on dominant trends in research. SETs function through tunneling a single electron and have unique advantages in low-power circuits and high-density circuits. Investigations have been aimed at enhancing SET performance and scaling up through the application of advanced materials and innovative circuit geometries.

Techniques such as cryogenic electron transport measurements are being employed to study the performance limits and viability of SETs at ultralow temperatures. Even as researchers refine the operating conditions of SETs, their application in quantum information technology and sensing devices increases to provide unprecedented sensitivity and accuracy. This desirable performance characteristic makes SETs an essential member of the quantum electronic device palette, setting them up for future advances. Together with the development of SETs and qubits, advancements in novel two-dimensional materials are reshaping the environment within which quantum devices can operate. The 2D material physics with no long-range order and unique electronic properties offers a rich substrate on which to build novel heterostructure devices, including tunneling transistors and resonant tunneling diodes. Researchers are endeavoring to study how the unique properties of each 2D crystal can be blended to achieve functionalities unattainable with traditional materials. Such materials have already shown promise to enhance the performance of quantum devices, particularly in cryogenic electronics where the control of temperature is more critical than anything else to maintain quantum coherence. As such materials are subjected to further investigation, their potential to transform quantum and single-electron devices becomes more evident, which has the potential to bring significant breakthroughs to both technology and applied physics (Brun-Picard et al., 2016; Novoselov et al., 2016).

9.6.1. Innovations in Quantum Device Technology

The research on developments in quantum device technology has positioned the creation of qubits at the forefront of research currently. Various kinds of qubits, such as those made of superconducting materials, semiconductors, and topological systems, each possess their own strengths and characteristics. Superconducting qubits, for example, have rapid gate speeds and scalability, making them suitable to construct extensive quantum circuits. Semiconductor-based qubits, on the other hand, incorporate matured fabrication techniques and potential integration with traditional electronics, hence easing the path towards hybrid quantum-classical devices. Topological qubits, while still in the experimental stage, promise more decoherence resistance, and further pushing the boundary of qubit technology. Synergies of these classes of qubits are at the heart of current research for fault-tolerant quantum computing architectures potentially able to far outperform their classical equivalents, so it is essential to continue studying their properties and interdependencies. Single-electron transistors (SETs) have been one of the most revolutionary technologies in quantum device technology, enabling the manipulation of individual electrons with high precision. This potential is central to future quantum circuits because SETs can act at the quantum level at which quantized charge is evident (Figure 9.11).

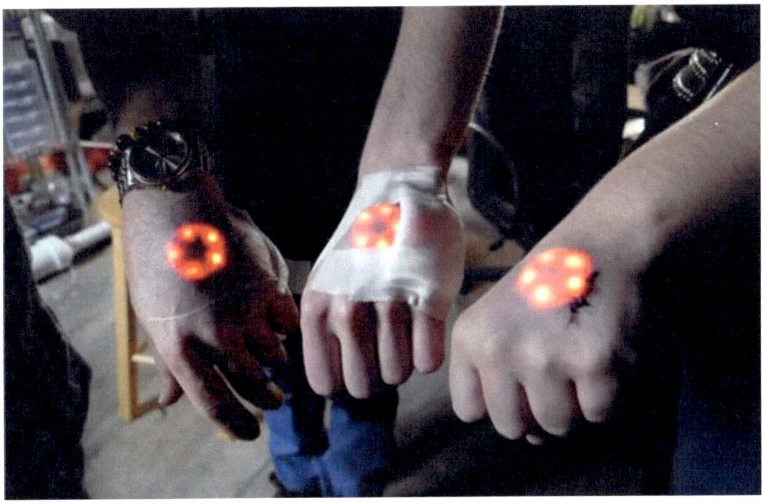

Figure 9.11. *The future advancement in quantum device technology.*

Source: https://thequantuminsider.com/wp-content/uploads/Quantum-Devices-article.png.

By confining electrons into a limited space, researchers are able to manipulate their quantum states, paving the way to implement designs of qubits using single electron spins. Developments in SET technology have combined advances in fabrication techniques and materials to facilitate higher temperature performance and reduced power usage. In addition, advances in cryogenic considerations in electronic design are essential to maximize the operation of SETs. The introduction of new materials and cooling designs will improve efficiencies in quantum information processing operations and permit more reliable quantum networks, which remain essential for meaningful applications to quantum computing. Analysis of quantum device technology also demands research into the challenging environment in which these devices operate, i.e., cryogenic systems. The functioning of quantum devices, including qubits and SETs, typically requires extreme refrigeration to minimize thermal noise and maintain quantum state coherence. Hence, in design work for cryogenic electronics, research focuses on the optimization of materials, architecture, and thermal management methods in order to realize improved device performance. Improvements in terms of dilution refrigerators and thermal couplers with high performance overcome such constraints so that quantum devices can be stably operated at temperatures close to absolute zero. Moreover, innovation in the development of flexible control electronics for use at cryogenic temperatures is a significant field, which allows for control and measurement of quantum states. Ultimately, the effort towards advancements in cryogenic technology is essential to advancing quantum devices and achieving scalable quantum computation models, which inform the future direction of quantum innovation.

9.6.2. Recent Breakthroughs in Single-Electron Devices

Single-electron devices have changed the landscape of quantum computing and opened the doors to a new class of efficient qubits. Superconducting qubits in recent years have harnessed the strange behavior of single-electron transistors (SETs) to achieve long coherence times and gate fidelities. SETs capitalize on the ability to address a single electron within a potential well and operate at high frequency, which are very crucial in quantum applications. Furthermore, when SETs are integrated with superconducting materials, scientists are now able to realistically minimize energy losses, a fundamental drawback in standard electronics. Device engineering breakthrough using next-generation materials and production techniques has explored electron temperature and coherence interactions and raised both operating reliability and scalability within quantum systems. With the frontier of single-electron devices further improving, their influence on qubit performance can potentially redefine the quantum computing paradigm. In the past decade, the technology of semiconductor qubits also made groundbreaking strides, namely in cryogenic electronics development to be used with single-electron devices. The conditions to operate qubit systems at cryogenic temperatures require pioneering electronic architectures with the capability to provide stability and efficiency in extreme situations. Researchers have developed sophisticated architectures with lower thermal noise and increased signal integrity so that SETs can function with more accuracy and robustness. The introduction of advanced insulating materials and tailored substrates created new opportunities for enhancing performance metrics such as fidelity and speed, breaking through key challenges in qubit design. Beyond augmenting our understanding of quantum mechanics, these developments also create the gateway to practical quantum processors that are capable of performing multi-step calculations with unprecedented accuracy. Furthermore, the study of topological qubits is also becoming an emerging area of growth within the field of single-electron devices. As they are decoherence-resistance, topological qubits offer tremendous possibilities for fault-tolerant quantum computing and are ideal candidates for future quantum systems. Recent studies have demonstrated how topological states can be formed by using single-electron devices and manipulating electron spins and braiding operations required for quantum error correction. The convergence of condensed matter physics, materials science, and quantum information theory in such developments speaks to the potential for single-electron devices to transform computational paradigms. As researchers delve deeper into the intricate relationship between qubit types and single-electron physics, the resulting technological innovations have the potential to make quantum computing an ordinary technological reality.

9.6.3. Future Directions in Quantum Research

The quantum research environment is changing at a rapid speed driven by the need for more efficient computational models and electronic device miniaturization. Revolutionary progress in qubit architecture, such as superconducting, semiconductor, and topological qubits, will contribute significantly to quantum computing's architecture evolution. Every qubit possesses certain operational potential and limitation with varied paths to achieve quantum algorithms. Choice of qubit can have an effect not only on computational speed, but error rates and scalability potential as well. As quantum processors grow larger with more qubits, supporting technology architectures also need to adapt, particularly with regards to cryogenic electronics. Designing electronic components to function at ultra-low temperatures is a focus of academic research, in attempting to minimize power usage and maximize signal fidelity, both of which are secrets to maintaining coherence in complex quantum systems. With advances in qubit technologies, single-electron transistors (SETs) are set to revolutionize nanoscale quantum devices. The ability of SETs to control the movement of single electrons makes their operation capability inherent in controlling single-electron movement, hence making them unpredictably precise in electronic signal processing. This property makes SETs even more appealing to applications in quantum information systems where fidelity of state and noise tolerance are priorities above all else. With scientists exploring integration of SETs with other qubit architectures, there is a potential for synergies that enhance quantum gate design and computational speed. Future efforts will likely seek to improve the scalability of SETs and also address fabrication and reliability issues. The convergence of SET technology and quantum computing is set to be a fertile ground for innovative solutions that will lead the discipline further into practical implementation. The next generation of quantum science is also set to invest heavily into the confluence of quantum materials and new electronic systems. For example, the recent findings of perovskite materials are a breakthrough area for energy storage and conversion technologies and quantum technologies. The materials exhibit outstanding properties that can be utilized in the development of robust energy-efficient quantum devices using piezoelectric, pyroelectric, and thermoelectric effects in combination. Multifunctionality in features like these has the potential to speed up quantum circuits in various environmental conditions. As quantum research evolves, eco-friendly material emphasis—like perovskites—can be in synergy with global initiatives towards sustainable technology. Such contemplation will surely entrap the adoption of interdisciplinary strategies as the quantum research horizon unfolds, with focus not only on technological advancements but also on sustainability-orientated innovation processes (Akyildiz et al., 2020; Pandey et al., 2019).

SUMMARY

The advancements in single-electron and quantum devices are a significant tradeoff between performance enhancement and complexity of design for computing ability. With the development of system integration of qubits ranging from superconducting to topological qubits, the importance of maintaining fidelity in quantum computations cannot be emphasized enough. The generation of coherent control of such qubits under cryogenic conditions poses unique challenges, particularly the control of noise and thermal fluctuations. Consequently, cryogenic electronics will therefore need to adapt to this, and the delicate quantum states be made available and manipulable in a stable fashion. Device structure and environmental influence interact in determining the optimum qubit performance, and minimizing decoherence, suggesting the need for continuous research into materials and fabrication technologies to create robust quantum systems. In contrast to traditional binary systems, SETs take advantage of quantum effects in a bid to achieve functionality at the ground level. Through the manipulation of individual electron charge states, SETs are a form of ultra-low power operation and are therefore pivotal towards the realization of scalable quantum circuits. However, their operation is heavily reliant on the accuracy of fabrication as well as the local electromagnetic setting. Nanoscale fabrication technology developments unique to SETs have indicated that electron tunneling can be controlled reliably such that quantum logic gates could potentially be realized. And with the requirement for ever-smaller and low-energy devices, exploring the synergism between SETs and qubit architectures could result in hybrid systems maximizing computation efficiency at minimal power consumption. As this technology moves towards practical application, the exploration of hybrid device architectures grows increasingly significant. Mixing quantum and classical components may bridge present performance gaps and counteract challenges in making theoretical ideas functional systems. For instance, the application of SETs with qubit technology can facilitate quantum computing advancement, possibly leading to architectures that are capable of executing complex algorithms. An intuitive understanding of such correspondences between device types and use context facilitates enhanced design strategies. The increasing number of research articles on quantum and single-electron devices indicates an age when systems integrated would harness the potential of quantum mechanics, in due course to revolutionize models of computing and advance technology frontiers, as evident from advances in materials like organometal halide perovskites (Hoye et al., 2015) and applications (Conings et al., 2013).

While nanoscale technologies continue to advance, research on quantum systems has evoked a profound curiosity in designing qubits, the fundamental quantum computational units. Among the different types of qubits, superconducting qubits have been of particular interest due to their relatively long coherence times and suitability with existing semiconductor technologies. These qubits implement Josephson junctions to produce nonlinear inductance so that states in quantum can be controlled at the microwave frequency. There are also semiconductor qubits, particularly quantum dots and donor atoms-based qubits, which have unique characteristics such as integrability with classical electronics and scalability, and therefore are extremely appealing for hybrid quantum-classical systems. As this occurs, however, topological qubits, grounded in exotic states of matter, offer the prospect of fault-tolerance to decoherence, and more robust quantum computation systems in general. Turning to SETs now, the devices embody the concepts of quantum transport at single-electron scales, with good sensitivity and minimal power requirement. SETs operate on the basis of quantum tunneling effects controlling the movement of single electrons and are therefore highly appropriate for quantum information processing and ultra-small circuitry design. Merging SETs and qubit technology can lead to more efficient quantum gates, which are vital in high-fidelity quantum computations. The new work concerns enhancing stability and operating margins of SETs in cryogenic conditions, a critical area since quantum systems are sensitive to heat. Optimization of nanoscale architecture and the selection of materials are crucial in overcoming such issues while enabling the possibility of coupling with superconducting qubits as well as supporting advanced quantum algorithms. Lastly, cryogenic electronics presents a unique set of design constraints for the effective functioning of quantum and single-electron devices. Operation at temperatures near absolute zero significantly reduces thermal noise, an important advantage in maintaining qubit coherence and enhancing SET performance. The cryogenic environment does impose stringent demands upon device-design materials and fabrication processes, however. A deep understanding of thermal management and signal integrity is then essential, as ordinary semiconductor behaviors are entirely changed under these conditions. Along with this, new packaging and interconnect technologies must ensure reliable scalability of such devices in quantum computer systems. As such design issues are addressed by researchers, quantum and single-electron devices are steadily becoming more realizable in the context of operating systems, with the promise of revolutionary technology shifts in both information technology and computation.

REFERENCES

1. Adachi, C., & Sandanayaka, A. S. D. (2020). The leap from organic light-emitting diodes to organic semiconductor laser diodes. *CCS Chemistry, 2*, 1203–1216. https://doi.org/10.31635/ccschem.020.202000327 (accessed on 27th September 2025).
2. Akyildiz, I. F., Kak, A. C., & Nie, S. (2020). 6G and beyond: The future of wireless communications systems. *IEEE Access, 8*, 133995–134030. https://doi.org/10.1109/access.2020.3010896 (accessed on 27th September 2025).
3. An, C., & Hou, J. (2022). Benzo[1,2-*b*:4,5-*b*′]dithiophene-based conjugated polymers for highly efficient organic photovoltaics. *Accounts of Materials Research, 3*, 540–551. https://doi.org/10.1021/accountsmr.2c00020 (accessed on 27th September 2025).
4. Ansaloni, F., Chatterjee, A., Bohuslavskyi, H., Bertrand, B., Hutin, L., Vinet, M., & Kuemmeth, F. (2020). Single-electron operations in a foundry-fabricated array of quantum dots. *Nature Communications, 11*. https://doi.org/10.1038/s41467-020-20280-3 (accessed on 27th September 2025).
5. Bauer, B., Bravyi, S., Motta, M., & Chan, G. K. L. (2020). Quantum algorithms for quantum chemistry and quantum materials science. *Chemical Reviews, 120*, 12685–12717. https://doi.org/10.1021/acs.chemrev.9b00829 (accessed on 27th September 2025).
6. Bauer, D., Khazdozian, H., Mehta, J., Nguyen, R. T., Severson, M., Vaagensmith, B., Toba, A. L., et al. (2023). 2023 critical materials strategy. https://doi.org/10.2172/1998242 (accessed on 27th September 2025).
7. Bharti, K., Cervera-Lierta, A., Kyaw, T. H., Haug, T., Alperin-Lea, S., Anand, A., Degroote, M., et al. (2022). Noisy intermediate-scale quantum algorithms. *Reviews of Modern Physics, 94*. https://doi.org/10.1103/revmodphys.94.015004 (accessed on 27th September 2025).
8. Bitton, O., Gupta, S. N., Houben, L., Kvapil, M., Křápek, V., Šikola, T., & Haran, G. (2020). Vacuum Rabi splitting of a dark plasmonic cavity mode revealed by fast electrons. *Nature Communications, 11*. https://doi.org/10.1038/s41467-020-14364-3 (accessed on 27th September 2025).
9. Blais, A., Grimsmo, A. L., Girvin, S. M., & Wallraff, A. (2021). Circuit quantum electrodynamics. *Reviews of Modern Physics, 93*. https://doi.org/10.1103/revmodphys.93.025005 (accessed on 27th September 2025).
10. Bokare, A., Chinnusamy, S., & Erogbogbo, F. (2021). TiO_2-graphene quantum dots nanocomposites for photocatalysis in energy and biomedical applications. *Catalysts, 11*, 319–319. https://doi.org/10.3390/catal11030319 (accessed on 27th September 2025).

11. Brun-Picard, J., Djordjevic, S., Leprat, D., Schopfer, F., & Poirier, W. (2016). Practical quantum realization of the ampere from the elementary charge. *Physical Review X, 6.* https://doi.org/10.1103/physrevx.6.041051 (accessed on 27th September 2025).

12. Bruzewicz, C., Chiaverini, J., McConnell, R., & Sage, J. (2019). Trapped-ion quantum computing: Progress and challenges. *Applied Physics Reviews, 6.* https://doi.org/10.1063/1.5088164 (accessed on 27th September 2025).

13. Conings, B., Baeten, L., De Dobbelaere, C., D'Haen, J., Manca, J., & Boyen, H. G. (2013). Perovskite-based hybrid solar cells exceeding 10% efficiency with high reproducibility using a thin film sandwich approach. *Advanced Materials, 26*, 2041–2046. https://doi.org/10.1002/adma.201304803 (accessed on 27th September 2025).

14. Cossarizza, A., Chang, H. D., Radbruch, A., Akdiş, M., Andrä, I., Annunziato, F., Bächer, P., et al. (2017). Guidelines for the use of flow cytometry and cell sorting in immunological studies. *European Journal of Immunology, 47*, 1584–1797. https://doi.org/10.1002/eji.201646632 (accessed on 27th September 2025).

15. Cui, Y., Yao, H., Zhang, J., Zhang, T., Wang, Y., Hong, L., Xian, K., et al. (2019). Over 16% efficiency organic photovoltaic cells enabled by a chlorinated acceptor with increased open-circuit voltages. *Nature Communications, 10.* https://doi.org/10.1038/s41467-019-10351-5 (accessed on 27th September 2025).

16. Cui, Z. Y., Abdurahman, A., Ai, X., & Li, F. (2020). Stable luminescent radicals and radical-based LEDs with doublet emission. *CCS Chemistry, 2*, 1129–1145. https://doi.org/10.31635/ccschem.020.202000210 (accessed on 27th September 2025).

17. Cuoco, M., & Di Bernardo, A. (2022). Materials challenges for $SrRuO_3$: From conventional to quantum electronics. *APL Materials, 10.* https://doi.org/10.1063/5.0100912 (accessed on 27th September 2025).

18. Degen, C. L., Reinhard, F., & Cappellaro, P. (2017). Quantum sensing. *Reviews of Modern Physics, 89.* https://doi.org/10.1103/revmodphys.89.035002 (accessed on 27th September 2025).

19. Farrugia, L. J. (2012). WinGX and ORTEP for Windows: An update. *Journal of Applied Crystallography, 45*, 849–854. https://doi.org/10.1107/s0021889812029111 (accessed on 27th September 2025).

20. Harrison, P., & Valavanis, A. (2016). *Quantum Wells, Wires and Dots: Theoretical and Computational Physics of Semiconductor Nanostructures.* https://doi.org/10.1201/9781315221670 (accessed on 27th September 2025).

21. He, F., & Li, Y. (2022). Advances on theory and experiments of the energy applications in graphing. *CCS Chemistry, 5*, 72–94. https://doi.org/10.31635/ccschem.022.202202328 (accessed on 27th September 2025).
22. Hirohata, A., Yamada, K., Nakatani, Y., Prejbeanu, L., Diény, B., Pirro, P., & Hillebrands, B. (2020). Review on spintronics: Principles and device applications. *Journal of Magnetism and Magnetic Materials, 509*, 166711–166711. https://doi.org/10.1016/j.jmmm.2020.166711 (accessed on 27th September 2025).
23. Hoye, R. L. Z., Chua, M. R., Musselman, K. P., Li, G., Lai, M. L., Tan, Z. K., Greenham, N. C., et al. (2015). Enhanced performance in fluorene-free organometal halide perovskite light-emitting diodes using tunable, low electron affinity oxide electron injectors. *Advanced Materials, 27*, 1414–1419. https://doi.org/10.1002/adma.201405044 (accessed on 27th September 2025).
24. Hu, Z., Deibert, B. J., & Li, J. (2014). Luminescent metal–organic frameworks for chemical sensing and explosive detection. *Chemical Society Reviews, 43*, 5815–5840. https://doi.org/10.1039/c4cs00010b (accessed on 27th September 2025).
25. Khan, I., Saeed, K., & Khan, I. (2017). Nanoparticles: Properties, applications and toxicities. *Arabian Journal of Chemistry, 12*, 908–931. https://doi.org/10.1016/j.arabjc.2017.05.011 (accessed on 27th September 2025).
26. Kühne, T. D., Iannuzzi, M., Del Ben, M., Rybkin, V. V., Seewald, P., Stein, F., Laino, T., et al. (2020). CP2K: An electronic structure and molecular dynamics software package–Quickstep: Efficient and accurate electronic structure calculations. *The Journal of Chemical Physics, 152.* https://doi.org/10.1063/5.0007045 (accessed on 27th September 2025).
27. Kuznetsov, A. I., Brongersma, M. L., Yao, J., Chen, M. K., Levy, U., Tsai, D. P., Zheludev, N. I., et al. (2024). Roadmap for optical meta surfaces. *ACS Photonics, 11*, 816–865. https://doi.org/10.1021/acsphotonics.3c00457 (accessed on 27th September 2025).
28. Liu, S., Srinivasan, K., & Liu, J. (2021). Nanoscale positioning approaches for integrating single solid-state quantum emitters with photonic nanostructures. *Laser & Photonics Review, 15.* https://doi.org/10.1002/lpor.202100223 (accessed on 27th September 2025).
29. Lyshevski, S. E. (2018). *Nano and Molecular Electronics Handbook.* CRC Press eBooks. https://doi.org/10.1201/9781315221670 (accessed on 27th September 2025).

30. Moody, G., Sorger, V. J., Blumenthal, D. J., Juodawlkis, P., Loh, W., Sorace-Agaskar, C., Jones, A. E., et al. (2021). 2022 roadmap on integrated quantum photonics. *Journal of Physics: Photonics, 4*, 012501–012501. https://doi.org/10.1088/2515-7647/ac1ef4 (accessed on 27th September 2025).

31. Naresh, V., & Lee, N. (2021). A review on biosensors and recent development of nanostructured materials-enabled biosensors. *Sensors, 21*, 1109–1109. https://doi.org/10.3390/s21041109 (accessed on 27th September 2025).

32. Novoselov, K. S., Mishchenko, A., Carvalho, A., & Castro Neto, A. H. (2016). 2D materials and van der Waals heterostructures. *Science, 353.* https://doi.org/10.1126/science.aac9439 (accessed on 27th September 2025).

33. Ozawa, T., Price, H. M., Amo, A., Goldman, N., Hafezi, M., Lü, L., Rechtsman, M. C., et al. (2019). Topological photonics. *Reviews of Modern Physics, 91.* https://doi.org/10.1103/revmodphys.91.015006 (accessed on 27th September 2025).

34. Pandey, R., Vats, G., Yun, J. S., Bowen, C., Ho-Baillie, A., Seidel, J., Butler, K. T., et al. (2019). Mutual insight on ferroelectrics and hybrid halide perovskites: A platform for future multifunctional energy conversion. *Advanced Materials, 31.* https://doi.org/10.1002/adma.201807376 (accessed on 27th September 2025).

35. Piot, N., Brun, B., Schmitt, V., Zihlmann, S., Michal, V. P., Aprá, A., Abadillo-Uriel, J. C., et al. (2022). A single hole spin with enhanced coherence in natural silicon. *Nature Nanotechnology, 17*, 1072–1077. https://doi.org/10.1038/s41565-022-01196-z (accessed on 27th September 2025).

36. Pirandola, S., Andersen, U. L., Banchi, L., Berta, M., Bunandar, D., Colbeck, R., Englund, D., et al. (2020). Advances in quantum cryptography. *Advances in Optics and Photonics, 12*, 1012–1012. https://doi.org/10.1364/aop.361502 (accessed on 27th September 2025).

37. Wang, G., Chernikov, A., Glazov, M. M., Heinz, T. F., Marie, X., Amand, T., & Urbaszek, B. (2018). Colloquium: Excitons in atomically thin transition metal dichalcogenides. *Reviews of Modern Physics, 90.* https://doi.org/10.1103/revmodphys.90.021001 (accessed on 27th September 2025).

38. Wang, Z., Cheng, X., Xie, Y., Liu, S., Dong, M., Zhao, J., Liang, F., et al. (2022). Recent advances in organic room-temperature phosphorescence of heteroatom (B/S/P)-containing chromophores. *CCS Chemistry, 5*, 292–

309. https://doi.org/10.31635/ccschem.022.202202242 (accessed on 27th September 2025).

39. Wehner, S., Elkouss, D., & Hanson, R. (2018). Quantum internet: A vision for the road ahead. *Science, 362.* https://doi.org/10.1126/science.aam9288 (accessed on 27th September 2025).
40. Wyrick, J., Wang, X., Kashid, R. V., Namboodiri, P., Schmucker, S., Hagmann, J. A., Liu, K., et al. (2019). Atom-by-atom fabrication of single and few dopant quantum devices. *Advanced Functional Materials, 29.* https://doi.org/10.1002/adfm.201903475 (accessed on 27th September 2025).
41. Zhang, Q., Hu, W., Sirringhaus, H., & Müllen, K. (2022). Recent progress in emerging organic semiconductors. *Advanced Materials, 34.* https://doi.org/10.1002/adma.202108701 (accessed on 27th September 2025).
42. Zhao, C., Zhang, D., & Qin, C. (2020). Perovskite light-emitting diodes. *CCS Chemistry, 2*, 859–869. https://doi.org/10.31635/ccschem.020.202000216 (accessed on 27th September 2025).
43. Zhao, Y., Liu, W., Zhao, J., Wang, Y., Zheng, J., Liu, J., Hong, W., et al. (2022). The fabrication, characterization and functionalization in molecular electronics. *International Journal of Extreme Manufacturing, 4*, 022003–022003. https://doi.org/10.1088/2631-7990/ac5f78 (accessed on 27th September 2025).
44. Zou, X., Xu, Y., & Duan, W. (2021). 2D materials: Rising star for future applications. *The Innovation, 2*, 100115–100115. https://doi.org/10.1016/j.xinn.2021.100115 (accessed on 27th September 2025).
45. Zwolak, J. P., & Taylor, J. M. (2023). Colloquium: Advances in automation of quantum dot devices control. *Reviews of Modern Physics, 95.* https://doi.org/10.1103/revmodphys.95.011006 (accessed on 27th September 2025).

CHAPTER 10

FABRICATION TECHNIQUES FOR NEXT-GEN DEVICES

Contents

Introduction 395

10.1. Advanced Lithography Techniques 400

10.2. Atomic Layer Deposition (ALD) 405

10.3. Integration of Heterogeneous Materials 409

10.4. 3D Printing in Device Fabrication 414

10.5. Nanofabrication Techniques 419

10.6. Quality Control and Testing Methods 424

Summary..429

References...431

INTRODUCTION

Over the past few years, the need for next-generation devices has increased exponentially, and with that came the requirement of new methods of fabrication that are able to keep up with the fast rate of technological change. An integral part of this revolution is the inclusion of cutting-edge lithography tools, including extreme ultraviolet (EUV) lithography and nanoimprint lithography, that enable high-fidelity patterning of materials at the nanoscale. Such techniques are fundamental to the production of high-performance parts to enable a variety of applications, ranging from semiconductors to flexible electronics. The ability to print complex designs with record-breaking resolution significantly enhances device performance, facilitating further miniaturization and improved performance. Deep insight into these lithographic techniques reveals their ability to meet current limits in device fabrication, opening the door to transforming the field. Moreover, future device fabrication relies more and more on atomic layer deposition (ALD), an approach that is best known for delivering atomic-level precision. Layer by layer, depositing thin films, ALD enables material properties to be controlled like never before, something critical to the creation of new coatings and interfaces inside electronic devices. This process is particularly valuable when integrating disparate materials requiring disparate properties, thereby achieving optimal device performance for diverse functions. In the face of industries seeking energy conservation and miniaturization, the role played by ALD becomes more critical. Merging ALD with emerging lithography techniques is the movement toward hybrid approaches combining the most attractive characteristics of various methodologies, eventually to next-generation devices that reflect both performance and flexibility. Finally, nanoscience and engineering integration propel the development of innovative fabrication techniques that fully benefit materials science. With the emergence of nanoscale material structure manipulation methods, researchers are focusing their efforts on creating methods permitting large-scale manufacturing with great precision. For instance, regulation of organic electronic device nanoscale structure and electroactive properties by emerging types of printing technology shows significant commercial potential. Techniques such as those discussed in comprehensive reviews highlight the importance of thorough understanding of material behavior and synergetic interaction of mixed manufacturing methods. Tailoring strategies for nanoscale properties during fabrication are essential, as highlighted in the literature, not only to improve device performance but also to enhance scalability within fast-evolving markets (Andersen et al., 2020). This research emphasizes the point that the future of next-generation devices will be at the intersection of advanced fabrication techniques and new material science (Al-Muhtaseb et al., 2010).

Definition of Fabrication Techniques

The sophisticated landscape of fabrication techniques comprises a vast range of methodologies required for the fabrication of next-generation devices. Of these, advanced lithography is a key strategy, particularly due to the evolution of EUV and nanoimprint lithography. EUV lithography enables nanometer-scale feature resolution, remedies to the challenges that arise from diminishing transistor size.

This is a valuable capability to ensure that semiconductor devices of current technology are able to meet the needs of increased performance in tighter power constraints. At the same time, nanoimprint lithography gives a low-cost approach to patterning high-resolution structures, provoking novel device architectures outside those available before. All these sophisticated techniques are complementary to one another in continually enhancing fabrication processes, pushing the boundaries of what is possible in electronic and photonic device production. In addition to advanced lithography, ALD is a key fabrication technique that is instrumental in achieving highly uniform thin films with highly controlled thickness at an atomic scale. The technique facilitates the integration of numerous layers of materials needed for semiconductors to enhance performance on the basis of improved electrical properties. ALD is remarkable in high-k dielectric and nanoscale transistor production, where atomic-level precision equates directly to device efficiency. As electronic system complexity increases, ALD's capability to deposit conformal films and coatings on intricate geometries becomes paramount in enabling the next generation of devices. Its significance is evidenced through use in applications across energy storage, sensors, and optoelectronics, where precise material properties delivered by ALD drive performance improvement and innovation. The integration of dissimilar materials also defines the scale of contemporary fabrication techniques, with the possibility of combining dissimilar properties and functionality in a single device. The multi-material approach provides routes for the design of multifunctional systems that exploit each material's unique properties, advancing device performance and increasing application scope. For instance, integration of semiconductors with dielectrics, insulators, and metals can produce devices that exploit advantages such as improved electrical conductivity, thermal characteristics, and optical characteristics. Techniques like selective area growth and wafer bonding are crucial in realizing such integrations so that one can have complex structures previously impossible to obtain. Technique innovations like those implemented at George C. Marshall Space Flight Center (MSFC), where P-MOS integrated circuits are fabricated, reflect the impact of such innovations towards technology and education in the microelectronics field (Bouldin et al., 2025). Continued improvement of these fabrication techniques is important to enabling continuing research and development in next-generation devices and leading the way to innovation in various fields.

Importance of Advanced Fabrication in Technology

The evolution of the art of fabrication has revolutionized the world of technology, particularly in next-generation device development and manufacturing. The employment of cutting-edge lithography techniques such as EUV and nanoimprint lithography has significantly enhanced the resolution and precision achievable in semiconductor processing. These techniques enable one to make smaller and more effective components that operate at unprecedented speeds. In addition, the unification of hetero-materials by means of methods like ALD enables the fabrication of sophisticated multi-material structures. This ability not only enhances device performance but also results in new functionalities that were not feasible before. As technology continues to evolve, the importance of advanced fabrication techniques in offering device miniaturization and efficiency cannot be overstated, with ongoing research promising even greater developments on this aspect (Rentmeesters et al., 1990). Apart from lithography, the advent of additive manufacturing is a major breakthrough in advanced fabrication techniques. The development of 3D printing technologies has created novel paradigms for the production of intricate geometries with intricate internal structures, which play critical roles in enhancing the performance of future devices. Several geometric algorithms have been presented according to recent research studies to facilitate process planning of additive manufacturing.

These algorithms help streamline the production process by pre-digesting three-dimensional models and optimizing machine instructions dedicated to specific objectives. The ability to manipulate geometry adaptively has several advantages, such as minimizing waste and facilitating rapid prototyping. Therefore, the role of advanced fabrication processes in achieving an elastic manufacturing environment that allows for rapid innovation cannot be downplayed, as noted in contemporary literature (Attene et al., 2017). Equally critical is addressing heterogeneity integration in the design of large devices, a challenge that remains a significant aspect of advanced fabrication. Integration of contrasting materials tends to create more robust device performance and functionality. The integration process is not problem-free, with a primary concern being maintaining material compatibility and their consistent interaction at interfaces. The counteracting demands of analog and digital designs present unique challenges in automated fabrication techniques, as emphasized in research studies. Nevertheless, ongoing technological advancements in automation technologies are bound to further counter these problems, providing the platform for simpler integration processes and more sophisticated functional devices. Last but not least, the ability to employ enhanced fabrication techniques to combine different materials will play an essential role in shaping future technologies, as companies struggle to satisfy improved performance requirements and innovative functionalities (Rentmeesters et al., 1990).

Overview of Next-Gen Devices

The manufacturing of next-generation devices is significantly influenced by innovative fabrication techniques that advance the performance of typical semiconductor processing. Amongst the new developments are advanced lithography processes such as EUV and nanoimprint lithography, which have emerged as primary enablers in achieving high-resolution patterns needed in small electronic devices. These processes enable the creation of intricate microstructures necessary in increasing the performance of devices, as reducing size generally means faster speed and efficiency. Also, working with the material at the nanoscale enables the formation of novel device structures, which leverage the unique properties of materials like silicon nanomembranes. As the industry continues to embrace these advanced fabrication techniques, it is obvious that the addition of advanced lithography will be the factor of determination in achieving the potential of next-generation technologies (Zhang et al., 2025) (Figure 10.1).

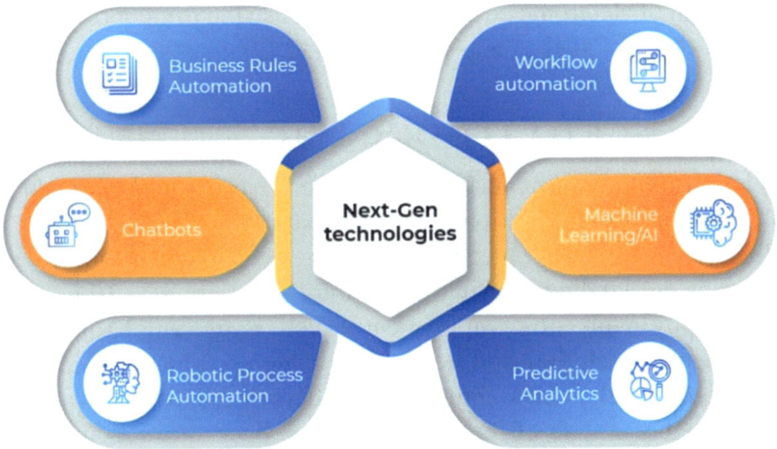

Figure 10.1. *Characteristics of next-gen technology.*

Source:https://www.speridian.com/wp-content/uploads/2019/07/next-gen.jpg.

Even though advanced lithography is a building block for next-gen device fabrication, the employment of atomic layer deposition (ALD) is equally revolutionary. This technique enables the deposition of thin films with atomic-level precision to create high-performance material for applications that vary from transistors to sensors. The ability of ALD to provide conformal coatings uniformly on complex geometries enhances the performance and reliability of devices. Notably, heterogeneous integration of materials through ALD makes it possible for device properties to be optimized by the combination of the functionality of different substances. For instance, in flexible electronics, inorganic and organic materials can be employed for realizing devices that are both lightweight yet strong with great functionality. The synergy in the integration of materials plays a major role in enhancing future device prospects by filling up performance and pragmatism gaps (Pathak et al., 2025). As technology continues to change, incorporating new materials in next-gen devices is one of the great opportunities and challenges. Multifunctional system design capabilities, particularly with respect to free-standing ferroelectric films, are under investigation. Free-standing films possess unique properties that enhance energy harvesting and sensor capabilities, a milestone shift towards green technologies. Complexity of processing capabilities allowing these combinations is at the core of scale-up and commercialization of new devices. Although promising, leakage currents and polarization stability issues have to be solved to enable maximal performance. With the knowledge of such issues and leveraging sophisticated techniques in material processing and engineering, the next generation of electronic devices will reestablish the frontiers of technology while addressing global challenges (Pathak et al., 2025).

10.1. ADVANCED LITHOGRAPHY TECHNIQUES

The evolution of lithography techniques has made nanoscale device fabrication possible, enabling the production of intricate patterns that were not previously possible. Among the advanced techniques, Extreme Ultraviolet (EUV) lithography is outstanding in patterning feature sizes less than 10 nanometers. EUV lithography involves the employment of shorter wavelengths of light, which provides greater resolution and accuracy for high-resolution semiconductor patterns. Significantly, EUV lithography offers the integration of several layers required by state-of-the-art microchips over diffraction limitations typical in traditional photolithography. Coupled with methods like nanoimprint lithography, centered on physical templates to create high-resolution designs, the diversity of advanced lithography occupies a focal position in pursuing next-generation devices with enhanced functionality capabilities. Along with the drive towards sophisticated lithography, Atomic Layer Deposition (ALD) becomes a critical complementary technique to device fabrication. ALD provides precise thickness control at the atomic scale, offering uniform coverage even in highly complex three-dimensional geometries. This is a critical characteristic required for the deposition of high-quality metal and dielectric layers required in modern electronic devices. The precision layering technique not only enhances the electrical and thermal management of components but also enables the integration of heterogeneous materials, a trend that becomes more crucial in the manufacture of multifunctional devices. Complementary synergy between ALD and advanced lithography ensures that devices are of superior performance and reliability, as every layer is engineered to meet specific functional requirements without adversely affecting the underlying infrastructure.

In addition, novel implementation of structured light technique in combination with cutting-edge lithography presents potential avenues for future breakthrough. As defined in landmark studies, structured light is capable of generating custom-designed light fields to enhance nanoscale manipulation of materials, opening up applications in photonic engineering. Specifically, the precise manipulation of light's characteristics is able to provide dynamic interaction at the nanoscale with the subsequent production of ultracompact optical circuits and advanced lab-on-a-chip systems. The major advances in studying the interaction of matter and light, such as the excitation of gap-surface plasmons and channeled plasmon polaritons as highlighted in recent studies, point out how these combined techniques are able to offer innovative solutions for current challenges in nanofabrication. The promise of semiconductor collaborative technologies sheds a bright light on a clear pathway to the achievement of next-generation device fabrication techniques.

10.1.1. Overview of Lithography in Device Fabrication

The past history of developments in lithography techniques has had a huge impact on the capability of manufacturing next-generation devices with increasingly smaller feature dimensions. Of the contemporary techniques, extreme ultraviolet (EUV) lithography is one such technology employed by the semiconductor sector, having resolutions as small as a few nanometers. This innovation allows for the production of advanced integrated circuits capable of meeting the growing demand for more computational power and efficiency. Using short wavelengths, EUV can create highly complex patterns on silicon wafers required for advanced applications like those found in quantum computing, which presents new challenges and possibilities for materials usage.

This move to next-generation lithography is in harmony with investigations of advanced materials with superior electronic properties, such as graphene nanoribbons (GNRs), which have been of promise in advancing device function and performance under stressful operating conditions (Bostwick et al., 2010). The use of different materials is becoming more salient in the design of modern devices, which enables their best characteristics. Atomic layer deposition (ALD) has emerged as a critical process here, facilitating the precise control of thin film thickness and composition at the atomic level (Figure 10.2).

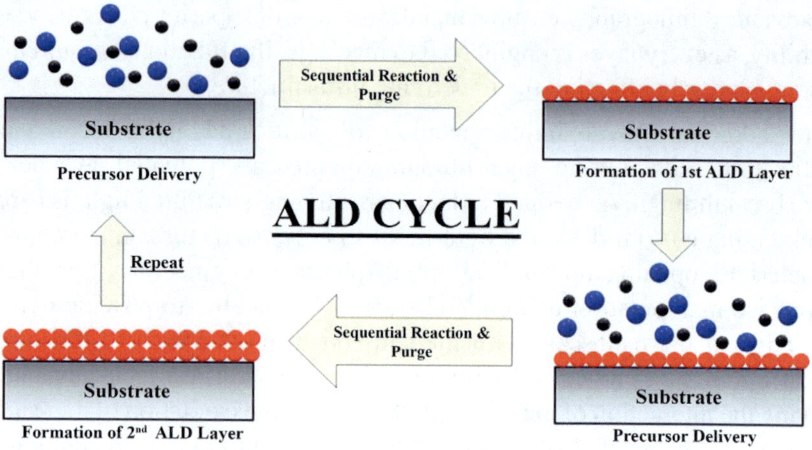

Figure 10.2. *Working cycle of atomic layer deposition (ALD).*

Source:https://ars.els-cdn.com/content/image/1-s2.0-S0300944024000444-gr11.jpg.

The process not only optimizes the performance of existing materials but also makes it possible for unlike substances to be effectively deposited on top of each other to create new functionalities. For example, ALD can be used in the manufacturing of high-k dielectric materials that are essential for transistors in next-generation logic circuits. The capability to change surface properties at a molecular scale significantly increases possibilities for innovation in device architecture, particularly as the company aims to condense multiple functions into progressively smaller form factors, addressing the immediate demands for energy efficiency and reduced physical size in device production (Armstrong et al., 2015). With the evolution of the electronic device landscape, the integration of sophisticated fabrication techniques such as lithography and ALD becomes crucial in meeting the growing computing and efficiency requirements. With the interaction between these techniques not only making next-generation devices possible but also giving a momentum boost to innovation in material science. The relentless pressure to miniaturize and to high performance is emphasized by the utilization of emerging technologies such as quantum computing, drawing heavily on the subtile device structures built from the advanced lithographic processes. Thus, subsequent research will likely be in extending further these processes, investigating the potential of novel materials, and transcending the constraints imposed by traditional methods. As such advancements continue to occur, the coupling of new material development and precise fabrication procedures will be a determining aspect in deciding the fate of electronic devices and their applications.

10.1.2. Extreme Ultraviolet Lithography (EUV)

In the ever-evolving semiconductor fabrication arena, which lithography technique to adopt is going to be a determining factor in choosing the resolution and effectiveness of device processing. EUV has become a leading-edge technology that is capable of printing complex patterns at sizes hitherto impossible. Having the wavelength of light fixed at around 13.5 nm, EUV allows the reduction of circuit elements with space for the inevitable need for higher integration and performance in coming devices. With EUV, the semiconductor industry has been able to go below the 7-nanometer node, where conventional optical lithography techniques begin to become constricted. This shift is reflective of an important response to the challenge posed by modern electronics, where processing power and power efficiency are most critical. EUV lithography materials play a significant role to improve its performance and efficacy. Research indicates that hybrid organic-inorganic resists, specifically designed for EUV application, have significant potential to improve pattern fidelity and resolution.

These resists, with components such as silicon, zirconium, and titanium oxides, are synthesized using sol-gel routes and show different photochemical responses upon exposure to EUV radiation. For instance, it has been demonstrated through research that the chemical composition of such materials influences their degradation processes and, in turn, the performance and yield of lithographic process (Brigo et al., 2018). Hence, understanding the nuances of these resists and their interaction with EUV light is essential to propel lithographic techniques further and achieve the required level of accuracy in the manufacturing of future devices. In addition, the combination of various materials is a sophisticated yet challenging issue in the realm of EUV lithography. With progressively complicated device architectures, the need for compatible materials that are capable of coexisting at nanoscale dimensions is more compelling than ever. The ability to pattern various materials efficiently not only enhances the functionality of semiconductor products but also facilitates innovations in chip design and architecture. With EUV, there is higher resolution and reduced feature size, leading to better device performance while, simultaneously, permitting multi-material system integration (Antonsen et al., 2019). The study of such new materials and their usability with EUV processes is also at the vanguard of pushing the capabilities of semiconductor production, further underlining the importance of EUV lithography in the creation of next-generation technologies.

10.1.3. Nanoimprint Lithography (NIL)

Fabrication processes for next-generation devices have seen tremendous progress, particularly through the inclusion of NIL. This technique is singular in its ability to make high-accuracy, relatively cheap fine nanoscale patterns. By utilizing the application of a stamping technique transferring nanoscale patterns to a broad spectrum of substrates, NIL permits mass production scaling without the commonly-enormous expenses normally associated with traditional photolithography methods. Specifically, multi-Si nanoridge fabrication advancements have established NILs potential for the fabrication of complex geometries, such as 120-nm-high and 40-nm-wide structures that future applications in electronics and photonics require (Berenschot et al., 2009). These efficiencies in fabrication are central to the mass production of next-generation devices because they promote cost-effectiveness and high throughput of the manufacturing process.

This capability of NIL to facilitate innovations in biochips is an excellent example of its applicability across various uses of technology and medicine. With biochip platforms being utilized more and more in the biomedical industry for conducting various biological tests, accuracy and cost-effectiveness during the manufacturing process have become essential conditions. Efficient replication of microelectrode arrays on solid substrates is possible with the application of NIL, significantly enhancing the functionality and reliability of the devices. In fact, the successful creation of a PDMS mold through NIL has demonstrated a mere 13% size variation in total, highlighting the technique's ability to be consistent between batches while utilizing cost-efficient production methods (Fahmi et al., 2016). This ability to create tiny-scale details at a macro scale not only meets market demands but is also critical to the growth of treatment and diagnostic technologies, signifying the site of intersection between innovation and cost-effectiveness. In addition, integrating heterogeneous materials with NIL brings new possibilities and advances device engineering. With the evolution of technology, materials science plays a vital role in designing next-generation devices requiring the convergence of unrelated functionalities. NIL facilitates the stacking of various materials, which enables one to create structures with integrated organic and inorganic elements required for multifunctional application. By facilitating nanoscale positioning of these materials with such fine control, NIL can potentially enhance optical, electrical, and mechanical properties and result in devices that are not only more capable but also more conformable to specific needs. This capacity to consolidate dissimilar materials with the help of sophisticated lithography proves NILs potential to drive the boundaries of technology in harmony with future smart devices and new opportunities in several industries.

10.2. ATOMIC LAYER DEPOSITION (ALD)

The need of nanotechnology requires precise fabrication techniques that can ensure atomic-level control of material properties. ALD stands out for its unique ability to deposit thin films with excellent uniformity and conformality, a prime requirement in the design of future devices. In contrast to other deposition techniques, ALD functions based on self-limiting surface reactions, enabling control at the atomic level of thickness and being well suited to coating geometries that are complex. This is critical in the manufacture of transistors, capacitors, and sensors, where film thickness can significantly impact both performance and reliability. By bypassing the constraints of conventional deposition techniques, ALD ensures that future technologies can meet the stringent demands of advanced electronic applications, leading to advancements in photonics as well as flexible electronics. Besides, ALD has been successfully used to create refractory metal thin films, a building block component in diverse new nanophotonic applications (Figure 10.3).

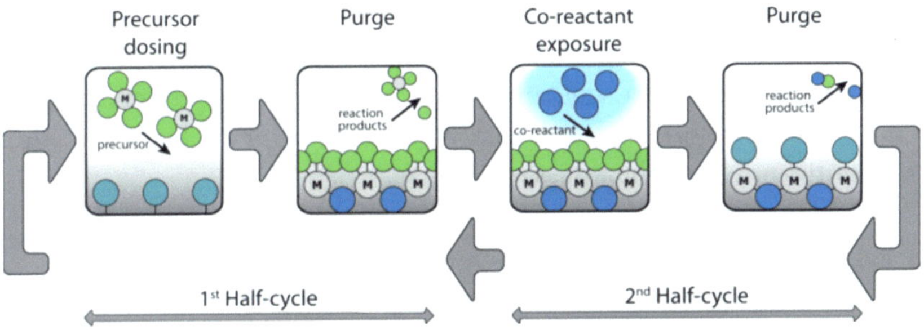

Figure 10.3. *Characteristics and process of atomic layer deposition (ALD).*

Source: https://i0.wp.com/www.atomiclimits.com/wp-content/uploads/2019/02/Fig.1_ALD-cycle.png?resize=738%2C251&ssl=1.

Research has established that NbN, NbTiN, and TiN materials are also crucial in high-temperature plasmonic devices and infrared superconducting single-photon detectors applied in optical communication and quantum computing technology. The materials have also demonstrated superior optical absorption characteristics when deposited using ALD and are thus perfectly suited for use in applications with a focus on performance. Moreover, the integration of ALD-deposited thin films into mainstream semiconductor technologies creates the possibility for the manufacturing of hybrid devices capitalizing on the optimum features of refractory materials and silicon-based platforms. Hence, not only does the versatility of ALD enhance material performance, but also allow one to devise new devices beyond the current technology. One of the impressive applications of ALD can be observed from the technique of making pressure sensors using self-assembled ZnO nanowires. In this technique, the application of ALD has been witnessed with other fabrication methods for creating devices with greater sensitivity and responsiveness. Here, the deposition of ZnO nanowires onto MOSFET gates suggests the potential for atomic layer deposition to increase device functionality at the expense of little to no loss of compatibility with standard CMOS processing. The ability of ALD to deposit nanostructures having well-defined electrical and mechanical properties means the creation of advanced sensors that can strongly sense subtle changes in pressure. Thus, the integration of ALD with other complementary fabrication processes dictates its significance in relentless progression towards next-generation devices across every industry like sensor technology and microelectronics.

10.2.1. Principles of Atomic Layer Deposition (ALD)

Evolution in the fabrication of next-generation devices is most frequently focused on ALD's significance owing to the fact that it has the special ability to form ultra-thin, conformal atomic layer films. This chemical vapor deposition process employs a cycle of sequential exposure to two or more precursors that each react with the surface, producing controlled, layer-by-layer growth with amazing uniformity and thickness control.

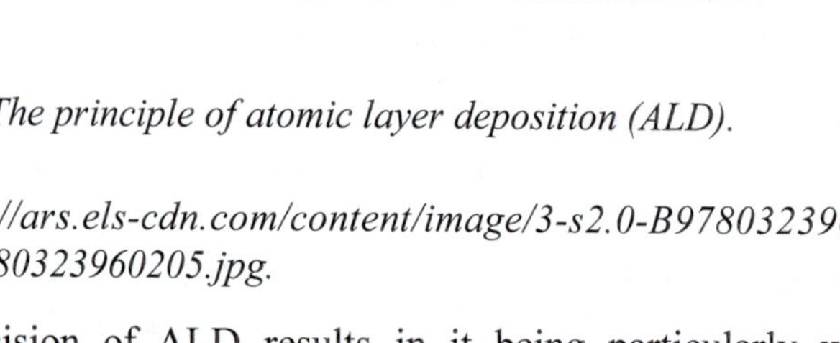

Figure 10.4. *The principle of atomic layer deposition (ALD).*

Source:https://ars.els-cdn.com/content/image/3-s2.0-B9780323960205000637-f00063-01-9780323960205.jpg.

The precision of ALD results in it being particularly well-suited for applications involving thin films of high quality, e.g., semiconductor production and nanoscale devices. These abilities have stimulated interest in integrating ALD with other high-end lithography technologies, thus raising overall device fabrication versatility. In exploring the broader implications of atomic layer deposition, one can see its fundamental role in the realization of new materials for next-generation electronic and photonic devices. For example, refractory metal thin films like NbN and TiN have been deposited using ALD because of their unique optical and electrical characteristics (Figure 10.4).

This capability is important in various applications ranging from high-temperature plasmonic structures to superconducting single-photon detectors required for quantum technology development. Atomic level control via ALD allows researchers to design material characteristics for a specified operational requirement. By achieving the desired film properties, designers can solve problems presented by conventional materials, contributing further to innovation in device miniaturization and capability. Atomic layer deposition also presents new prospects for integrating complex device architectures by enabling electrochemically deposited or grown heterogeneous materials. The deposition of materials such as ZnO nanowires using ALD over pre-existing semiconductor structures enhances the functioning of devices such as pressure sensors in Metal-Oxide-Semiconductor Field-Effect Transistors (MOSFETs). The combination of ALD with traditional fabrication methods such as photolithography and anodic alumina oxide templates leads to the creation of structures with excellent aspect ratios and sensitivity. These innovations not only improve the performance of the devices but also the CMOS technology compatibility, which renders them particularly desirable in the context of advanced electronic circuits. Given the growing necessity of multifunctional devices, atomic layer deposition concepts will remain the best fabrication techniques, which will drive innovation across various applications in electronics and photonics.

10.2.2. Applications of Atomic Layer Deposition (ALD) in Semiconductor Manufacturing

The complex geometry of future semiconductors requires new processing techniques, one of which is ALD. With its ability to deposit thin films layer by layer with very good control, ALD is extremely useful in the fabrication of high-quality materials that are essential to leading-edge lithography applications, including Extreme Ultraviolet (EUV). These advancements require materials of superior uniformity and adhesion to complicated surface topographies, both of which are very well supported by ALD techniques. With increasing demand for smaller yet more powerful devices, combining materials with different properties and yet achieving performance requirements is becoming a make-or-break necessity. ALDs capability of material depositing like dielectric oxides and high-k dielectrics directly influences device efficiency and running speed, paving the way for improved nanoscale fabrication techniques that enable the path of Moores Law and further, shaping the future of the semiconductor industry. The use of ALD in semiconductor processing extends beyond mere material deposition; it is essential for hetero material integration. The technological shift towards multi-material devices demands quality junctions between different types of semiconductors, such as silicon and gallium nitride (GaN). ALD techniques enable the growth of thin films that are able to create effective interfaces between these materials, and this can lead to new circuit topologies that exploit each semiconductor's unique attributes. GaN, for instance, has thermal stability and high electron mobility that provide advantages in power conversion and high-frequency applications, thus addressing the requirements of next-generation devices in various markets (Amano et al., 2018). The ability to deposit heterostructures with atomic control not only enhances device performance but also hastens innovation in system architectures needed for further advancements in power electronics applications.

In addition, ALD helps to overcome the physical constraints of traditional silicon-based technologies as scaling of CMOS continues toward physical boundaries. As the semiconductor industry turns to novel materials and processes, ALD seeks to unlock the promise of novel oxide films and heterostructures, which have been stimulating alternatives. The oxides exhibit unique functional behaviors that range across electronic applications as varied as energy conversion and information storage (Althammer et al., 2019). The collective efforts of researchers in this arena have led to innovations in thin film technology needed for the development of next-generation devices that are more efficient and flexible. As industries make the transition towards these newer fabrication techniques, ALD will be at the forefront, contributing significantly to the redesigning of the future of semiconductors.

10.2.3. Advantages of Atomic Layer Deposition (ALD) Over Traditional Deposition Methods

The quest for newer fabrication techniques has driven the semiconductor industry to invent many techniques that enhance precision and efficiency. Among them, ALD was one of the serious contenders for the traditional deposition techniques like chemical vapor deposition (CVD) and physical vapor deposition (PVD). ALD allows the deposition of thin films at the atomic scale with full control over film thickness and composition. This control at the atomic level is particularly useful in constructing next-generation devices, where small differences can be crucial to performance. Reducing constraint that is inherent with non-conformal coating and maximizing uniformity on intricate geometries, ALD is firmly ahead, particularly critical in silicon photonics and other semiconductor materials, as highlighted in recent studies (Lee et al., 2025). Another huge advantage of ALD is its unprecedented material flexibility, enabling the use of numerous different materials in one process without incurring significant contamination or defects. While traditional methods suffer from film adhesion and evenness in many layers, ALD processes can deposit effectively high-quality metal films, oxides, and nitrides on a vast array of substrates, including two-dimensional materials. This capability is vital for the design of innovative devices that require heterogeneous material integration, a technique that continues to play a significant role in optoelectronics and integrated circuits. Further, the low-temperature deposition characteristics of ALD maintain the integrity of temperature-sensitive structures and substrates and thus make it an essential technology to the production of complex layers required for advanced electronic functionality (Gamiz et al., 2025). The level of precise control that ALD makes available also makes possible the minimization of material loss, something that is especially important in the current emphasis on sustainability. Traditional processes are thicker and require extensive etching or masking, which is based on huge material loss. ALD's layer-by-layer growth, by contrast, only deposits the deposited material, resulting in minimal loss and production cost on large-scale production. This cost-effectiveness is also aligned with the mission of the industry to meet high-functionality devices in terms of concurrent resolution of environmental concerns around material use and disposal. As the semiconductor world keeps on evolving, the integration of ALD surfaces as a leading methodology further highlights its performance and environmentally friendly advantage toward next-generation device development (Lee et al., 2025).

10.3. INTEGRATION OF HETEROGENEOUS MATERIALS

The requirement for optimum device performance in next-generation applications necessitates innovative integration strategies of disparate materials. Future lithography, such as extreme ultraviolet (EUV) and nanoimprint lithography, has played a pivotal role to define fine details on substrate surfaces that support numerous materials. The technology enables the formation of intricate patterns of high precision, which is essential to integrate various functional materials without disturbance. For instance, III-V-based photonic device production on silicon substrates illustrates such integration whereby manipulation of diverse material properties is used to meet specific operating requirements and create devices beneficial in optical communication and sensing (Abbasi et al., 2015). Precise combination of dissimilar materials allows engineers to combine individual strengths while overcoming traditionally intolerable incompatibility problems of heterogeneous constituents. In addition, the application of atomic layer deposition (ALD) in the thin-film fabrication process is critical to the successful integration of the heterogeneous material. ALD provides greater control over the film thickness as well as the composition, which is critical in the formation of the desired electrical and optical properties in multilayer devices (Figure 10.5).

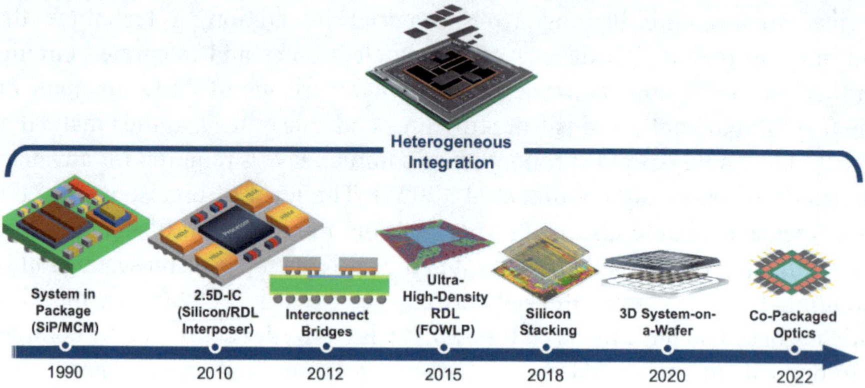

Figure 10.5. *The implementation of heterogeneous integration.*

Source:https://i0.wp.com/semiengineering.com/wp-content/uploads/Screenshot-2023-03-31-at-12.13.52-PM.png?fit=2140%2C948&ssl=1.

In next-generation device production, the method can be utilized to deposit precise materials in layer-by-layer form, thereby enhancing the functionality and performance of integrated systems. Highly customized layers like these can be engineered to maximize interactions among different materials and minimize defects and maximize efficiency. Therefore, the overall behaviour of integrated layers allows for enhanced performance parameters of electronics and photonics (Abate et al., 2017). This layered structure provides new grounds for multifunctional integration, driving innovation across a wide range of technological fields. Beyond simple functional extension, the integration of disparate materials is also important for addressing contemporary challenges such as miniaturization and energy efficiency. Integration technologies facilitate devices to take advantage of special properties of different materials and develop synergies leading to more efficient and compact systems. For example, microfluidic integration into molecular diagnostics is a novel approach in which materials of varying chemical properties come together to enhance the detection capabilities (Abate et al., 2017). These platforms not only accelerate diagnostic processes but also reduce costs involved in them, making them more manageable to be used on a mass scale in major industries like healthcare and the environment. With continuous innovations in research, the phenomenon of heterogeneous integration will remain at the center of next-generation fabrication technique development, solidifying it as a driving force in shaping future technologies.

10.3.1. Definition and Significance of Heterogeneous Integration

The evolution of electronic equipment towards enhanced performance and functionality needs advanced fabrication techniques with the ability to combine several materials and components. One major paradigm in this context is heterogeneous integration, where different materials are combined to make their inherent properties useful. By layering multiple materials, each specially designed for single functionalities—e.g., semiconductors, conductors, and insulators—manufacturers can develop multi-faceted devices that outperform their homogeneous counterparts.

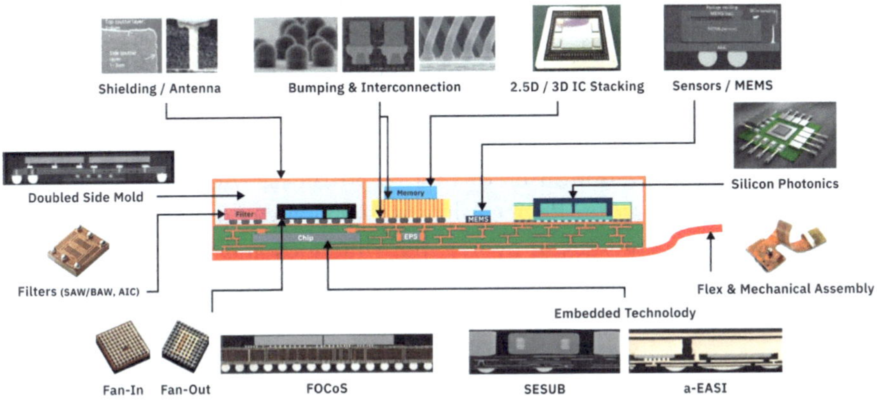

Figure 10.6. *Technology building blocks for heterogeneous integration.*

Source:https://ase.aseglobal.com/wp-content/themes/ase/img/img-hi-tech.png.

This complement can particularly enhance performance in photonic circuits, where silicon may be combined with III-V compounds to create efficient light-emitting devices. Its significance is more and more manifest in applications as diverse as optical communication and sensing, as in the advancement towards heterogeneous III-V-on-silicon photonic devices with a wide wavelength range from 850 nm to 3.85 μm, demonstrating versatility and efficacy in contemporary device realization (Abbasi et al., 2015). As the need for such small and sophisticated devices grows, new technologies such as advanced lithography and atomic layer deposition (ALD) have emerged as enabling technologies for heterogeneous integration. Both methods offer nanoscale control over material deposition, enabling the creation of complex structures required for future applications. For instance, second-generation lithography techniques, including extreme ultraviolet (EUV) lithography and nanoimprint lithography, are crucial in defining intricate patterns that integrate dissimilar materials. Simultaneously,

ALD provides unmatched uniformity and thickness control and offers for the deposition of conformal films and ultrathin films that are essential for attaining highest device performance (Figure 10.6).

The combination of the processes makes it possible to produce devices having unique electronic, optical, and mechanical properties, facilitating more efficient energy control and signal processing (Bonvin et al., 2016). Heterogeneous integration is more than just mixing materials; it motivates the formation of novel device architectures that can address evolving challenges in electronics. By using methods that stimulate spatial positioning and directed development of nanostructures, the process is made efficient. Guided growth methods, for instance, can help assemble complex nanowires into pre-defined architectures, offering new opportunities for the manufacture of solar cells, lasers, and light-emitting diodes. The technologies enhance performance and lower production costs with increased material efficiency and reduced waste. Thus, heterogeneous integration drives technological innovation by offering new avenues for innovation in fabrication approaches, which are required for developing next-generation devices that are sophisticated and cost-reducing. As the technology continues to evolve, the strategies and methods associated with heterogeneous integration will play a critical role in shaping the future of next-generation electronics.

10.3.2. Challenges in Integrating Different Materials

In the changing scenario of next-generation device manufacturing technology, integration of hetero-materials is a multifaceted challenge to be resolved by engineers and researchers. While near-atomic resolution by EUV and nanoimprint lithography becomes increasingly available, the need for compatible properties of materials gains primacy. Materials of different thermal and electrical conductivities or expansion coefficients can create sites of stress when combined within a composite material. Such incompatibility is likely to lead to mechanical failure or reduced device performance. Additionally, uniform adhesion between layers also requires overcoming intrinsic material properties and potential contaminants that may serve as performance inhibitors. So, the development of methods for predicting and compensating for such issues is crucial to advancing the field without detracting from next-generation device functionality reliability and lifespan. It is another degree of sophistication because of the need for enhanced functionality within newly developed devices. Atomic layer deposition (ALD) technologies provide the ability to produce ultra-thin films with precision, which can be applied to bringing together various materials at a nanoscopic level. But this precision requires a complete understanding of chemical and physical material interaction under the deposition process. When combining materials for electronics or sensors, contaminants or interface reactions can undermine performance, often resulting in reduced signal integrity or reduced energy efficiency. Thus, scientists have to exert a lot of effort in studying each material's character and material compatibility with novel deposition techniques. Overcoming such challenges becomes increasingly important with the continually increasing demand for high-performance and multifunctional devices. Finally, the realization of model-based design tools is a potential route to making the handling of heterogeneous materials in fabrication easier. Existing assistance in laminate robot design, as evidenced in Aukes et al. (2018), necessitates the use of automated simulation and analysis to understand dynamic material interactions. Applications extend into robotics because such interfaces can be utilized to enhance microfluidic platforms and molecular diagnostic platforms, as discussed in Abate et al. (2017). Yet, the non-availability of design and analysis frameworks that can be used universally remains a severe limitation. Advanced software would allow engineers to simulate and visualize interactions before they make physical prototypes, saving trial-and-error cost and time efficiency in integrating materials. The development of such tools would be a step up in addressing the difficulties of integrating diverse materials into high-performing, coherent systems.

10.3.3. Case Studies of Successful Heterogeneous Integration

In the field of next-generation devices, the use of heterogeneous integration is at the core of enhanced device performance and capability. Using highly advanced lithography, such as extreme ultraviolet (EUV) lithography and nanoimprint lithography, researchers have been able to fabricate complex device architectures integrating different materials with varied properties. The methods facilitate the precise patterning required to align various semiconductor materials, thereby making the integration of novel devices that leverage the optimum features of each component feasible. Case studies also show that the application of these advanced techniques has made impressive improvements in operational efficiency and functionality, demonstrating that there is a requirement for tailored fabrication techniques in enhancing device performance. Furthermore, atomic layer deposition (ALD) is also critical in the integration of heterogeneous materials, allowing precise control over thickness and film composition. This strategy is central to improving interfacial properties in a range of materials such that loss of performance due to ill-defined interfaces is reduced. Examples use case studies to point out applications in microfluidics and biosensing where ALDs unique strengths have improved sensitivity and lowered noise in signal sensing, illustrating the revolutionary impact of this deposition technique. These advancements, in addition to improving device reliability, create windows of opportunity for microscopic diagnostic and targeted therapeutic innovations, underpinning the significance of ALD in integrated heterogeneous systems. The issues that come with heterogeneous integration require sustained research and development efforts, specifically in the process of optimizing the waveguide fabrication of integrated photonic devices. Technological innovations in waveguides have been motivated by the need for lower propagation losses, particularly for devices such as soliton microcombs. A successful case study demonstrated that silicon nitride ($\mathrm{Si_3N_4}$) waveguide integration used outstanding smoothness and dimensional control by using the photonic Damascene process together with a novel reflow technique. These innovations have also shown promise in breaking through traditional optical integration limitations, a return that defies the propagation losses, which are established through systematic investigation of scattering and absorption contributions. In general, these studies underscore the dynamic nature of heterogeneous integration in shaping fabrication process development for future devices (Abate et al., 2017; Ghadiani et al., 2018).

10.4. 3D PRINTING IN DEVICE FABRICATION

The latest advances in manufacturing technologies have significantly impacted the capacity of device fabrication, particularly through the integration of 3D printing systems. Through this process, complex geometries that are not achievable using traditional fabrication techniques can be designed and manufactured. Compared to other state-of-the-art lithography techniques, such as extreme ultraviolet (EUV) and nanoimprint, which are constrained by their processing limits, 3D printing provides unparalleled freedom of design. It enables the straightforward integration of disparate materials to create strongly customized components tailored for a specific application. 3D printing's flexibility not only simplifies production processes but also reduces material waste, which is a goal in increasing focus by manufacturers today. Such a synergy between design flexibility and waste reduction implies that a paradigm shift is imminent in how devices are envisioned and commercialized, which will open up many innovations across industries ranging from electronics to biomedical applications. Additionally, the innovation of geometric algorithms to enhance the additive manufacturing process has transformed the method of making devices. Due to manufacturers adopting these sophisticated algorithms, the demand for optimization of processes is on the rise, making it possible to improve efficiency in operations. There is evidence that there is a growth of techniques developed to prepare a 3D model for production, each with a unique objective essential to the production process (Attene et al., 2017) (Figure 10.7).

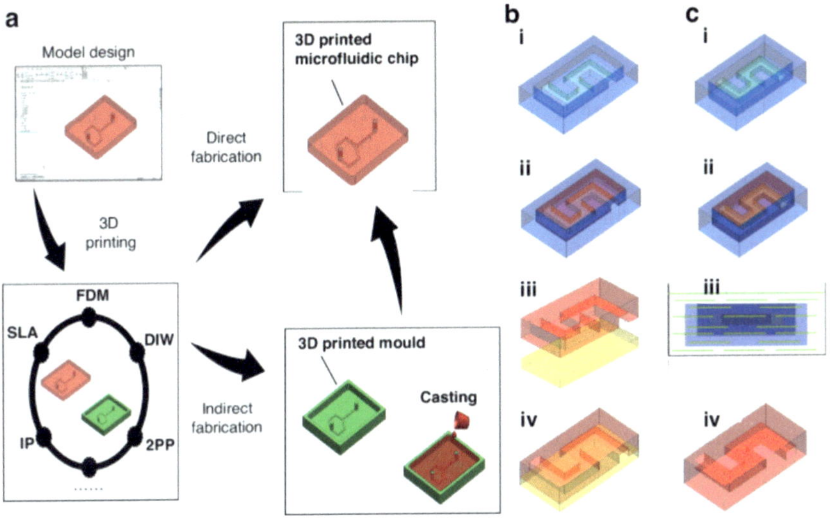

Figure 10.7. *The pattern of 3D printing.*

Source: https://media.springernature.com/lw685/springer-static/image/art%3A10.1038%2Fs41378-024-00812-3/MediaObjects/41378_2024_812_Fig6_HTML.png.

This trend of algorithmic advancements and the type of devices produced results in more intelligent, more responsive manufacturing conditions. This not only speeds up the design-to-manufacturing cycle but also facilitates iterative prototyping that greatly enhances product optimization. With the integration of new geometric algorithms, the discipline is now poised to overcome most challenges, further enhancing the usability and efficiency of 3D printing in device production. Finally, the use of 3D printing in the context of industry 4.0 is another breakthrough in the history of manufacturing. This fourth industrial revolution highlights the imperative for increased integration between information technologies and smart manufacturing systems, with additive manufacturing as a pioneer (Dehghanghadikolaei et al., 2019). The potential for producing complex components rapidly in a decentralized manner has benefits such as saving time and material, as well as enhancing overall productivity. As industries adopt these novel approaches, they not only respond to the demands of customization but also address greater sustainability issues within the manufacturing process. The economic, social, and environmental potential that 3D printing has already demonstrated ensures its position as an enabling technology for future device production. As we look into the evolution and applications of additive manufacturing, we can see that its continuous evolution is important to advance innovation and achieve sustainability in the rapidly changing nature of technology.

10.4.1. Overview of 3D Printing Technologies

In the recent past, advancements in 3D printing technologies have significantly enhanced new fields of manufacture, particularly in such sectors where a complex design and personalized functionalities are called for. Several additive manufacturing techniques, ranging from Fused Deposition Modeling (FDM) to Selective Laser Sintering (SLS), have different strengths more suited for specific applications. FDM uses thermoplastic filaments and is extensively employed due to the availability of easy access and affordability, thereby making it ideal for quick prototyping. On the other hand, SLS uses powdered materials to be treated with laser sintering to facilitate complex geometries that are not readily achievable with traditional subtractive methods.

Subsequently, these technologies are being integrated into other disciplines, including aerospace, automotive, and biomedical engineering, evidence of the versatility and room for innovation that 3D printing technologies possess. The progress of 3D printing may also be assessed from the lens of novel methods such as Direct Ink Writing (DIW) and bioprinting. DIW facilitates printing with viscous inks, e.g., made of biological material, thereby creating structures with new mechanical and biological properties that are attractive to contemporary engineering needs. Additionally, the discovery of bioprinting is a breakthrough area in biomedical devices that enables the generation of patient-specific organs and tissues through the printing of living cells in a highly precise fashion (Figure 10.8).

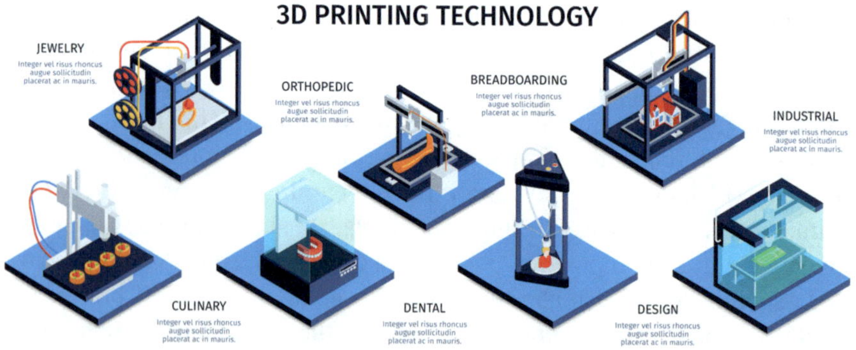

Figure 10.8. *Application of 3D printing technology.*

Source: https://cdn.prod.website-files.com/620126acdd13716fb8a8b3e3/647491dd7458229b2c7ab799_14742491_1908.i504.018_isometric%203d%20printing%20horizontal%20infographics.jpg.

As stated, these advancements hold open the gate for personalized medicine, addressing specific patient needs and achieving enhanced therapeutic results. These advances indicate the immense impact that the future 3D printing technologies will have on device production, as clearly demonstrated in the medical sector. Other than underlying challenges, issues with restrictions on material and process optimization are still there. Advancements in geometric algorithms used in Additive Manufacturing have been advancing at a rapid pace, making it possible to optimize production and manufacturing designs in a manner to pursue specific goals, according to recent studies (Attene et al., 2017). Additionally, creating the microarchitectural complexity needed within printed tissues poses additional challenges; technological limitations must be overcome in a successful attempt to regenerate complex anatomical structures (Chia et al., 2015). With the development of the technology, ongoing research will be crucial in surmounting the inadequacies, thus 3D printing will not only facilitate the production of future devices but also support innovations that transform human medicine and engineering solutions.

10.4.2. Benefits of 3D Printing for Rapid Prototyping

The revolution brought about by the 3D printing technology is a pivotal moment in design and manufacturing procedures, particularly in rapid prototyping. With additive manufacturing, complex three-dimensional models can be created very rapidly with astonishing accuracy. With such capability not only is the cycle of development shortened but creative freedom is also enhanced for designers and engineers. For instance, the flexibility to customize prototypes to demand cuts the iterative cycle to a fraction, allowing different designs to be developed and tested within a minute fraction of what would be possible through traditional means. The use of technologies such as inkjet printing in the area of additive manufacturing continues to enable quick turnaround prototypes, reducing time-to-market for subsequent devices by orders of magnitude. As noted, the steps are also possible to integrate with advanced manufacturing, including structural electronics, outlining the scalability and flexibility of 3D printing in current manufacturing practices (Algarni et al., 2016). 3D printing provides heterogeneous materials the ability to be integrated, a feature essential for the development of complex prototypes. This merging allows for the inclusion of electronic components or other mechanical attributes in a single print job, creating effectively multifunctional prototypes that are resilient against various working conditions. These functionalities not only enhance the ease with which prototyping can be carried out but also push the boundaries of what is possible within product design. By using layered manufacturing techniques, designers can control the distribution of material at the micro-level, so that strength, flexibility, and functionality are coordinated with specific design requirements. Also, by using biocompatible materials for rapid prototyping, considerable breakthroughs can be achieved in medical and dental fields, where fit and function require high accuracy. Consequently, 3D printing is revolutionary technology in which form and function are combined in perfect harmony in the production of next-generation devices (Bencherif et al., 2014). Not only does the prototyping speed advantage of 3D printing offer speed but also tremendous cost savings and minimization of resources. Traditional production involves high upfront costs and material wastage, together with the need for detailed tooling and fixtures. While traditional manufacturing produces excessive waste because it makes parts in batches, additive manufacturing mitigates this through its layer-based operation where only the materials needed are utilized in the areas specified, hence enhancing prototyping sustainability. Also, localized on-demand part manufacturing minimizes the inventory required, consequently lowering the cost of operations directly. Such economic advantages make 3D printing an attractive approach for both startups and corporations, encouraging product innovation. Lastly, 3D printing's transformative value in rapid prototyping is revolutionizing the face of fabrication technologies to offer unprecedented opportunities for efficiency, customization, and higher integration of materials to the quest for next-generation devices.

10.4.3. Limitations and Challenges of 3D Printing in Electronics

Even as the revolutionary potential of 3D printing in electronics cannot be overstated, the technology comes with severe shortcomings that discourage extensive applications. The first challenge is to achieve the resolution and precision necessary for accurate electronic components that are critical to today's device performance. The conventional processes of additive manufacturing fall far short of the level of detail attained with cutting-edge lithography equipment such as extreme ultraviolet (EUV) and nanoimprint lithography. The aged methods excel in generating micro-scale features critical to semiconductor devices, which 3D printing must complete to match on an equal footing. The inability to achieve such high-resolution needs restricts the types of applications for which 3D printing is possible within the electronics domain, suggesting a need for further advances in additive manufacturing technologies to increase the fidelity of printed electronics. Furthermore, the integration of heterogeneous materials is another significant issue in 3D printing electronics. Optimized electronic systems require a variety of materials with varying electrical, thermal, and mechanical properties. 3D printing technology is not as versatile as it must be, so the simultaneous deposition and bonding of numerous materials become challenging, and the creation of complex electronic structures with high levels of customization become complicated. Techniques such as atomic layer deposition (ALD) provide solutions towards the fabrication of thin films with excellent conformality, but integrating these advanced processes with 3D printing is a technological hindrance. This limitation inhibits new functional development in next-generation devices because flawless integration of different materials needs to be achieved to stimulate electronic applications and enhance performance characteristics. Lastly, despite the numerous benefits enumerated by 3D printing, such as quicker prototyping and material efficient use, scalability remains a primary challenge. While additive manufacturing allows rapid iterations in design, it remains challenging to create large numbers of parts for mass use in industry. With the industry moving toward Industry 4.0 principles, the need for seamless, consolidated integration of smart manufacturing systems becomes increasingly paramount but current 3D printing technologies lag behind when it comes to matching the needs of scalability and speed. In order to make a successful transition from low-volume production towards industrial adoption, improving the speed, processing capability, and cost benefits of 3D printing technologies becomes imperative. The existing limitations thus remain as a discouragement not just to the implementation of the technology within the electronics sector but even to its application as a pillar of sustainable manufacturing practices as intended by an exhaustive piece of research on the subject matter (Attene et al., 2017; Dehghanghadikolaei et al., 2019).

10.5. NANOFABRICATION TECHNIQUES

The progress in nanofabrication methods has played a pivotal role in the creation of next-generation devices, facilitating performance and miniaturization development. Of these technologies, the high-end lithography of EUV and nanoimprint lithography have gained attention with the potential to create intricate nanoscale patterns. EUV lithography leverages short light wavelengths to deliver high-resolution capabilities in response to the demand for reduced-size circuit elements in semiconductor manufacturing. Alternatively, nanoimprint lithography delivers a compelling option through mechanical nanostructure imprinting onto a substrate that holds promise for cost-effectiveness and enhanced throughput for manufacturing. These advanced lithographic technologies have emerged as indispensable instruments in the quest for accuracy and efficiency and finally lead to the robust manufacturing of future electronic devices that are able to respond to the complexities of modern applications. Complementing lithography, atomic layer deposition (ALD) has emerged as a critical technique for producing thin films and coatings with atomic-level accuracy. One of ALD's key strengths is its ability to provide conformal coverage of intricate high-aspect-ratio structures, making it the preferred technique in nanofabrication. By depositing films in layers, ALD offers the highest level of control over film thickness and composition, which is critical in optimizing device performance, particularly in the semiconductor industry (Figure 10.9).

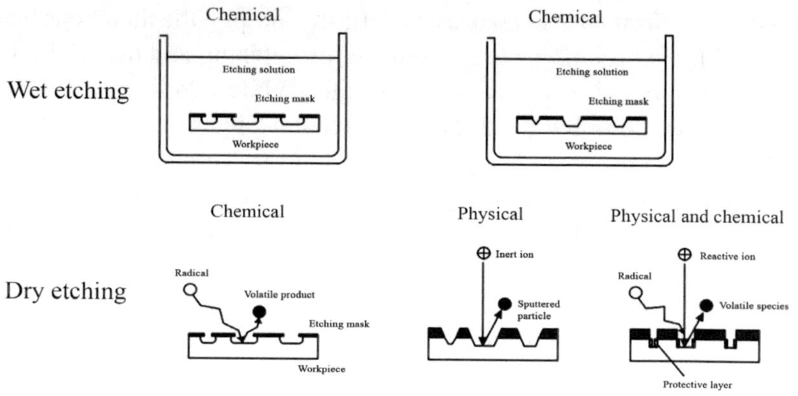

Figure 10.9. *The methods of nanofabrication techniques.*

Source:https://cdnintech.com/media/chapter/87904/1706266821/media/F5.png.

The extremely precise nature of ALD allows the introduction of varied materials, making possible innovations such as high-k dielectrics and novel conductive materials. These provide the ability to have more desirable electrical properties and reduced leakage currents for transistors, again making miniaturization and the operation of future electronic devices possible. In this sense, ALD is an important factor in the resolution of nanoelectronics' issues. Also, the bonding of disparate materials is increasingly important in the development of new nanofabrication techniques. With increasing complexity of devices, the ability to blend materials at the nanoscale allows for multi-functionality and performance improvement. Techniques such as hybrid nanostructuring, where disparate materials are blended in two or more, generate new properties that cannot be offered by a single material. For instance, silicon nanowires in a polymer matrix have the potential to integrate the merits of high conductivity and flexibility, which would bring new opportunities in electronics and optoelectronics. Much research provides the means through which integrating materials with techniques like roll-to-roll processing and solution-based approaches can lead to high-end multi-functional devices.

The coupling of different nanomaterials not only optimizes performance but also enables environmentally benign approaches in next-generation device manufacturing, showcasing the pivotal role of nanofabrication in technology frontier improvement.

10.5.1. Definition and Importance of Nanofabrication

The modern technological landscape is increasingly defined by the ability to fabricate nanoscale materials, which has brought into being a discipline known as nanofabrication. This discipline concerns the fabrication of structures, devices, and systems at the nanometer scale, which holds relevance to the manufacture of next-generation devices in various fields, e.g., electronics, biotechnology, and materials science. Among the many methods employed in nanofabrication, advanced lithography, i.e., extreme ultraviolet (EUV) lithography and nanoimprint lithography, is leading the way by enabling appropriate patterning of materials. These methods enhance the resolution and magnitude of manufacturing processes, making way for quicker, miniature, and more efficient devices. By allowing the creation of intricate nanoscale architectures, nanofabrication eventually enhances the performance and capabilities of future technologies. Nanofabrication is more than some manufacturing process improvements; it transforms our understanding of material properties and engineering. The ability to create structures at these high resolutions allows researchers to explore and harness quantum phenomena that manifest at the nanoscale. For instance, atomic layer deposition (ALD) allows ultrathin film deposition with atomic precision, which further makes it possible to build novel materials with distinct electronic, optical, and mechanical properties. This level of control is integral to dissimilar materials' fusion, wherein dissimilar substances can be combined in such a way as to leverage their relative strength. The interdisciplinarity of these technologies, which cross chemistry, physics, and engineering, is an ideal demonstration of the essential importance of nanofabrication in overcoming challenging problems and advancing technology in new ways. Nanofabrication is having a great impact on industries, bringing with it a new era of innovation driven by the unique properties of nanomaterials. An example is that focused electron beam induced deposition (FEBID) has been recognized as an innovative method for the creation of magnetic nanostructures, from information storage all the way through to nanoscale sensing technologies (Córdoba et al., 2016). The approach allows for the selective delivery of materials, i.e., cobalt nanodeposits, that have been designed with specific functional properties. Additionally, the mention of plasma-specific nanofabrication phenomena suggests the potential of self-organized nano-solids to acquire enhanced properties, depending on localized control of energy and matter (Meyyappan et al., 2013). These innovations not only point to the versatility of nanofabrication technologies but also speak to their ability to shape the future of technology by enabling the development of more efficient, compact devices that have the ability to carry out sophisticated functions.

10.5.2. Techniques Used in Nanofabrication (e.g., Etching and Deposition)

The nanofabrication environment is evolving with the integration of next-generation lithography technologies, which are significant for next-generation device development. EUV lithography is special through its capability to produce intricate patterns at a nanoscale, with significantly better resolution compared to traditional photolithography methods. Nanoimprint lithography also offers a cost-effective solution to creating high-fidelity nanostructures, allowing for the reproduction of complicated patterns onto other substrates. These lithography methods not only improve patterning fidelity but also play a pivotal role in shaping integrated photonics and quantum systems, where precision and miniaturization are key. As emphasized by recent publications, the effective combination of such methods is unveiling new fields of novel applications in optical communications, thereby representing a revolutionary shift in the ability of nanofabrication processes, particularly with regard to the realization of quantum photonic integrated circuits (Moody et al., 2021). Atomic layer deposition (ALD) lies at the forefront of achieving the desired material properties and functionality of future devices. The technique allows for the growth of thin films in a controlled manner with atomic precision, hence allowing for the deposition of various materials like metals, dielectrics, and semiconductors. The layer-by-layer technique ensures consistent coatings, a prerequisite for the development of devices with better performance parameters. The precision used in ALD not only enables devices to be miniaturized down to nanoscale dimensions but also ensures integration of disparate materials, hence facilitating the creation of complex structures that exhibit excellent electrical and optical properties. Additionally, conformal coating on complex geometries is possible, hence allowing advancing technologies, such as spintronics, to benefit from these advances. In such applications, the search for multifunctional devices is greatly driven by the synthesis capability of ALD in meeting the performance needs of future applications (Hirohata et al., 2020). Heterogeneous material integration is a milestone in nanofabrication. With different materials integrating at the nanoscale, scientists have been able to create interfaces with unique characteristics that homogenous materials cannot. This approach proves most beneficial in developing next-generation electronic and photonic devices, wherein the interaction between various types of material can lead to higher function and efficiency. For instance, it proves most crucial for quantum devices, wherein the elements have to talk to each other flawlessly to enable processes such as entanglement and superposition. In addition, the development of tailored materials through hybrids of organic, inorganic, and metal compounds can yield performance advantages, such as enhanced thermal and electrical conduction. While advances in fabrication processes continue, the focus will be on refining these integrations, the building blocks for the developments of devices that can leverage the advanced paradigms of next-generation technologies (Moody et al., 2021).

10.5.3. Future Trends in Nanofabrication

Progress in nanofabrication technology continues to advance innovation in various fields of technology. Emerging lithographic technologies, including extreme ultraviolet (EUV) lithography and nanoimprint lithography, are responsible for enhancing the resolution in future fabrication. EUV lithography, which can produce structures at the nanometer range, enables the creation of microchips with much smaller nodes, with enhanced performance and efficiency saved. Simultaneously, nanoimprint lithography offers a cost-effective path to patterning intricate nanostructures and hence to making advanced microfabrication equipment broadly accessible. The simultaneous advances, not only allow for the synthesis of advanced devices, but also provide the basis for more robust quantum computing environments and advanced sensors that are essential to future digital infrastructure. The synergistic combination of these advanced techniques promises tremendous advancement in nanofabrication, driving future device performance to unprecedented heights. Of these technologies, atomic layer deposition (ALD) is a pillar technique with extensive control over material layers down to the atomic level. The ability of ALD to deposit ultrathin films with superior uniformity and conformality holds considerable advantages in the fabrication of nanostructured devices. As heterogeneous materials play an increasingly larger role in the architecture of devices, ALD becomes essential to the creation of complex, layered structures that enhance device functionalities. The layering precision created by ALD not only improves material characteristics but also paves the way for hybrid devices, which combine diverse materials with contrasting electrical, optical, and thermal characteristics. Consequently, as industries are increasingly reliant on miniaturization and multi-functional devices, the combination of ALD with other nanofabrication techniques is destined to play a pivotal role in shaping future technological trends. Besides, the insertion of disparate materials in nanofabrication will revolutionize device technology to enable the production of even more sophisticated and efficient systems. It allows for the synthesis of materials with varying properties, such as the tensile property of metals and the semiconductive property of silicon. These hybrids can also lead to devices with improved efficiencies and tailored for specific applications, e.g., renewable energy to advanced communication systems. For instance, recent research on photonic integrated circuits (PICs), e.g., development toward terahertz communication applications, demonstrates the potential of these heterogeneous solutions to revolutionize data processing and transmission (Zhu et al., 2016). Similarly, 3D silicon detector technology shows the capabilities of such materials even in adverse environments with enhanced performance at the expense of reliability (Lange et al., 2015). As innovation in the area is taking place, synergy between these approaches will undoubtedly set the course of nanofabrication and its use in different disciplines.

10.6. QUALITY CONTROL AND TESTING METHODS

Next-generation device manufacturing calls for a robust quality control and test protocols platform as the fabrication processes evolve. With ongoing innovations in forms of advanced lithography, including extreme ultraviolet (EUV) and nanoimprint techniques, further pushing the envelope of device manufacturing, the importance of maintaining high precision cannot be overstated. Quality control practices implemented with precision ensure that the intricate patterns created throughout these advanced lithography processes are within tight specifications. They accomplish this by marrying real-time monitoring with statistical process controls, enabling manufacturers to spot anomalies early in production. With this inclusion of quality assurance protocols, not only are yield rates enhanced, but costly errors are reduced, ultimately creating a more trustworthy manufacturing environment for next-generation devices. In combination with advanced lithography, the application of atomic layer deposition (ALD) in semiconductor fabrication proves to require the need for rigorous testing protocols. ALD allows precise thin film deposition, a prerequisite for the achievement of desired electrical and thermal properties of modern devices. Owing to nanoscale material dimension used in ALD, any variation can influence device performance radically. Thus, quality control is integrated through employing various test methods—such as spectroscopic ellipsometry and atomic force microscopy—to ensure thickness and uniformity of films follow the set standards. These methods enable manufacturers to attain tight tolerances, which are absolutely essential for the proper functioning of more intricate and miniaturized electronic systems. Thus, incorporating a thorough examination strategy effectively minimizes the occurrence of defects while enhancing overall process efficiency in fabrication. The application of heterogeneous materials further complicates the quality control environment in device production. Future devices become more reliant on bringing together different materials to enhance performance, and those responsible for producing such devices experience unusual challenges in determining compatibility and performance. The quality control systems need to account for the varied properties of the materials and how they interact during the process of device manufacturing. To prevail over these challenges, a number of test methods such as thermal cycling and mechanical stress tests are used, which test the reliability and durability of integrated materials when exposed to working conditions. Additionally, sophisticated geometric algorithms used in additive manufacturing have a crucial role in improving the design and configuration of such heterogeneous structures, as highlighted in recent studies (Attene et al., 2017). In the end, effective quality control and testing processes assist in the mass production of future-generation devices to keep up with the demands of cutting-edge technology and to address academic and industrial problems (Gilman et al., 2025).

10.6.1. Importance of Quality Control in Fabrication

In the area of advanced fabrication processes, incorporation of advanced processes such as atomic layer deposition (ALD) and advanced lithography processes such as extreme ultraviolet (EUV) and nanoimprint technology require quality control systems of high standards. Each of these techniques plays a pivotal role in addressing next-generation devices whose tolerance for errors is greatly decreased. With advancements in fabrication technology, the risk of defects increases, and quality control is crucial not only to maintain product quality but also to ensure the reliability of new electronic applications. Since the nature of analog IC layouts is such that they must be different from digital layouts, quality control measures must be applied in a proper manner to address specific problems within the analog realm (Rentmeesters et al., 1990). This distinction emphasizes the necessity of specialized quality control measures that can specifically account for the intricate variations in these advanced manufacturing techniques. Moreover, the new technology of additive manufacturing poses additional challenges and variables for manufacturing quality control (Figure 10.10).

Geometric algorithms have emerged as 3D printing technologies expanded, to facilitate the optimizations required for efficient production processes (Attene et al., 2017). These algorithms rank chief objectives in the process planning phase, including the geometry analysis and optimization that are directly linked to the product function and integrity. Top standards in this diversified field are a necessity; fractured quality control not only stops the productivity of operations but jeopardizes the end device performance. Therefore, quality control systems must evolve to cover comprehensive assessments of both hardware and software components, in an effort to integrate the rapid innovations involved in manufacturing processes. Stringency in the quality assurance process offers the basis for innovative approaches to enhancing both dimensional accuracy and material integrity. The incorporation of heterogeneous materials in fabrication processes also underscores the need for interactive quality control procedures. When manufacturers develop next-generation devices that take advantage of the diversities of material properties, possibilities for incompatibility and performance issues are higher. Quality control in such circumstances includes not only assessing discrete parts but also understanding dissimilar material interaction behaviors, which is challenging. Through diligent current practices and by adopting good testing procedures, engineers can offset the challenges of such complexities. For instance, employing quality control processes to verify thermal and mechanical properties prevents failure in multi-component constructions. Therefore, maintaining high-quality control standards enables the manufacture of devices that not only meet technological specifications but also operate reliably with changing operating conditions, enabling innovations in material science to effectively gain implementation in devices of the future.

Figure 10.10. *Purpose of quality control.*

Source: https://resources.duralabel.com/hubfs/Imported_Blog_Media/quality-control-benefits-1-1-1-1-1-1.jpg.

10.6.2. Common Testing Methods for Fabricated Devices

The evaluation of devices synthesized requires a variety of test methods, each necessary in ascertaining functionality and reliability. Electrical characterization is one of the typical methods, where parameters such as threshold voltage, subthreshold slope, and leakage current are measured in semiconductor devices. Capacitance-voltage (C-V) measurements are among the methods employed to obtain information on device performance through revelation of doping profile and interface states. These electrical measurements are imperative to validating the performance of advanced fabrication processes, such as atomic layer deposition (ALD), which yields high-quality films with uniform thickness. As fabrication processes continue to improve, solid testing methodologies become integral to verifying that future devices can withstand the harsh requirements of many applications, from consumer electronics to space technology, thereby validating the efficacy of meticulous design, performance, and testing (Bouldin et al., 2025). Another vital test technique is reliability testing, whereby units are put under stress tests simulating real-world operating conditions. Such testing involves thermal cycling, mechanical stress, and electrical overstress to determine long-term stability and performance degradation. Results of these tests are crucial in determining the lifespan of integrated circuits, more so those fabricated using advanced lithography techniques, like extreme ultraviolet (EUV) lithography. These methods enable excellent accuracy at nanoscale dimensions, and it is therefore essential to conduct comprehensive testing protocols to confirm their endurance against changes in the environment. The information gained from reliability tests is applied to continuous optimization of manufacturing technologies, ultimately leading to more resilient and efficient devices, particularly in microelectronics and integrated systems (Bouldin et al., 2025). Ultimately, innovative methods from new Additive Manufacturing (AM) technologies are also gaining traction in the testing arena. Geometric algorithms allow for pre-fabrication simulations, including design optimization and defect detection before actual manufacturing. This innovative approach synergizes with modern fabrication technologies, complementing traditional testing processes while adapting to the unique challenges posed by heterogeneous material integration. The ability to study and enhance geometry with analog algorithms enhances the validity of the devices manufactured to perform as needed in various uses. The verification of the performance of devices that have been manufactured using AM technologies, for instance, by conducting impact and thermal tests, is vital with increasing industries relying on the technology to push the boundaries of the manufacturing and application fronts, thereby addressing academic and business problems.

10.6.3. Role of Automation in Quality Assurance

With device manufacturing today, the implementation of automated quality assurance techniques has been made mandatory to achieve the highest levels of production efficiency and product dependability. Advanced processes, such as extreme ultraviolet (EUV) lithography and atomic layer deposition (ALD), are increasingly reliant on automation to achieve precise material deposition and patterning at nanoscale thicknesses. Automation facilitates real-time monitoring and control, which allows for immediate responses to be implemented during the manufacturing process. This minimizes the risk of human error and ensures that it is far less likely for defective products to find their way to the market. The use of automation for quality control increases the operation's efficiency as well as allows for the joining of heterogeneous materials, a necessity for the development of next-generation devices that incorporate complex multi-material structures. As observed in NASA workshop findings, there is a need to invest in automation technology in order to advance the materials and methods of aerospace manufacturing and much more, and toward a long-term future of innovation and sustainability in fabrication processes (Glaessgen et al., 2025). Also, as the semiconductor world continues to evolve in terms of complexity, automation plays a vital role in safeguarding production integrity. Globalized manufacturing of integrated circuits (IC) exposes products to wide-ranging threats like theoretical supply chain attacks and manufacturing errors. To mitigate these threats, automated quality control processes employ sophisticated algorithms and data analysis to identify inconsistencies and patterns of defects. Implementing automated inspections not only improves accuracy but also makes systems more robust overall against potential hardware trojans. This is particularly critical as devices become ever more interdependent, because even small mismatches can result in large failures in cryptographic and data-sensitive applications. The use of automated quality assurance procedures is therefore the key to supporting high assurance within products while staying within manufacturing budgets, as recent industry studies have emphasized (Cerulli et al., 2017). Ultimately, automation in quality control transcends improving manufacturing processes; it also fuels innovation in the development of new fabrication processes. While scientists tinker with new procedures such as 3D printing and microfabrication, automated processes for quality control are key to trying out and perfecting these technologies. Automation allows for rapid prototype testing and creation, facilitating a faster response to market demand and technological advancements. In addition, by harvesting and processing large sets of data throughout the production cycle, organizations can derive insights that fuel ongoing improvement and optimization. As seen in the case of next-generation manufacturing technologies, automation embedded in quality control ensures that the new manufacturing processes not only achieve the aggressive requirements of quality but also enable the move towards the next level of leading-edge applications, thereby embedding the basics for future innovation in next-generation devices (Glaessgen et al., 2025).

SUMMARY

With a growing need for next-generation devices, continuous innovation in manufacturing techniques is needed to meet the sophisticated needs of modern technology. A key aspect of this innovation is the improvement of lithography techniques, such as EUV and nanoimprint lithography, that allow for the formation of significantly smaller features on semiconductor wafers. These advanced techniques enable patterning with higher resolution, and thus more effective device performance and simpler incorporation of new materials. Although the use of these advanced lithographic techniques has seen improved device functionality, it is also accompanied by humongous technical complexities and economic costs. The semiconductor world therefore has to tread with extreme caution over these matters to be able to harness advanced lithography towards creating more robust and efficient electronic devices. The heterogenic material combination is a powerful aspect of next-generation device fabrication techniques as well. The technique such as ALD provides the material interfaces control at atomic levels, allowing the formation of heterostructures that are crucial to the high-performance optoelectronic devices. With the use of ALD, businesses can deposit films with atomically precise thicknesses uniformly coated, a feat instrumental in bettering the performance of the devices utilizing semiconductors and insulators. Furthermore, this specific deposition method can optimize the scalability of devices, paving the way toward commercial viability for next-generation uses. These advancements point toward a greater movement in the market—enhancing functionality through targeted coupling of several materials instead of reverting to traditional practices. This innovation in new materials not only advances development but also dictates the direction for future next-generation device production. The culmination of these sophisticated fabrication techniques is not only in the imagination but also foretells profound alteration in the capabilities landscape for electronics. The increasing ubiquity of solid-state optical emitters, particularly those based on quantum mechanical properties, demonstrates the information technology and sensing application potential of such developments. The incorporation of GaAs photonic crystals and InAs self-assembled quantum dots is a move to include high-performance material and new methods of fabrication. The accomplishment brings to the fore the necessity to expand studies and utilization of advanced methods of fabrication in the manufacturing of devices that must operate on low power and process information at high speed. As industries work to exploit these technologies, there is a need to assess not just the technical but also the economic feasibility of new fabrication methods, specifically those that allow for transition to next-generation devices as society increasingly depends on advanced electronics for a wide range of uses.

In the context of the evolving world of fabrication processes for next-generation devices, very advanced lithography is a key element in the development of resolution and accuracy. Semiconductor firms have increasingly used methods such as EUV lithography and nanoimprint lithography, which allow for the development of miniaturized, high-efficiency electronic components. These processes allow for the manufacturing of highly intricate patterns at a nanoscale that is hard to achieve for conventional lithography. The advantages of EUV lithography are that it can reduce feature sizes much smaller, which is extremely critical as the devices become smaller and more advanced. Moreover, nanoimprint lithography offers an alternative by physically stamping an energy-sensitive polymer to create high-fidelity patterns with the advantage of low cost over traditional photolithography, and thus stimulates innovation in device fabrication. In addition to advanced lithography, ALD seems to be a significant technique for the building of thin films with precise dimensions on substrates at the atomic scale. ALD is remarkable for the capacity to coat uniform layers across intricate and deep geometries, being increasingly engineered into contemporary devices. Through alternate exposure to reactants, ALD enables controlled film material composition, morphology, and thickness. This is particularly valuable in the creation of gate dielectrics in transistors or protective coatings that enhance material characteristics for enhanced device reliability. The extent of precision offered by ALD is priceless in the creation of multifunctional devices required in future electronics so that scientists and engineers can tailor materials to application needs and operating conditions. Moreover, combining heterogeneous materials also has opportunities and challenges in the field of device making. The capability to integrate various materials—organic and inorganic semiconductors, for instance—is improved in terms of function and performance, as well as fostering device architecture innovation. Methods such as printed electronics are a good example of integrating different materials by facilitating the deposition of organic electronic materials on flexible substrates through roll-to-roll methods, thereby reducing the cost of production and speeding up production (Andersen et al., 2020). But the integration of different materials must fight against thermal, mechanical, and chemical compatibility issues. Ongoing research still attempts to find new solutions for these problems, with the development and use of next-generation devices standing at the forefront of scientific innovation (Attene et al., 2017).

REFERENCES

1. Abate, A., Aebersold, P., Ahlford, A., Almassian, R., Almoguera, L., Asiello, P., et al. (2017). Advances in microfluidics and lab-on-a-chip technologies. http://arxiv.org/abs/1703.08583 (accessed on 27th September 2025).
2. Abbasi, A., Baets, R., Bauwelinck, J., Cardile, P., et al. (2015). III–V-on-silicon photonic devices for optical communication and sensing. https://core.ac.uk/download/55743989.pdf (accessed on 27th September 2025).
3. Al-Muhtaseb, S. A., Verma, A., Asakawa, K., Sharma, A., Brooksby, P. A., Campo, A. D., Sharma, C. S., et al. (2010). Micro-fabrication of carbon structures by pattern miniaturization in resorcinol-formaldehyde gel. http://arxiv.org/abs/1009.3703 (accessed on 27th September 2025).
4. Algarni, A., Allen, M., Baldwin, R., Bergstrom, P., Bhardwaj, R., Bidoki, S. M., Bogy, D., et al. (2016). Integration of additive manufacturing and inkjet printer electronics: A potential route to parts with embedded multifunctionality. https://core.ac.uk/download/96767162.pdf (accessed on 27th September 2025).
5. Althammer, M., Bibes, M., Boschker, H., Calleja, F., et al. (2019). Towards oxide electronics: A roadmap. https://core.ac.uk/download/199214191.pdf (accessed on 27th September 2025).
6. Amano, H., Baines, Y., Borga, M., Bouchet, R., et al. (2018). The 2018 GaN power electronics roadmap. https://core.ac.uk/download/154431388.pdf (accessed on 27th September 2025).
7. Andersen, T. R., Cottam, S., Elkington, D. C., Griffith, M. J., et al. (2020). Manipulating nanoscale structure to control functionality in printed organic photovoltaic, transistor and bioelectronic devices. https://core.ac.uk/download/288432664.pdf (accessed on 27th September 2025).
8. Antonsen, T., Attwood, D., Bennett, B., Born, M., Carter, A., Choi, Y., Cummings, B., et al. (2019). Optical lithography. https://core.ac.uk/download/287622025.pdf (accessed on 27th September 2025).
9. Armstrong, C., Bryce, B. A., Copel, M., Elmegreen, B., et al. (2015). Pathway to the piezoelectronic transduction logic device. http://arxiv.org/abs/1503.07467 (accessed on 27th September 2025).
10. Attene, M., Ellero, S., Lefebvre, S., Livesu, M., et al. (2017). From 3D models to 3D prints: An overview of the processing pipeline. https://core.ac.uk/download/84977973.pdf (accessed on 27th September 2025).
11. Aukes, D. M., Khodambashi, R., & Sharifzadeh, M. (2018). An integrated design and simulation environment for rapid prototyping of laminate

robotic mechanisms. http://arxiv.org/abs/1808.03697 (accessed on 27th September 2025).

12. Bajcsy, M., Englund, D., Faraon, A., Kim, H., et al. (2011). Integrated quantum optical networks based on quantum dots and photonic crystals. https://core.ac.uk/download/10211147.pdf (accessed on 27th September 2025).
13. Banerjee, A., Hadfield, R. H., Heath, R. M., Hemakumara, G., et al. (2018). Optical properties of refractory metal-based thin films. https://core.ac.uk/download/159754713.pdf (accessed on 27th September 2025).
14. Belmonte Molina, A., Berry, M. V., Forbes, A., Pérez Torres, J., et al. (2017). Roadmap on structured light. https://core.ac.uk/download/87648764.pdf (accessed on 27th September 2025).
15. Bencherif, S. A., Braschler, T. M., & Renaud, P. (2014). Advances in the design of macroporous polymer scaffolds for potential applications in dentistry. https://core.ac.uk/download/28947444.pdf (accessed on 27th September 2025).
16. Berenschot, J. W., Elwenspoek, M. C., Huskens, J., Jansen, H. V., et al. (2009). Multi-silicon ridge nanofabrication by repeated edge lithography. https://core.ac.uk/download/pdf/11471245.pdf (accessed on 27th September 2025).
17. Bonvin, E., Forró, L., Horváth, E., Náfrádi, B., et al. (2016). Controlled growth of $CH_3NH_3PbI_3$ nanowires in arrays of open nanofluidic channels. http://arxiv.org/abs/1612.01891 (accessed on 27th September 2025).
18. Bostwick, A., Cresti, A., Dato, A., Castro Neto, A. H., Giesbers, A. J. M., Javey, A., Geim, A. K., et al. (2010). Towards graphene nanoribbon-based electronics. http://arxiv.org/abs/1002.4461 (accessed on 27th September 2025).
19. Bouldin, D. L., Feltner, W. R., Hollis, B. R., Routh, J. D., et al. (2025). High and low threshold P-channel metal oxide semiconductor process and description of microelectronics facility. https://core.ac.uk/download/pdf/42882473.pdf (accessed on 27th September 2025).
20. Bozhevolnyi, S. I., Kristensen, A., Mortensen, N. A., et al. (2015). Gap and channeled plasmons in tapered grooves: A review. https://core.ac.uk/download/43250842.pdf (accessed on 27th September 2025).
21. Brigo, L., Brusatin, G., Della Giustina, G., et al. (2018). Changes in the near edge X-ray absorption fine structure of hybrid organic-inorganic resists upon exposure. http://arxiv.org/abs/1809.09916 (accessed on 27th September 2025).

22. Castro López, R., Fernández Fernández, F. V., Rodríguez Vázquez, Á. B. (2005). On the suitability and development of layout templates for analog layout reuse and layout-aware synthesis. https://core.ac.uk/download/286564418.pdf (accessed on 27th September 2025).
23. Cerulli, A., Cvrcek, D., Danezis, G., et al. (2017). A touch of evil: High-assurance cryptographic hardware from untrusted components. https://core.ac.uk/download/111036080.pdf (accessed on 27th September 2025).
24. Chia, H. N., & Wu, B. M. (2015). Recent advances in 3D printing of biomaterials. https://core.ac.uk/download/323078409.pdf (accessed on 27th September 2025).
25. Clavijo, W. (2014). *Nanowire Zinc Oxide MOSFET Pressure Sensor.* https://core.ac.uk/download/51290965.pdf (accessed on 27th September 2025).
26. Córdoba, R., De Teresa, J. M., Fernández-Pacheco, A., et al. (2016). Review of magnetic nanostructures grown by focused electron beam induced deposition (FEBID). https://core.ac.uk/download/78917676.pdf (accessed on 27th September 2025).
27. Dehghanghadikolaei, A., Emamian, S. S., Fotovvati, B., et al. (2019). The potential of additive manufacturing in the smart factory industrial 4.0: A review. https://core.ac.uk/download/231748826.pdf (accessed on 27th September 2025).
28. Fahmi, S., Khairudin, M., Khi, K. B., et al. (2016). The fabrication of PDMS mould for microelectrode array biochip using NIL. https://core.ac.uk/download/159187087.pdf (accessed on 27th September 2025).
29. Francisco Gamiz, C., Márquez, M., Caño, M., Sampedro, C., Navarro, C., Ortega, R., Lorenzo, F., et al. (2025). (Invited) The game-changing potential of 2D materials in next-generation electronics and sensors. *ECS Meeting Abstracts.* https://www.semanticscholar.org/paper/4fe4a1810869d512c9456c9b55b61b8d5e8efc02 (accessed on 27th September 2025).
30. Galan Moody, Sorger, V. J., Blumenthal, D. J., Juodawlkis, P., Loh, W., Sorace-Agaskar, C., Jones, A. E., et al. (2021). 2022 roadmap on integrated quantum photonics. *Journal of Physics Photonics, 4*, 012501–012501. https://doi.org/10.1088/2515-7647/ac1ef4 (accessed on 27th September 2025).
31. Ghadiani, B., Kippenberg, T. J., Liu, J., et al. (2018). Probing the loss origins of ultra-smooth Si_3N_4 integrated photonic waveguides. http://arxiv.org/abs/1802.08315 (accessed on 27th September 2025).
32. Gilman, J. A., Paige, J. B., & Schwenk, F. C. (2025). *NASA SBIR Abstracts of 1991 Phase 1 Projects.* https://core.ac.uk/download/pdf/42806098.pdf (accessed on 27th September 2025).

33. Glaessgen, E. H., Jensen, B. J., & Ransom, J. B. (2025). *ARMD Workshop on Materials and Methods for Rapid Manufacturing for Commercial and Urban Aviation.* https://core.ac.uk/download/pdf/286392860.pdf (accessed on 27th September 2025).
34. Heath, J. R., Kurtin, J. N., Sheriff, B. A., et al. (2008). Complementary symmetry nanowire logic circuits: Experimental demonstrations and *in silico* optimizations. https://core.ac.uk/download/4879629.pdf (accessed on 27th September 2025).
35. Hirohata, A., Yamada, K., Nakatani, Y., Prejbeanu, L., Diény, B., Pirro, P., & Hillebrands, B. (2020). Review on spintronics: Principles and device applications. *Journal of Magnetism and Magnetic Materials, 509*, 166711–166711. https://doi.org/10.1016/j.jmmm.2020.166711 (accessed on 27th September 2025).
36. Lange, J. (2015). Recent progress on 3D silicon detectors. http://arxiv.org/abs/1511.02080 (accessed on 27th September 2025).
37. Lee, T. Y., Chen, P. T., Huang, C. C., Chen, H. C., Chen, L., Lee, P. T., Chen, F. C., et al. (2025). Advances in core technologies for semiconductor manufacturing: Applications and challenges of atomic layer etching, neutral beam etching and atomic layer deposition. *Nanoscale Advances, 7*, 2796–2817. https://www.semanticscholar.org/paper/9684a14f7361c98ef20416bd87d6b2c2ff28374f (accessed on 27th September 2025).
38. Meyyappan, M., Neyts, E. C., & Ostrikov, K. (2013). Plasma nanoscience: From nano-solids in plasmas to nano-plasmas in solids. http://arxiv.org/abs/1306.6711 (accessed on 27th September 2025).
39. N/A. (2013). *The Boston University Photonics Center Annual Report 2013–2014.* https://open.bu.edu/bitstream/2144/29752/1/Photonics-Center-Final-Annual-Report-2013-2014-v2.pdf (accessed on 27th September 2025).
40. Pathak, R., Anoop, G., & Samanta, S. (2025). Advancements in free-standing ferroelectric films: Paving the way for transparent flexible electronics. *Journal of Composites Science.* https://www.semanticscholar.org/paper/e42d9109117ff23958df0a18e2aa9e1b785e5d5e (accessed on 27th September 2025).
41. Rentmeesters, M. J. (1990). Automatic synthesis of analog layout: A survey. https://core.ac.uk/download/287623298.pdf (accessed on 27th September 2025).
42. Zhang, Z., Wang, Y., Cai, T., Wu, B., Hu, B., Li, X., Song, E., et al. (2025). Recent progress on flexible silicon nanomembranes

for advanced electronics and optoelectronics. *Advanced Functional Materials.* https://www.semanticscholar.org/paper/edf3101e0eddcf55a075b27972ba27fa9cf99eea (accessed on 27th September 2025).

43. Zhu, H., Marsh, J. H., Xu, J., Han, L., Hou, L., Deng, Q., & Liang, S. (2016). Fully integrated multi-optoelectronic synthesizer for THz pumping source in wireless communications with rich backup redundancy and wide tuning range. https://core.ac.uk/download/42372375.pdf (accessed on 27th September 2025).

INDEX

A

Active-Bridge-Active-Clamp (ABAC) 227
Adaptability 183
Additive manufacturing 398, 414, 415, 416, 417, 418, 419, 425, 431, 433
Amplification 1
Angular momentum 268, 274, 280, 285
Artificial intelligence 2, 6, 8, 9, 10, 11, 14, 16, 17, 18, 19, 21, 22, 25, 26, 27, 28, 29, 30, 33, 34, 35
Atomic layer deposition (ALD) 396, 399, 402, 405, 406, 409, 411, 413, 419, 421, 423, 424, 425, 427, 428
Atomic scale 397, 400, 408, 430
Autonomous systems 2

B

Ballistic transport 90, 91, 92, 93, 99, 102, 108, 109, 110, 113, 114, 115, 116, 117, 121, 123, 124, 126
Bandgap engineering 48, 50, 54, 55, 57, 60, 62, 64, 66, 67, 68, 69, 70, 71, 76, 78, 79, 81
Bandwidth 312, 315, 316, 317, 318, 320, 323, 325, 326, 330, 332, 334, 335, 336, 339, 340, 341, 342, 343, 347
Big data-processing systems 136
Bimolecular recombination 92, 93, 108
Biomedical imaging 312
Bipolar transistors 223, 226, 228, 233, 235, 243, 251, 253, 255

C

Capacitance devices 180, 195
Carrier energy 94
Carrier mobility 90, 91, 92, 93, 95, 99, 102, 104, 107, 109, 110, 112, 114, 116, 119, 120, 121, 123, 124
Carrier transport 89, 90, 91, 92, 93, 94, 95, 96, 97, 99, 100, 101, 102, 103, 104, 105, 106, 107, 108, 109, 110, 111, 112, 113, 115, 116, 117, 119, 120, 121, 122, 123, 124, 125, 126, 127, 128
Carrier wavelengths 94
Channel charge carriers 184
Channel conduction 179
Charge carriers 89, 90, 92, 93, 95, 96, 97, 102, 104, 105, 106, 107, 108, 111, 113, 114, 115, 116, 120, 121, 123, 124, 125
Charging stations 225
Circuit designs 184, 188
Classical physics 358
Cloud computing 316, 334, 336, 341
Commercialization 180, 354
Communication networks 314, 316, 319, 320, 324, 331, 335, 336, 339, 340, 343

Communication system 223, 263
Compatibility 316, 318, 319, 320, 322, 326, 327, 328, 336, 339, 342, 343
Complexity 400
Component miniaturization 2
Computational density 271
Computational devices 270
Computational requirements 179
Computing environments 271, 290, 300
Computing systems 134
Concentration regions 91, 95
Condensed matter physics 90
Conduction model 91
Conductivity control 140
Conformal films 397, 412
Consumer electronics 134, 136
Conventional designs 1
Conventional memory technologies 269, 280, 290
Conventional transistors 179, 188, 198, 212
Cryogenic temperatures 355, 357, 359, 361, 367, 368, 372, 377, 378, 379, 381, 384, 385
Crystal lattice 51, 53, 54, 59, 63, 65, 68
Crystalline structure 93
Crystal structure 47, 49, 50, 51, 52, 53, 54, 55, 59, 69, 71, 78
Customer needs 137
Cutting-edge lithography 395, 398, 401, 418

D

Data communication 311, 317, 319, 321, 335, 338, 339
Data connections 315
Data processing 2, 6, 15, 23, 26, 34
Data rates 313, 316, 321, 339, 340, 343
Data retention 270, 272, 288, 299, 304
Data storage 313
Data transfer 312, 321, 326, 328, 330, 341, 345, 346
Data transmission rates 315, 322, 333
de Broglie wavelength 90
Decoherence 354, 355, 357, 358, 359, 362, 364, 366, 367, 370, 372, 375, 377, 383, 385, 386, 387
Department of Energy (DOE) 182
Deterioration 137
Determination 399
Device fabrication 396, 399, 400, 401, 406, 414, 429, 430
Dielectrics 397, 407, 420, 422, 430
Diffraction 318
Digital transformation 2
Digital world 271
Digitization 179
Drift mechanism 93
Dual-Active-Bridge (DAB) 227
Durability 3, 225, 236

E

Electrical component integrity 224
Electrical conductivity 49, 50, 51, 57, 59, 63, 65, 78, 81
Electrical machinery 224
Electrical performance 134, 145, 147, 169
Electric car technology 50
Electric fields 90, 91, 93, 95, 96, 99, 101, 108, 109, 110, 111, 112, 113, 115, 121, 124
Electric vehicle 225, 226, 244, 253, 259
Electromagnetism 312
Electron conduction 358
Electron flow 355
Electronic devices 1, 5, 6, 42
Electronic function 66
Electronic functioning 178
Electronic property 66
Electronics innovation 267
Electronics sector 269
Electron mobility 222, 224, 225, 228, 233, 235, 243, 251, 253, 254, 255, 262
Electrostatic channel control 138
Electrostatic confinement 137, 146, 148, 158

Electrostatic control 134, 135, 136, 137, 141, 144, 145, 147, 149, 151, 152, 153, 154, 156, 157, 158, 159, 161, 162, 166, 167, 169, 170, 179, 181, 185, 191, 192, 193, 200, 202, 205, 212
Energy efficiency 1, 9, 15, 19, 25, 32, 33, 37
Energy gap reduction 48
Energy loss 50
Engineering processes 47
Environmental conditions 3
Environmental monitoring 312, 325, 333
Environmental observation 313
Error tolerance 355
Exponential technology 180
Extreme ultraviolet (EUV) 395, 401, 409, 411, 413, 414, 418, 421, 423, 424, 425, 427, 428
Extreme Ultraviolet (EUV) lithography 400

F

Fabrication processes 397, 398, 406, 423, 424, 425, 426, 427, 428, 430
Fascinating synergy 272
Ferroelectric 178, 179, 181, 185, 187, 191, 193, 194, 196, 197, 198, 199, 203, 205, 206, 207, 213, 215, 218
Finite Element Analysis (FEA) 3
Flexible electronics 396, 400, 405, 434
Fusion 268

G

Gate-All-Around Field-Effect Transistors (GAAFETs) 134, 154, 167
Gate-All-Around (GAA) 137
Gate terminal 135, 138
Geometric algorithms 398, 414, 415, 417, 425
Geometries 134, 138, 141, 142, 143, 144, 146, 147, 148, 149, 150, 154, 167, 169, 170

H

Heat dissipation 225, 227, 228, 232, 235, 236, 241, 246, 247, 251, 253, 254, 256
Heat management 223
Heterogeneous integration 3, 5, 8, 10, 32, 33
High Electron Mobility Transistors (HEMTs) 222, 226, 227, 234, 237, 238, 240, 241, 246, 247, 248, 250, 252
High-end computing 2
High-frequency magnetic design 222
High power processing 224
High-speed computation 139
Human brains process 269

I

Imaging systems 313, 330
Impurity concentration 183
Industrial applications 223, 235, 254
Industrial power systems 225
Information capture 316
Information processing 269, 271, 272, 275, 277, 287, 290, 294, 301, 304, 311, 316, 324, 353, 355, 359, 362, 363, 366, 370, 373, 374, 375, 381, 384, 388
Information technology 268, 269, 302
Instrumentation 355
Insulated Gate Bipolar Transistors (IGBTs) 222, 226, 230, 234, 238, 240, 244, 246, 247, 249
Insulators 397, 411, 429
Integrated circuit design 134, 173
Interconnectivity 2
Internet of Things (IoT) 2, 6, 8, 10, 12, 13, 14, 16, 17, 19, 21, 22, 23, 25, 26, 27, 28, 29, 30, 31, 33, 34, 35

L

Laser technology 312, 327, 329
Lateral Diffused Metal Oxide Semicon-

ductors (LDMOS) 222
Lateral Diffusion Metal-Oxide-Semiconductor (LDMOS) transistors 226
Leakage control 138, 141, 142, 145, 146, 153, 159, 161, 162, 163, 169
Light emissions 314
Light-emitting devices 92, 123
Light-emitting diodes (LEDs) 49, 79
Lightweight 400
Lithographic techniques 396, 403, 429
Longevity 51

M

Machine learning 268, 269, 272, 274, 275, 279, 290, 292, 293, 295, 299, 304
Magnetic field detection 356
Magnetic moments 268, 272, 280, 300
Magnetic ordering 268
Magnetic random-access memory (MRAM) 267, 272, 275, 278, 290, 293, 296, 298, 300, 301
Magnetic technology 270, 299
Magnetoresistive Random Access Memory (MRAM) 269, 277, 279, 281, 286, 288, 302
Manufacturing technology 2
Marketplace 182
Marshall Space Flight Center (MSFC) 397
Material development 136
Material science 90, 99, 103, 109, 113, 125
Material science innovation 356
Mechanical stress tolerance 3
Medical diagnosis 313
Medical imaging 315, 317, 330, 333
Metal dichalcogenides 49, 57, 67, 68, 69, 76, 78, 80, 81, 85, 86
Metalorganic vapor phase 66
Metal-Oxide-Semiconductor Field-Effect Transistor (MOSFET) 134, 138, 140, 150, 154, 164
Microelectronics 51, 140, 173, 174
Microelectronics field 397
Microfabrication technology 357
Microprocessors 135
Microstructures 399
Microwave technology 355
Miniaturization 133, 136, 137, 145, 146, 150, 151, 152, 157, 158, 159, 170
Mobile communication 139
Mobility manipulation 90
Modern electronic systems 221, 236, 253
Moores Law 1, 3, 6, 8, 17, 22, 26, 35, 36
More Electric Aircraft (MEA) systems 227

N

Nanoimprint lithography 395, 396, 397, 398, 399, 400, 409, 411, 412, 413, 418, 419, 421, 423, 429, 430
Nanoscience 396, 434
Nanosheet (NS) transistors 137
Nanosheet transistors 134, 135, 136, 138, 140, 142, 145, 147, 148, 149, 151, 152, 154, 155, 156, 157, 158, 159, 161, 162, 163, 164, 166, 167, 169, 170, 176
Nanotechnology 89, 116
Neumann architectures 268
Neural networks 270, 272, 273, 275, 282, 286, 289, 290, 291, 292, 294, 296, 297, 303, 305, 308
Neuromorphic computing 268, 269, 270, 272, 273, 275, 276, 277, 279, 280, 281, 282, 285, 286, 287, 289, 290, 291, 292, 293, 294, 295, 296, 297, 299, 300, 301, 302, 303, 304
Next-generation solutions 1
Noise reduction 355

O

Operational capacity 312
Operational integrity 224
Optical technology 315
Optoelectronic 47, 49, 50, 51, 52, 53, 55, 57, 60, 65, 66, 68, 70, 71, 78
Optoelectronic devices 91, 93, 106, 107, 108, 119, 123, 127

Optoelectronic innovations 47

P

Performance enhancement 223, 264
Phase transitions 47
Photodetector development 317
Photodetectors 312, 313, 315, 316, 317, 319, 320, 321, 323, 325, 326, 327, 328, 329, 330, 331, 332, 333, 334, 335, 336, 338, 339, 340, 341, 342, 343, 344, 345, 347, 348
Photoluminescence spectroscopy 55, 66, 87
Photonic device production 397, 409
Photonic elements 315, 324, 330
Photonic integrated circuits (PICs) 311, 334, 337
Photovoltaic cells 90
Physical realizations 354, 358
Planar geometries 178, 197
Power consumption 133, 136, 138, 139, 142, 145, 148, 151, 153, 154, 162, 170, 179, 180, 181, 182, 185, 186, 187, 189, 190, 191, 192, 194, 195, 196, 197, 201, 202, 203, 205, 206, 209, 211, 212, 213
Power density 3, 15, 222, 226, 228, 233, 249, 252, 253, 261
Power dissipation 141, 159, 161, 162, 169, 176, 354, 369
Power efficiency 178, 186, 187, 188, 190, 194, 195, 199, 201, 207, 212
Power generation 223
Power management system 225
Power reduction 178
Power regulation systems 135
Power utilization 136
Practical engineering 355, 357
Pragmatism gaps 400
Precision metrology 356
Prototyping 398, 415, 416, 417, 418, 419, 431

Q

Quantum computer 2
Quantum-confined systems 90
Quantum confinement 91, 97, 104, 113, 118, 119, 125, 131
Quantum cryptography 354, 359, 363, 392
Quantum dots 92, 94, 97, 98, 119, 122, 123, 124, 126, 129, 130
Quantum effects 3, 4, 17, 32, 36
Quantum electronics 354, 368, 370, 373, 390
Quantum mechanics 90, 102, 125, 353, 357, 358, 359, 360, 365, 385, 387
Quantum regime 355
Quantum research 355, 370, 385, 386
Quantum sensors 356, 379
Quantum tunneling 177, 179, 185, 186, 189, 190, 191, 195, 196, 203, 206

R

Radio frequency (RF) devices 224
Rectification 1
Reflection 318, 319
Refraction 318
Reliability 89, 98, 106, 109, 117
Revolutionary evolution 1

S

Satellite earth observation 67
Scalability 180, 182, 185, 190, 191, 207, 208, 211, 215, 269, 274, 275, 276, 277, 281, 283, 285, 288, 301
Scattering 90, 91, 92, 93, 96, 108, 109, 110, 113, 115, 116, 120, 121, 123, 126
Self-assembled systems 48
Self-organization 48, 56, 80
Semiconductor alloys 48
Semiconductor chips 3, 29
Semiconductor device 1, 2, 3, 5, 6, 8, 9, 10, 11, 16, 17, 18, 20, 21, 24, 25, 26, 27, 28, 30, 35, 36, 37, 38, 43
Semiconductor materials 47, 48, 49, 50, 51, 53, 56, 57, 60, 61, 62, 63, 64, 65, 66, 67, 68, 69, 70, 71, 82, 83,

84, 86
Semiconductor physics 178
Semiconductor production processes 180
Semiconductor qubits 353, 356, 358, 360, 362, 364, 365, 370, 381, 384, 387
Semiconductor technology 2, 4, 5, 6, 7, 8, 9, 10, 12, 13, 15, 16, 19, 20, 23, 25, 26, 28, 30, 31, 32, 34, 36, 37
Sensor technology 316
Signal amplification 225, 240, 254
Silicon-based technologies 177, 205
Silicon fabrication processes 318
Silicon nanomembranes 399, 434
Silicon photonics 312, 316, 317, 321, 322, 323, 328, 329, 337, 340
Silicon technology 222
Single-electron devices 353, 354, 355, 356, 357, 359, 360, 369, 370, 371, 373, 374, 375, 376, 378, 379, 380, 381, 382, 384, 385, 387, 388
Single-electron transistors (SETs) 354, 357, 358, 360, 362, 365, 368, 369, 371, 374, 378, 381, 382, 383, 384, 385
Smart devices 2, 25
Solar cells 50, 56, 57, 58, 59, 62, 66, 70, 75, 81, 83, 84
Solid-state systems 90
Special attention 134
Spin-based neural networks 269
Spin-orbit torque (SOT) 270, 290
Spin-qubit systems 2
Spin transport 268, 270, 271, 272, 274, 275, 276, 277, 278, 279, 280, 281, 282, 283, 285, 287, 288, 289, 290, 291, 292, 295, 296, 297, 298, 299, 301, 302, 303, 304
Spintronic devices 268, 269, 270, 272, 275, 276, 277, 278, 280, 286, 287, 289, 290, 291, 292, 294, 295, 296, 297, 298, 299, 300, 301, 302, 304, 310
Spintronics 267, 268, 269, 272, 273, 277, 286, 296, 303, 307, 309
Strategic engineering 90
Structural complexity 140, 167
Superconducting qubits 353, 355, 358, 360, 361, 362, 363, 364, 370, 376, 378, 381, 387, 388
Sustainability 178, 185, 188, 202, 205, 213
Sustainable development 224, 256
Sustainable energy 90
Synergetic interaction 396

T

Technological development 311
Technology advancement 3, 25
Technology converge 312
Telecommunication 48, 50
Thermal conductivity 61, 66, 71, 73, 74
Thermal dissipation 2, 3, 29
Thermal emissions 314
Thermal management 2, 3, 12, 15, 27, 29, 30, 36, 222, 224, 225, 226, 227, 228, 230, 231, 232, 234, 235, 236, 237, 239, 241, 242, 243, 244, 245, 246, 247, 251, 252, 254, 256, 257, 258, 259, 264, 400
Thermal noise 354, 357, 359, 364, 368, 375, 376, 377, 378, 379, 380, 384, 385, 388
Thermal performance 224, 227, 231, 234, 242, 245, 257
Thermal resistances 223
Thermal stability 48, 50, 51, 59, 63, 68, 79, 82
Thermionic emission 181, 206
Topological media 356
Transformational character 181
Transformations 177
Transistor design 134, 149, 151, 156, 163
Transistor development 178
Transistor size 396
Transistor technology 134, 151, 170, 180, 184, 185, 186, 194, 199, 202, 205, 206, 207, 208
Tunability 316, 326, 328, 333, 342

Tunneling mechanism 180, 189, 190, 202, 213

U

Ultra-scaled transistors 135
Ultrasonic saw 3, 41
Ultra-wide bandgap (UWBG) 2

V

van der Waals magnetic materials 268, 307
Vertical orientation 179
Vertical transistors 182
Vertical transport 179, 180, 182, 186, 191, 192, 196, 197, 201, 202, 204, 205, 207
Vertical transport devices 179, 180, 182, 186, 191, 192, 196, 197, 207
Volatile memory 268, 273, 280, 281, 282, 284, 285, 286, 292

W

Wavelengths 48, 49, 66, 68, 72, 82
Welding 3
Wide bandgap (WBG) 2